Lecture Notes on Data Engineering and Communications Technologies

Volume 20

Series editor

Fatos Xhafa, Technical University of Catalonia, Barcelona, Spain
e-mail: fatos@cs.upc.edu

The aim of the book series is to present cutting edge engineering approaches to data technologies and communications. It will publish latest advances on the engineering task of building and deploying distributed, scalable and reliable data infrastructures and communication systems.

The series will have a prominent applied focus on data technologies and communications with aim to promote the bridging from fundamental research on data science and networking to data engineering and communications that lead to industry products, business knowledge and standardisation.

More information about this series at http://www.springer.com/series/15362

Natalia Kryvinska · Michal Greguš

Editors

Data-Centric Business and Applications

Evolvements in Business Information Processing and Management—Volume 1

 Springer

Editors
Natalia Kryvinska
Department of e-Business,
  Faculty of Business, Economics
  and Statistics
University of Vienna
Vienna
Austria

Michal Greguš
Department of Information Systems,
  Faculty of Management
Comenius University
Bratislava
Slovakia

ISSN 2367-4512   ISSN 2367-4520   (electronic)
Lecture Notes on Data Engineering and Communications Technologies
ISBN 978-3-319-94116-5   ISBN 978-3-319-94117-2   (eBook)
https://doi.org/10.1007/978-3-319-94117-2

Library of Congress Control Number: 2018945071

Printed on acid-free paper

This Springer imprint is published by the registered company Springer International Publishing AG part of Springer Nature
The registered company address is: Gewerbestrasse 11, 6330 Cham, Switzerland

Preface

Since global market is becoming more and more complex with the increased availability of data and information, doing business with information is getting more popular impacting modern society immensely. Accordingly, there is an increasing need for a common understanding of how to create, access, use, and manage business information.

For that reason, we explore in this book different aspects of data and information processing. Relevant themes covered include the information generation, representation, structuring, organization, storage, retrieval, navigation, human factors in information systems, and use of information. The book provides also an experience in processes and procedures in information/data processing and management.

With this volume, we start to analyze challenges and opportunities for doing business with information. It considers research outputs on varied perspectives of business information managing.

In particular, the chapter entitled "Digital Communications and a Smart World" examines turbo codes that are currently introduced in many international standards and implemented in numerous advanced communication systems, applied in a smart world, and evaluates the process of extrinsic information transfer (EXIT). The convergence properties of the iterative decoding process, associated with a given turbo-coding scheme, are estimated using the analysis technique based on so-called EXIT charts. This approach provides a possibility to predict the bit-error rate (BER) of a turbo code system with only the extrinsic information transfer chart. The idea behind this research was to consider the associated soft-input soft-output (SISO) stages as information processors, which map input a priori log likelihood ratios (LLRs) onto output extrinsic LLRs, the information content being obviously assumed to increase from input to output and introduce them to the design of turbo systems without the reliance on extensive simulation. Compared with other methods, the suggested approach provides insight into the iterative behavior of linear turbo systems with substantial reduction in numerical complexity.

Next Chapter "A Qualitative Evaluation of IoT-Driven eHealth: Knowledge Management, Business Models and Opportunities, Deployment and Evolution" gives a reasonable, qualitative evaluation of IoT-driven eHealth from theoretical and practical viewpoints. The authors look at associated knowledge management issues and contributions of IoT to eHealth, along with requirements, benefits, limitations, and entry barriers. Important attention is given to security and privacy issues. Finally, the conditions for business plans and accompanying value chains are realistically analyzed. The resulting implementation issues and required commitments are also discussed. The authors approve that IoT-driven eHealth can happen and will happen; however, much more needs to be addressed to bring it back in sync with medical and general technological developments in an industrial state-of-the-art perspective and to recognize and the get timely benefits.

Further, the research performed on Chapter "WLAN Planning and Performance Evaluation for Commercial Applications" recognizes that capacity is equally important as coverage for the WLAN planning, as the number of users with more high-throughput applications, supported by sophisticated mobile computing devices, continues to grow rapidly on the one hand, and on the other hand, the new high-speed IEEE 802.11-based WLANs continue to have a significant wireless communication market spread. Subsequently, WLANs capacity planning and performance evaluation focusing on the impact of the applications throughput requirements, high-speed IEEE 802.11 WLAN standards, and user terminal capabilities are discussed. To investigate the impact of the high-speed WLANs and user terminal capabilities on the network throughput, performance evaluation is carried out by means of measurements, using IEEE 802.11ac and IEEE 802.11n standards.

In Chapter "Traffic Fluctuations Optimization for Telecommunication SDP Segment Based on Forecasting Using Ateb-Functions", the authors propose a network traffic prediction method based on a theory of differential equations with them solutions can be written as Ateb-functions. The proposed method first uses the hypothesis about cyclic nature of network traffic pulsations. Then, description of the traffic intensity fluctuations into the computer network builds a prediction model, as nonlinear oscillating system with single degree-of-freedom under the conditions of small-scale disturbances. From the simulation, the proposed prediction method in distinction from existing methods gives more qualitative solution. Thereafter, the obtained values of the predicted traffic intensity are used as one of the key indicators in solving the problem of optimal resource allocation in heterogeneous network platforms by correcting the metrics or priorities of traffic flows routing. Decisions taking on a basis the procedures for evaluating alternatives, by mutually related and mutually controversial criteria, are using fuzzy logic approach with triangular fuzzy numbers aggregation.

The chapter authored by Maria Ivanova "IP-Telephony—Comparative Analysis of Applications" deals with the Internet Protocol Telephony. It explains the meaning of this technology, its conveniences and disadvantages, evolution and history. Next, it informs about the most popular and widely used programs which provide IP Telephony—Skype as a pioneer in a modern IP Telephony market with its peer-to-peer system; and Whatsapp, a relative newcomer with a client–server system, which has gained much on the popularity among both young and elder generations in the last few years. Finally, a comparison matrix is given—briefly displaying the main differences and similarities between other popular VoIP applications (Viber, Telegram, Facebook Messenger, Google Hangouts, including Skype and Whatsapp).

Chapter "Product Usage Data Collection and Challenges of Data Anonymization", authored by Ümit G. Peköz, demonstrates the importance of the product usage data collection and processing for the future enterprise and highlights the technical challenges in data analyzing. The Big Data processing and analyzing can bring also risks to the individuals, such as privacy threads. Thus, the aim of the paper is to show briefly the Big Data consequences for a private person and to review existing technical privacy protection methods. It is worth to compare state-of-the-art data anonymization techniques with each other to show the privacy challenges in Big Data. The limitation in the findings is that each technique can have its special application area and running a full comparison can be not possible.

In the work headed "Challenges in Management of Software Testing", Martina Halás Vančová and Zuzana Kovačičová focus on the test management within projects of software development. The high quality of test management is a part of the success of IT projects. There exists a high quantity of recommended practices and techniques for test management. However, it is necessary to complete these recommendations by actual trends and practical experience, since every IT project is different in its nature and it is managed in a specific environment. The first part of the chapter deals with the theoretical background of IT testing. The second part specifies the research objective, methodology, and research procedures. The last part is empirical, and it describes activities of test management carried out at the observed IT project.

The chapter authored by Raoul Gorka, Christine Strauss, Claus Ebster "The Role of Variety Engineering in the Co-creation of Value" explores that by integrating methods and models from systems science, the interdisciplinary field of service science has provided the basis for a possible radical shift in understanding the coordination mechanisms underlying global market dynamics. The variety engineering model is found to support and extend the evolving framework of service-dominant logic by shedding light on the relational nature of interacting social agents, who co-create value by steering their behavior toward shared meanings through conversation. As the authors claim—in our highly complex present-day world the main driver of balancing agents' complexity asymmetries is self-organization. Without adequate management, this self-regulation seldom produces socially desirable outcomes. Thus, the authors conceive the proposed systemic methodology as an effective guideline for supporting managers to coordinate this process toward common policies, which may foster sustainable structures.

Chapter "Enterprises Servitization for an Efficient Industrial Management a Survey-Based Analysis", authored by S. Kaczor, strives to explain the concept of servitization in detail in order to declare its capabilities in being a powerful strategy to provide a competitive advantage in manufacturing industries. This is conducted by initially elaborating its notion and origins in the literature as well as the characteristics and processes regarding value creation, which is consolidated as fundamentals of servitization. Subsequently, the driving characteristics separated by general environmental trends, financial-, strategic-, and marketing-drivers are discussed in detail. The final section refers to the classifications of servitization and the actual options of implementation.

In Chapter "The Future of Industrial Supply Chains: A Heterarchic System Architecture for Digital Manufacturing?", Corinna Engelhardt-Nowitzki and Erich Markl claim that economic value networks are occasionally described as heterarchic systems. Hence, the application of supply chain management (SCM) practices means to be subjected to the fundamental principles of such systems—whether knowing and managing them consciously or not. However, this kind of knowledge has predominantly remained abstract and far from practical application. Besides, successful SCM relies on a fast and flexible information flow between the involved parties. Modern IT extends previous, at most hierarchical data structures and software systems by means of ubiquitous random access facilities. Comparable to the heterarchically negotiated structure of supply chain processes, these information systems are increasingly able to process data in reticulate structures, eventually even based on software agents with negotiation skills. Thus, the chapter characterizes basic principles of heterarchic systems. Accordingly, heterarchic, network-like IT-approaches are shortly discussed regarding the application in an SCM-context. Subsequently, business implications for practical application are deduced regarding two questions: (1) 'What is heterarchy and why does it matter in SCM?' and (2) 'Could hierarchical layer models, as frequently used in SCM, informatics and automation, still serve for the purpose of achieving a holistic model of heterarchic systems in the context of digital manufacturing?'.

Next chapter authored by Kitty Klacsánová, Mária Bohdalová "Analysis of GDP and Inflation Drivers in the European Union" describes different effects of inflation and GDP on the European Union countries pointing out the costs attendant on high and low rate of inflation. To improve the understanding of GDP and inflation differences, there were applied statistical methods. They verified some of the big economic crises in the history. Concerning inflation, the authors have investigated co-movements and the heterogeneity in inflation dynamics across the analyzed countries over two periods (1967–2015 and 1994–2015). The findings indicate that there are three substantial common principal components explaining 99.72% of the total variance in the consumer price indexes in the period from 1994 to 2015 that can be related to the common monetary policy in the euro area.

In the chapter called "Office 365 as a Tool for Effective Internal Communication in Corporate Environment", the author finds that despite massive expansion of social media sites, email communication keeps its position and importance within organizational communication process and tends to increase. Thus, the author aims in this chapter to analyze Office 365 and its components with a focus on SharePoint Online in corporate environment. The chapter also describes a partial implementation of services within selected workplaces. Office 365 components, components of SharePoint Online are implemented to improve an effective communication within an organization and to reduce the number of emails sent or received only to necessary minimum.

As a final theme, Chapter "Simulating and Reengineering Stress Management System—Analysis of Undesirable Deviations" proposes a method for diagnosing the level of criticality of an undesirable deviation in the supply–production–marketing chain. It is based on simulating the impact of a potential or actual incident that causes such a deviation on each link in the chain using a number of representative parameters (i.e., reliability of the supply chain, inventory management system, negative response from external marketing stakeholders, competitiveness of the marketing complex, quality of the production process, flexibility of the production process, level of planned processes violation). Also, according to the results of the performed research, a set of heuristic rules for diagnosing incidents in the supply–production–marketing chain was proposed.

Vienna, Austria   Natalia Kryvinska
Bratislava, Slovakia   Michal Greguš

Contents

Digital Communications and a Smart World

Izabella Lokshina and Hua Zhong

Abstract This chapter is devoted to digital communications in a smart world. The authors examine turbo codes that are currently introduced in many international standards and implemented in numerous advanced communication systems, applied in a smart world, and evaluate the process of Extrinsic Information Transfer (EXIT). The convergence properties of the iterative decoding process, associated with a given turbo-coding scheme, are estimated using the analysis technique based on so-called EXIT charts. This approach provides a possibility to predict the Bit-Error Rate (BER) of a turbo code system with only the extrinsic information transfer chart. The idea is to consider the associated Soft-Input Soft-Output (SISO) stages as information processors, which map input a priori Log Likelihood Ratios (LLRs) onto output extrinsic LLRs, the information content being obviously assumed to increase from input to output and introduce them to the design of turbo systems without the reliance on extensive simulation. Compared with other methods, the suggested approach provides insight into the iterative behavior of linear turbo systems with substantial reduction in numerical complexity.

Keywords Smart world · Digital communications · Reliability and performance Error control and correction · Turbo codes · Extrinsic information transfer (EXIT) EXIT chart · Reduced numerical complexity

I. Lokshina (✉) · H. Zhong
Management, Marketing and Information Systems, SUNY Oneonta,
Oneonta, NY 13820, USA
e-mail: Izabella.Lokshina@oneonta.edu

H. Zhong
e-mail: Hua.Zhong@oneonta.edu

© Springer International Publishing AG, part of Springer Nature 2019
N. Kryvinska and M. Greguš (eds.), *Data-Centric Business and Applications*,
Lecture Notes on Data Engineering and Communications Technologies 20,
https://doi.org/10.1007/978-3-319-94117-2_1

1 Introduction

Digital communications (or, data communication) is defined as data transfer over a communication channel. In a smart world, digital communications support a wide range of multimedia applications, such as audio, video and computer data that differ significantly in their traffic characteristics and performance requirements [15–17, 23].

Turbo codes have been adopted for providing error correction in a number of advanced communication systems, such as the 3rd-Generation Wideband Code Division Multiple Access (3G WCDMA) and 4th-generation Long Term Evolution (4G LTE) systems [9, 18, 20]. Turbo codes comprise a parallel concatenation of two constituent convolutional codes. By iteratively exchanging information between the two corresponding constituent decoders, turbo codes facilitate reliable communication at transmission throughputs that approach the channel capacity [3, 5, 7, 10, 16, 17, 19, 20].

Turbo codes represent a great advancement in the coding theory. Their excellent performance, especially at low and medium signal-to-noise ratios, has attracted an enormous interest for applications in digital communications. Historically, turbo codes were first deployed for satellite links and deep-space missions, where they offered impressive Bit-Error Rate (BER) performance beyond existing levels with no additional power requirement (a premium resource for satellites). Since then, they have made their way in the 3rd-generation wireless phones, Digital Video Broadcasting (DVB) systems, Wireless Metropolitan Area Networks (WMAN) and Wi-Fi networks.

Currently in a smart world, even if several research issues are still open, the success of turbo codes is growing, and their introduction in many international standards is in progress; the 3rd-Generation Partnership Project (3G PP) Universal Mobile Telecommunications System (UMTS) standard for 3rd-Generation Personal Communications and the European Telecommunications Standards Institute (ETSI) Digital Video Broadcasting—Terrestrial (DVB-T) standard for terrestrial digital video broadcasting are among them [1, 2, 5, 11, 13–15]. In turbo processes the channel decoder and the demodulator at the receiver exchange extrinsic information. Such turbo processes are frequently analyzed using Extrinsic Information Transfer (EXIT) charts [4, 6, 8, 11, 12, 16, 17, 20]. This approach provides a possibility to predict the bit-error rate of a turbo code system using only the extrinsic information transfer chart [16, 17].

Extrinsic information transfer charts are powerful tools to analyze and optimize the convergence behavior of iterative systems applying the turbo principle, i.e., systems exchanging and refining extrinsic information, without dependence on extensive simulations [16, 17]. Compared with the other known methods for generating extrinsic information transfer functions, the proposed analytical method provides insight into the iterative behavior of linear turbo systems with substantial reduction in numerical complexity.

This chapter is comprised of ten sections and is organized as follows. Section 2 reviews performance of error-control codes, also called error-correcting codes or channel codes, in digital communications. Section 3 focuses on application of Reed-Solomon (RS) codes in digital communication systems. Section 4 describes advantages of Berlekamp–Massey (BM) and Euclidean algorithm used for time-domain decoder. Section 5 focuses on Internet Protocol (IP) interfacing and enhanced Forward Error Correction (FEC). Section 6 describes Symbol Error Rate (SER) simulation and provides suggestions on performance analysis. Section 7 consider extrinsic information transfer in digital communications. Section 8 provides analysis of the turbo code behavior with extrinsic information transfer charts. Section 9 discusses numerical results. Section 10 presents summary and conclusions, which are followed by acknowledgment and references.

2 Overview of Error-Control Codes in Digital Communications

Error-control codes, also referred to error-correcting codes or channel codes, are fundamental components of nearly all digital transmission systems in use today. Channel coding is accomplished by inserting controlled redundancy into the transmitted digital sequence, thus allowing the receiver to perform a more accurate decision on the received symbols and even correct some of the errors made during the transmission.

In his landmark 1948 paper that pioneered the field of Information Theory [22], Claude E. Shannon proved the theoretical existence of good error-correcting codes that allow data to be transmitted practically error-free at rates up to the absolute maximum capacity (usually measured in bits per second) of a communication channel, and with surprisingly low transmitted power (in contrast to common belief at that time). However, Shannon's work left unanswered the problem of constructing such capacity-approaching channel codes [15–17, 24, 25].

This problem has motivated intensive research efforts during the following four decades and has led to the discovery of very good codes, frequently obtained from sophisticated algebraic constructions. However, 3 dB or more stood between what the theory promised, and the practical performance offered by error-correcting codes in the early 90's [23–25].

The introduction of Convolutional Turbo Codes (CTC) in 1993 quickly followed by the invention of Block Turbo Codes (BTC) in 1994, closed much of the remaining gap to capacity. Classical convolutional turbo codes, also called Parallel Concatenated Convolutional Codes (PCCC), resulted from a pragmatic construction conducted by researchers, who, in the late 80's, highlighted the interest of introducing probabilistic processing in digital communications receivers. Previously, other researchers had already imagined coding and decoding systems closely related to the principles of turbo codes [3–5, 21, 23–25].

Later, Block Turbo Codes (BTC), also called Turbo Product Codes (TPC), offered an interesting alternative to CTC for applications requiring either high code rates ($R > 0.8$), very low error floors, or low-complexity decoders able to operate at several hundreds of megabits per second (and even higher) [6, 8, 11, 12, 23, 25].

Nowadays, advanced Forward Error Correction (FEC) systems employing turbo codes commonly approach Shannon's theoretical limit within a few tenths of a decibel. Practical implications are numerous [11, 12, 15–17, 20]. Using turbo codes, system designers can, for example, achieve a higher throughput (by a factor 2 or more) for a given transmitted power, or, alternatively, achieve a given data rate with reduced transmitted energy.

Accordingly, while reviewing this advanced error control and correction technology in existing standards, a decade after the discovery of turbo codes, as well as discussing future applications of this technology in a smart world, this chapter analyzes turbo processes using extrinsic information transfer charts.

3 Application of Reed-Solomon Codes in Digital Communication Systems

For example, in a smart world, a Reed–Solomon (RS) code for control and correcting errors and erasures can be used to improve performance in a variety of systems, including a space communication link, a compact disc audio system, High-Definition Television (HDTV) and a Digital Versatile Disk (DVD) [15].

The error-correcting ability of any Reed-Solomon code is determined by n − k, the measure of redundancy in the block. If the locations of the error symbols are not known in advance, then a Reed–Solomon code can correct up to (n − k)/2 erroneous symbols, i.e., it can correct half as many errors as there are redundant symbols added to the block. Sometimes error locations are known in advance (e.g., "side information" in demodulator signal-to-noise ratios)—these are called erasures.

A Reed–Solomon code (like any linear code) is able to correct twice as many erasures as errors, and any combination of errors and erasures can be corrected as long as the inequality 2E + S < n − k is satisfied, where E is the number of errors and S is the number of erasures in the block. It is well known that the Berlekamp–Massey (BM) algorithm or the Euclidean algorithm can be used to solve Berlekamp's key equation for decoding RS codes [11, 15].

Forney defined an errata locator polynomial, using what are now called Forney syndromes, to correct both errors and erasures for RS codes. Based on the ideas of Forney, various authors have developed simplified algorithms for correcting both errors and erasures for RS codes [11, 15].

Recently, an inverse-free method was proposed to simplify the BM algorithm for finding the error locator polynomial in an RS decoder for correcting both errors and erasures [11, 15]. When digital data is transmitted or stored, it is common to be concerned about both errors and erasures.

An error occurs in a data stream when some data element is corrupted; an erasure occurs when a data element is detected missing or known to be faulty, and it defines the relation of errors with respect to erasures that themselves are the consequence of errors. Consequently, erasures describe part of the errors in the communications systems, however, the errors the authors refer to in this paper are the errors left after removing the errors included in the erasures. Shannon's basic theorems on coding [21] tell us that to correct e errors, a data stream needs to guarantee a minimum distance between correct streams of at least 2e + 1; to detect e errors, or to correct e erasures, valid entries in the data streams need to be separated by at least e + 1 bits.

Reed and Solomon [11, 15] devised a method for constructing linear codes over finite fields, where the individual data items can be represented as elements of the finite field. For digital data, it is most convenient to consider the Galois fields GF (2^m), for m = 8, 16, or 32, where the underlying data words are correspondingly, single octets, pairs of octets, or quads of octets.

4 Berlekamp–Massey and Euclidean Algorithm for Time-Domain Decoder

First, the authors let GF (2^m) be a finite field of 2^m elements. Also, the authors let C be a (n, h) RS code over GF (2^m), with minimum distance d, where $n = 2^m - 1$ is the block length, h is the number of m-bit message symbols and $d - 1$ is the number of parity symbols. The minimum distance d of this RS code is related to n and by h by $d = n - h + 1$.

Next, the authors define the errata vector as $\tilde{\mu} = \mu + e = (\tilde{\mu}_0, \tilde{\mu}_1, \ldots, \tilde{\mu}_{n-1})$, where $e = (e_0, e_1, \ldots, e_{n-1})$ and $\mu = (\mu_0, \mu_1, \ldots, \mu_{n-1})$ are the error and erasure vectors, respectively.

Finally, the authors represent the obtained vector $r = (r_0, r_1, \ldots, r_{n-1}) = c + \tilde{\mu}$, where $c = (c_0, c_1, \ldots, c_{n-1})$ is the codeword vector. Suppose that v errors and s erasures occur in the received vector r an assume that $s + 2v \leq d - 1$.

At this time, the authors let α be a primitive element in GF (2^m). The syndromes of the code computed from the received vector r are given by (1):

$$S_k = \sum_{i=0}^{n-1} r_i \alpha^{ik} = \sum_{i=1}^{s} W_i Z_i^k + \sum_{i=1}^{v} Y_i X_i^k. \tag{1}$$

for $1 \leq k \leq d - 1$, where W_i and Z_i are the ith erasure amplitude and the ith erasure location, respectively. Y_i and X_i are the ith error amplitude, and the ith error location, respectively.

The authors let $\sigma(x)$ be the erasure locator polynomial with zeros at inverse erasure locations. That is shown in (2):

$$\sigma(x) \prod_{j=1}^{s} \left(1 - Z_j x\right) = \sum_{j=0}^{s} \sigma_j x^j. \tag{2}$$

where $\sigma_0 = 1$ and $\deg\{\sigma(x)\} = s$. Similarly, the authors let $\tau(x)$ be the error locator polynomial with zeros at the inverse error locations. That is given in (3):

$$\tau(x) \prod_{j=1}^{s} \left(1 - X_j x\right) = \sum_{j=0}^{v} \tau_j x^j. \tag{3}$$

where $\tau_0 = 1$ and $\deg\{\tau(x)\} = v$. As the error locations X_i's are computed, the last two terms in (1) can be computed and expressed as (4):

$$S_k = \sum_{i=1}^{s+v} \hat{W}_i \hat{Z}_i^k, \text{ for } 1 \leq k \leq d - 1 \tag{4}$$

where $\hat{W}_i$ and $\hat{Z}_i$ are the ith errata amplitude and errata location, respectively. Consequently, the syndrome polynomial $S(x)$ can be derived as (5):

$$S_k = \sum_{k=1}^{d-1} S_k x^k = \sum_{k-1}^{d-1} \left(\sum_{i=1}^{s+v} \hat{W}_i \hat{Z}_i^k\right) x^k = \sum_{i=1}^{s+v} \hat{W}_i \frac{\hat{Z}_i x - \left(\hat{Z}_i x\right)^d}{1 - \hat{Z}_i x}. \tag{5}$$

Now, the authors define the errata locator polynomial with zeros at the inverse errata locations as (6):

$$\Lambda(x) = \sigma(x)\tau(x) = \prod_{j=1}^{s+v} \left(1 - \hat{Z}_j x\right) = \sum_{j=0}^{s+v} \Lambda_j x^j. \tag{6}$$

where $\Lambda_0 = 1$ and $\deg\{\Lambda(x)\} = s + v$. Finally, the authors define the errata evaluator polynomial $\Omega(x)$ as (7):

$$\Omega(x) = x \sum_{i=1}^{s+v} \hat{W}_i \hat{Z}_i \prod_{l \neq i} \left(1 - \hat{Z}_l x\right). \tag{7}$$

where $\deg\{\Omega(x)\} = s + v$ and $\Omega(0) = 0$.

Therefore, Berlekamp's key equation for erasures and errors [11, 15] is given by (8):

$$S(x)\Lambda(x) \equiv \Omega(x) \bmod x^d \tag{8}$$

or shown in (9)

$$\Omega(x) = \Theta(x) \cdot x^d + \Lambda(x) \cdot S(x).\tag{9}$$

where $\Theta(x)$ serves as an auxiliary polynomial which is not actually computed. The errata magnitudes are given in (10):

$$\hat{W}_l = \frac{\Omega(\hat{Z}_l^{-1})}{\hat{Z}_l^{-1}\Lambda'(\hat{Z}_l^{-1})}, \text{ for } 1 \le l \le s+v\tag{10}$$

where $\Lambda'(Z_l^{-1})$ is the derivative of $\Lambda(x)$ with respect to x evaluated at $x = \hat{Z}_l^{-1}$ and the errata locations can be computed from $\Lambda(x)$ with use of the Chien search method [11, 15], or the Berlekamp algorithm combined with the Chien search method [15]. The problem of decoding algorithms is to solve the Eqs. (8) or (9) for the errata locator polynomial $\Lambda(x)$ and the errata evaluator polynomial $\Omega(x)$ satisfying the following inequalities (11):

$$\deg\{\Omega(x)\} \le s+t \text{ and } \deg\{\Lambda(x)\} \le s+t.\tag{11}$$

where $t = \lfloor (d-1-s)/(2) \rfloor$ denotes the maximum number of errors that can be corrected and $\lfloor x \rfloor$ denotes the greatest integer less than or equal to x. Obviously, one has $v \le t$. The errata locator polynomial and the errata evaluator polynomial in (9) can be obtained with use of either the BM algorithm or the Euclidean algorithm with the stop criteria in (11).

5   IP Interfacing and Enhanced Forward Error Correction

Contrary to other Digital Video Broadcasting (DVB) transmission systems that are based on the DVB transport stream [15], adopted from the Moving Picture Experts Group (MPEG-2) standard, the Digital Video Broadcasting—Handheld (DVB-H) system is based exclusively on the Internet Protocol (IP). Using IP allows the coding to be decoupled from the transport, then opening the door to a number of features benefiting handheld mobile terminals including a variety of encoding methods, which require lower power from a decoder. Therefore, IP is the Open System Interconnection (OSI-layer 3) protocol used in the mobile handheld convergence terminals. In addition, IP itself is relatively insensitive to any buffering or delays within the transmission (unlike MPEG-2). It is appropriate for handheld packet mode as well as for time-division multiplexed transmission. Actually, IPv6 could be more appropriate in mobile environments in comparison with IPv4.

Thus, IPv6 may be the preferred option on the broadcast interface. However, both packet mode and time division multiplexing may be used with both IPv4 and IPv6. DVB-H is an extension of DVB-T with some backward compatibility, i.e., it can share the same multiplex frame with DVB-T. It may use a mechanism called the Multi-Protocol Encapsulation (MPE), making it possible to transport data network protocols on top of MPEG-2 transport streams: the IP protocol units are embedded into the transport stream by means of the MPE, an adaptation protocol defined in the DVB standards.

On the level of MPE, an additional stage of Forward Error Correction (FEC) is added. This scheme is used to improve the robustness of the information stream.

The technique, called the Multi-Protocol Encapsulation—Forward Error Correction (MPE-FEC) is the main innovation of DVB-H. MPE-FEC complements the physical layer FEC of the underlying DVB-T standard. It is intended to reduce the signal-to-noise S/N requirements for reception by a handheld device. The MPE-FEC processing is located on the link layer at the level of the IP input streams before they are encapsulated by means of the MPE. The MPE-FEC, the MPE, and the time division multiplexing technique were defined jointly and directly aligned with each other [15].

All three elements together form the DVB-H codec which contains the essential DVB-H functionality as shown in Fig. 1. The IP input streams provided by different sources as individual elementary streams are multiplexed according to the time division multiplexing method. The MPE-FEC error protection is calculated separately for each individual elementary stream. Afterwards encapsulation of IP packets and embedding into the transport stream follow. All relevant data processing is carried out before the transport stream interface in order to guarantee compatibility to a DVB-T transmission network. All three elements together form the DVB-H codec which contains the essential DVB-H functionality.

The new MPE-FEC scheme contains a RS code in conjunction with a block interleaver [12, 15]. Interleaving is a technique to arrange data in a non-contiguous way to increase performance: the MPE-FEC encoder creates a specific frame structure, the FEC frame, incorporating the incoming data of the DVB-H codec as demonstrated in Fig. 2.

The FEC frame consists of a maximum of 1024 rows and a constant number of 255 columns; every frame cell corresponds to one byte. The frame is separated into two parts, the application data table on the left (191 columns) and the RS data table

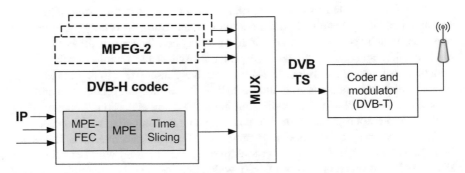

Fig. 1 Diagram of DVB-H codec and transmitter

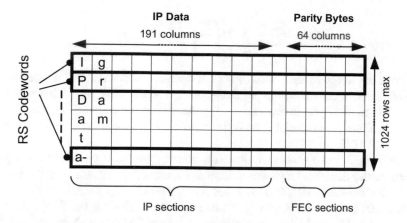

Fig. 2 MPE–FEC frame structure

on the right (64 columns). The application data table is filled with the IP packets of the service to be transported. After applying the RS (255,191) code to the application data row-by row, the RS data table contains the parity bytes of the RS code.

After the coding, the IP packets are read out of the application data table and are encapsulated in IP sections in a way which is well known from the MPE method. These application data are followed by the parity data which are read out of the RS data table column-by-column and are encapsulated in separate FEC sections. The FEC frame structure also contains a "virtual" block interleaving effect in addition to the coding.

6 Simulation and Performance Analysis

To evaluate the proposed RS codes, the authors have performed simulations with the use of MATLAB. The figures below show the input Symbol Error Rate (SER) versus the output SER for the RS code used for the MPE-FEC block in DVB-H. The code takes 191 bytes, adds 64 bytes of redundancy and by using the Cyclic Redundancy Check (CRC) in the MPE-FEC packet header, it allows verifying the correctness of the MPE-FEC packet contents, identifying a packet as unreliable if the CRC check fails.

As soon as the MPE-FEC packet contents are unreliable, the bytes in the packet are termed "erasure" symbols, and using erasures allows the decoder to correct twice the number of bytes that could be corrected when erasures were not used. Currently, erasures are used, and this code can correct up to 64 bytes out of the 255-byte codeword. The Eq. (12) that used to calculate the simulation graphs is as follows:

$$P_E \approx \frac{1}{2^m - 1} \sum_{j=t+1}^{2^m-1} j\left(\frac{2^m - 1}{j}\right) p^j (1 - p)^{2^m - 1 - j} \tag{12}$$

which is the weighted-sum of the binomial distribution function for all possibilities where the code cannot correct all the errors, i.e. when there are 65 bytes in error (t + 1) up to 255 bytes in error ($2^m - 1$), where P_E is the output SER, m = 8 is the number of bits per symbol, t = 64 is the number of correctable symbols and p is the input SER [12].

The Figs. 3 and 4 show only the SER transfer function (the output for a given input) of the decoder, they demonstrate the astounding strength of RS coding—especially the (255, 191, 64) code for the MPE-FEC that used in DVB-H. The Figs. 5 and 6 show the output SER versus the input SER for the 204, 188, 8 RS code used in Digital Video Broadcasting—Terrestrial/Handheld/Satellite (DVB-T/H/S) and Digital Multimedia Broadcasting (DMB). The code takes 188 bytes, encodes them by adding 16 parity bytes, does not use erasures, and can correct any 8 bytes of the 204-byte codeword.

To sum up, the authors must note that for DVB-H block any 64 bytes can be corrected, regardless whether errors exist in one bit or all eight bits. The figures received by the simulations confirm the symbol error rate (one symbol in the case of RS codes used for Digital Video Broadcasting (DVB) and Digital Multimedia Broadcasting (DMB) consist of one byte of data), as opposed to the Bit Error Rate (BER).

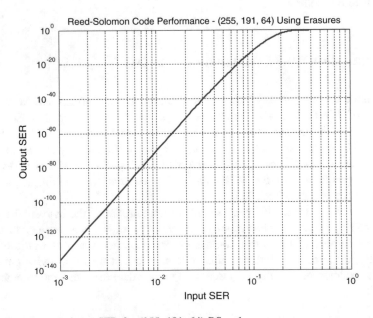

Fig. 3 Output versus input SER for (255, 191, 64) RS code

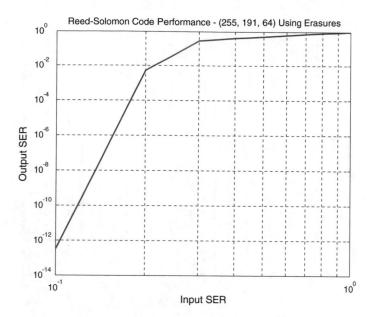

Fig. 4 Output versus input SER for (255, 191, 64) RS code at High SER

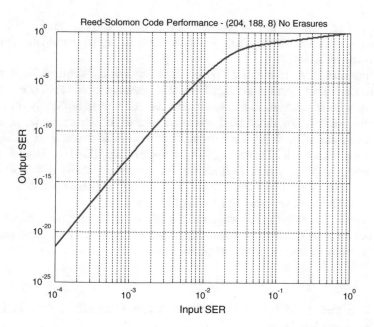

Fig. 5 Output versus input SER for (204, 188, 8) RS code

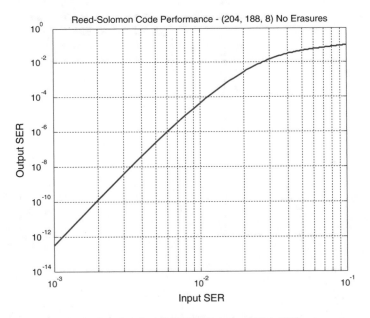

Fig. 6 Output versus input SER for (204, 188, 8) RS code at High SER

Also, the results are calculated from an Eq. (12) that deals with uniformly-distributed errors. The errors that occur in digital communications typically occur in bursts.

On the one hand, if bursts of errors exist, the peak short-term BER will be significantly higher than the average BER. This means that the RS code would be more likely to make errors than in case when the errors were uniformly-distributed.

On the other hand, these high short-term bursts of errors create significant problems to use of the Viterbi algorithm for the inner convolutional coding. The interleaving, used in Digital Audio Broadcasting (DAB), Digital Video Broadcasting (DVB), Digital Multimedia Broadcasting (DMB) and other digital communication systems in a smart world, is an attempt to make the errors to appear as much uniformly-distributed as possible [11, 12, 23–25].

In combination with the RS codes and the applied error correction, this is shown to provide in most situations a reduced residual bit error rate, less retransmissions, and as a result an improved overall performance [15].

7 The Process of Extrinsic Information Transfer in Digital Communications

In a smart world, the process of extrinsic information transfer in digital communications can be described with analysis of iterative turbo decoder.

The block diagram of an iterative turbo decoder is shown in Fig. 7, where each APP decoder corresponds to a constituent encoder and generates the corresponding extrinsic information $\Lambda_{e,r}^{(m)}$ for $m = 1$, 2, using the corresponding received sequences.

The interleavers are identical to the turbo encoder's interleavers, and they are used to reorder the sequences so that the sequences at each decoder are properly aligned. The algorithm is iterated several times through the two decoders; each time the constituent decoder uses the currently calculated a posteriori probability as input.

The a posteriori probabilities produced by the first decoder are shown as (13),

$$L(u_r) = \Lambda_{e,r}^{(2)} + \Lambda_{e,r}^{(1)} + \Lambda_s. \tag{13}$$

where (14) is the a posteriori probability of the systematic bits, which are conditionally independently distributed:

$$\Lambda_s = \log \frac{P\left(y_r^{(0)} | u_r = 1\right)}{P\left(y_r^{(0)} | u_r = 0\right)}. \tag{14}$$

Direct use of (13) would lead to an accumulation of "old" extrinsic information by calculating the a posteriori probabilities given as (15):

$$L_r^{(1)'} = \Lambda_{e,r}^{(1)'} + \underbrace{\Lambda_{e,r}^{(2)} + \Lambda_{e,r}^{(1)} + \Lambda_s}_{L_r^{(1)}}. \tag{15}$$

As a result, the decoders are constrained to exchange extrinsic information only, which is achieved by subtracting the input values to the APP decoders from the output values as indicated in Fig. 7.

The extrinsic information is a reliability measure of each constituent decoder's estimate of the transmitted information symbols based on the corresponding

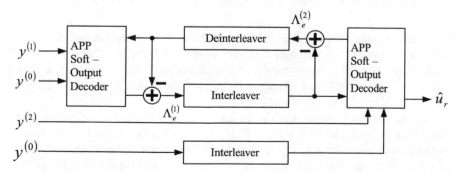

Fig. 7 Block diagram of turbo decoder with two constituent decoders

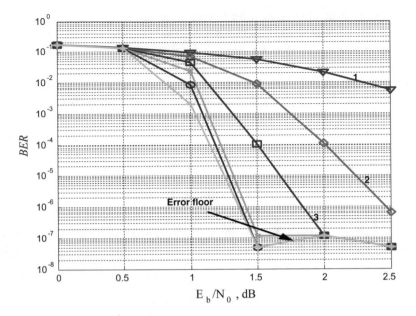

Fig. 8 Performance of the (37, 21, 65536) turbo code as function of decoder iterations

received constituent parity sequence only. Since each constituent decoder uses the received systematic sequence directly, the extrinsic information allows the decoders to share information without biasing.

The effectiveness of this technique can be seen in Fig. 8, which shows the performance of the original (37, 21, 65536) turbo code as a function of the decoder iterations. It is remarkable that the performance of the code with iterative decoding continues to improve with increasing number iterations, but frequently, after 6 iterations a sufficient convergence has already been reached.

8   Constituent Decoder as Extrinsic LLR Transformer

In digital communications, the distance spectrum and the minimum distance of turbo codes can explain the error floor and contribute to the threshold behavior of these codes. The behavior of turbo codes at the onset of the turbo cliff is better understood by a statistical analysis, called the Extrinsic Information Transfer (EXIT) analysis [4, 10, 12, 16, 17, 20, 21].

The extrinsic information transfer analysis is related to similar methods that are based on the transfer of variances [6, 7, 11, 12, 19]. The basic philosophy of these methods is to view the constituent decoders of the turbo decoder, as a statistical processor that transforms an input value—that is, the extrinsic Log Likelihood

Ratio (LLR) of the information symbols—into an output value, the recomputed extrinsic LLR, as is illustrated in Fig. 7.

The constituent decoder is after seen as a nonlinear LLR transformer, as is shown in Fig. 9. The input LLR is denoted by $\Lambda_e^{(A)}$ since it takes on the function of a priori probability, and the output extrinsic LLR is denoted by $\Lambda_e^{(E)}$.

Evidently, the authors expect the constituent decoder to improve the extrinsic LLR in the course of the iterations, so that $\Lambda_e^{(E)}$ is better than $\Lambda_e^{(A)}$, in some sense, since otherwise the iterations will lead nowhere. Measurements of $\Lambda_e^{(m)}$ show that it has an approximately Gaussian distribution, presented in (16) and (17),

$$\Lambda_e^{(m)} = \mu_\lambda u + n_\lambda. \tag{16}$$

$$n_\lambda \approx N\left(0, \sigma_\lambda^2\right). \tag{17}$$

where $u \in \{-1, 1\}$ is the information bit whose LLR is expressed by $\Lambda_e^{(m)}$, μ_λ is the mean, and n_λ is a zero-mean independent Gaussian random variable with variance σ_λ^2.

One of the main questions is how to measure the reliability of $\Lambda_e^{(m)}$. The mutual information $I\left(u, \Lambda_e^{(m)}\right)$ between $\Lambda_e^{(m)}$ and u has proven to be the most accurate and convenient measure, which has the following advantages, defined by means of (18), (19) and (20):

- The measure is bounded

$$0 \le I\left(u, \Lambda_e^{(m)}\right) \le 1; \tag{18}$$

- The upper limit indicates high reliability that is, if $\Lambda_e^{(m)}$ has variance σ_λ^2 then

$$\sigma_\lambda^2 \to 0; \tag{19}$$

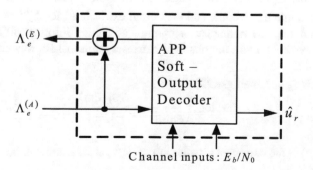

Channel inputs : E_b/N_0

Fig. 9 Constituent decoder considered as extrinsic LLR transformer

- The lower bound indicates low reliability, and

$$\sigma_\lambda^2 \to \infty; \tag{20}$$

- The measure is monotonically increasing.

In order to capture the input–output behavior of the constituent decoder, simulated extrinsic values $\Lambda_e^{(A)}$ are generated at its input, according to the Gaussian distribution (16) with independent realizations. This independence models the effect of the interleaver, which, ideally, destroys any dependence between successive LLR values.

In the case of Gaussian model input LLR values, the mutual information measure, in bits, can be calculated as (21):

$$I_A = I\left(u; \Lambda_e^{(A)}\right) = \frac{1}{\sqrt{2\pi}\sigma_\lambda} \int_{-\infty}^{\infty} \exp\left(-\frac{(\lambda - \mu_\lambda)^2}{2\sigma_\lambda^2}\right)(1 - \log_2[1 + \exp(-\lambda)])d\lambda. \tag{21}$$

The mutual information of the output extrinsic LLR and the information bit u is more complex to evaluate since $\Lambda_e^{(E)}$ is not exactly Gaussian. Additionally, the values of $\Lambda_e^{(E)}$ corresponding to $\Lambda_e^{(A)}$ must be found when simulating the constituent decoder. Other than in very simple cases, such as the rate $R = 1/2$ repetition code, there is neither closed form nor even analytical formulas for the output extrinsic mutual information that exist to date.

Then, the output extrinsic information I_E is an empirical function of the constituent decoder, the input I_A and the channel signal-to-noise ratio at which the decoder operates, formally provided by the extrinsic information transfer function (22):

$$I_E = T(I_A, E_b/N_0). \tag{22}$$

To evaluate this function the following steps were executed. At the first step a Gaussian model input LLR with mutual extrinsic information value I_A was generated and sent into the decoder and the output extrinsic LLRs $\Lambda_e^{(E)}$ were achieved. At the second step the numerical estimation of mutual information between the information symbols u and the obtained extrinsic output LLRs was completed.

$\Lambda_e^{(E)}$ yields I_E specified as (23):

$$I_E = \frac{1}{2} \sum_{u \pm 1} \int_{-\infty}^{\infty} p_E(\xi|u) \log_2\left(\frac{2p_E(\xi|u)}{p_E(\xi|U = -1) + p_E(\xi|U = 1)}\right) d\xi. \tag{23}$$

where $p_E(\xi|u)$ is the empirical distribution of $\Lambda_e^{(E)}$, as measured at the output of the APP decoder.

9 Numerical Results

The recommended method to evaluate the output extrinsic information is demonstrated in Fig. 10. The extrinsic information transfer (EXIT) function $T(I_A, E_b/N_0)$ is considered for a memory 4, rate $R = 1/2$ constituent convolutional code with $h_0(D)$ and $h_1(D)$ given as (24) and (25), respectively:

$$h_0(D) = D^4 + D^3 + 1. \tag{24}$$

$$h_1(D) = D^4 + D^3 + D^2 + D + 1. \tag{25}$$

The numbers on the trajectories are correspondent to the E_b/N_0 calculated in dB. In a full turbo decoder this extrinsic information is exchanged between the decoders, and the output extrinsic LLRs become input extrinsic LLRs for the next decoder. In suggested approach these iterations are captured by a sequence of applications of constituent decoder EXIT functions, presented as (26),

$$0 \xrightarrow{T_1} I_E^{(1)} = I_A^{(2)} \xrightarrow{T_2} I_E^{(2)} = I_A^{(2)} \xrightarrow{T_1} I_E^{(1)} = I_A^{(2)} \xrightarrow{T_2} \dots, \tag{26}$$

where T_1 is the extrinsic information transfer (EXIT) function of decoder 1 and T_2 is that of decoder 2.

The behavior of this exchange can be visualized in the extrinsic information transfer chart, where the extrinsic information transfer functions of both constituent decoders are plotted as shown in Fig. 11.

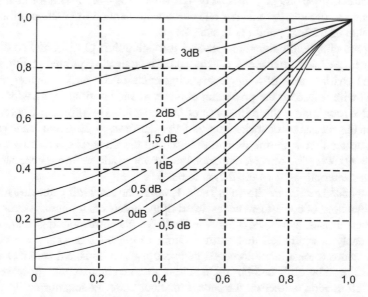

Fig. 10 EXIT function of a rate 1/2, 16-state convolutional code

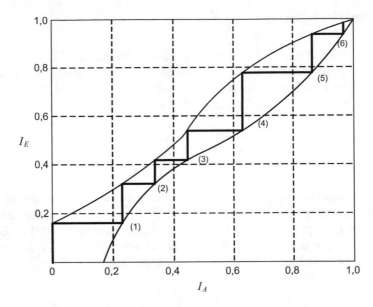

Fig. 11 EXIT chart combining EXIT functions of two 16-state decoders involved in turbo decoder for $E_b/N_0 = 0.8$ dB

Since the iterations start with zero LLRs—that is, with zero mutual information at point (0, 0) in the chart—the iterations progress by bouncing from one curve to the other as illustrated in Fig. 11. The first iteration through decoder 1 takes an input value of $I_A= 0$ and produces an output value of $I_E= 0.18$. This value then acts as the input to decoder 2, and after one completed iteration the mutual information has a value of 0.22, at point (1) in the plot.

Correspondingly, the iterations proceed through points (2), (3), and so on, until they reach $I_A= I_E= 1$ after about 6 iterations. The trajectory plotted in Fig. 11 is a measured characteristic of a single decoding cycle of the turbo decoder, and the accuracy with respect to the prediction of the extrinsic information transfer chart is impressive, confirming the robustness of this method. The growing discrepancies between the measured characteristic and the extrinsic information transfer chart starting at iteration 7 are the outcomes based on the finite-size interleaver, in this case $N = 60,000$. The selected superior interleaver produces more accurate EXIT charts that later are used as prediction tools.

The obtained results are shown in Table 1, where the pinch-off values are given for all combinations of constituent codes. Most of the processed iterations occur at very high bit-error rates, that is, $P_b > 10^{-2}$, in fact $I_E \approx 0.9$ corresponding to $P_b = 10^{-2}$.

The analysis confirmed that extrinsic information transfer charts are effective tools to evaluate constituent decoders, their cooperative statistical behavior such as onset of the turbo cliff, number of iterations to convergence, etc. However, they deliver information neither on the code error floor performance, nor on its ultimate operational performance.

Table 1 Pinch-off signal-to-noise ratios for various combinations of constituent codes

v	C_1/C_2	(3, 2) (dB)	(7, 5) (dB)	(13, 15) (dB)	(23, 37) (dB)	(67, 45) (dB)	(147, 117) (dB)
1	(3, 2)	>2					
2	(7, 5)	1.49	0.69				
3	(13, 15)	1.14	0.62	0.62			
4	(23, 37)	1.08	0.65	0.64	0.68		
5	(67, 45)	0.86	0.62	0.66	0.70	0.77	
6	(147, 117)	0.84	0.63	0.67	0.72	0.81	0.84

In order to demonstrate the use of extrinsic information transfer charts to predict the start of turbo cliff behavior, the signal-to-noise ratio E_b/N_0 for the codes in Fig. 11 was reduced until the open channel was closed at $E_b/N_0 = 0.53$ dB.

This suggests the pinch-off signal-to-noise ratio of the original turbo code combinations for encoders. The sharpness of this pinch-off signal-to-noise ratio depends on the size of the interleaver, as seen in Table 1, since finite-size interleavers always leave some residual correlation, in particular for larger iteration numbers, which violates the assumption of independence used to generate the extrinsic information transfer functions of the constituent decoders.

10   Conclusion

This chapter describes digital communications and a smart world. In a smart world, digital communications support a wide range of multimedia applications, such as audio, video and computer data that differ significantly in their traffic characteristics and performance requirements. In digital communications, one of the most important parameters to evaluate the channel coding methods is the Bit-Error Rate (BER) versus E_b/N_0, where E_b denotes the energy per binary information symbol, and N_0 denotes the noise power spectral density, that is the probability for bit error at a given ratio E_b/N_0. In digital communications, during transmission, the digital data (e.g., the encoded signal) is affected by different factors, including interference, fading, and channel noise, which distort the signal.

Normally, the BER charts of iterative decoding schemes are divided into three regions: the region of low with negligible iterative BER reduction; the turbo cliff region with persistent iterative BER reduction over many iterations; and the BER floor region for moderate to high, in which a rather low BER can be reached after just a few number of iterations.

If the receiver is to reach a given BER (that is a residual bit-error rate as the bit-error rate after enhancement, error detection and recovery is considered), for example, 10^{-5}, then it is necessary either to increase the transmitter power, or to use proper methods for channel coding. As of all practical error correction methods known to date, turbo codes and Low-Density Parity-Check (LDPC) codes come closest to approaching the Shannon limit ($\approx$0.5 dB), the theoretical limit of maximum information transfer rate over a noisy channel [3, 11, 17].

This is possible through iterative decoding process, which allows correction of larger number of error bits during the transmission in the channel, but also increases decoding time. After 18 iterations, a BER of 10^{-5} is reached at an E_b/N_0 ratio at only 0.5 dB from the theoretical limit predicted by the Shannon capacity.

A new analytical approach to analysis and optimization of the convergence behavior of iterative systems utilizing the turbo principle, i.e., systems exchanging and refining extrinsic information, with Extrinsic Information Transfer (EXIT) charts was proposed. The suggested approach didn't rely on extensive simulations.

The implementation of the suggested approach for determining the pinch-off signal-to-noise ratios for the turbo codes, using the constituent codes and their extrinsic information transfer functions was demonstrated. Compared with the other known methods for generating extrinsic information transfer functions, the proposed analytical method provided insight into the iterative behavior of linear turbo systems with substantial reduction in numerical complexity.

The authors considered results presented in Table 1 as pinch-off values for all possible combinations of constituent codes. As can be seen in the numerical examples, the use of larger codes, e.g. v > 4 did not produce stronger turbo codes since the pinch-off was not improved. A combination of two strong codes provided a worse pinch-off value when to compare with one of the best combinations that had a pinch-off signal-to-noise ratio of 0.53 dB, as shown in Table 1. That has demonstrated the likelihood for a superior choice of constituent codes with v = 4, (37, 21).

Acknowledgements The authors would like to thank Cees J. M. Lanting from DATSA Belgium, who contributed his time, insights and review, and kindly shared his views on ongoing development during the preparation of this chapter. In addition, the authors wish to extend their gratification to the anonymous reviewers for their time, assistance, and patience.

References

1. Auer L, Kryvinska N, Strauss C (2009) Service-oriented mobility architecture provides highly-configurable wireless services. IEEE, pp 1–1
2. Bashah NSK, Kryvinska N, Do TV (2012) Quality-driven service discovery techniques for open mobile environments and their business applications. J Serv Sci Res 4(1):71–96
3. Berrou C, Glavieux A, Thitimajshima P (1993) Near Shannon limit error-correcting coding and decoding: turbo-codes. In: Proceedings of the IEEE international conference on communications (ICC), vol 2, Geneva, Switzerland, pp 1064–1070
4. Brink S (2000) Design of serially concatenated codes based on iterative decoding convergence. In: Proceedings of the second international symposium on turbo codes and related topics, Brest, France, 2000, pp 319–322
5. Brink S (2001) Convergence behavior of iteratively decoded parallel concatenated codes. IEEE Trans Commun 49(10):1727–1737
6. Clevorn T, Godtmann S, Vary P (2006) BER prediction using EXIT charts for BICM with iterative decoding. IEEE Commun Lett 10(1):49–51

7. Colavolpe G, Ferrari G, Raheli R (2001) Extrinsic information in iterative decoding: a unified view. IEEE Trans Commun 49(12):2088–2094
8. Divsalar D, Dolinar S, Pollara P (2000) Serial turbo trellis coded modulation with rate-1 inner code. In: Proceedings of the international symposium on information theory, IEEE Globecom, San Francisco, CA, USA, pp 194–200
9. ETSI TS 136 212 LTE (2013) Evolved universal terrestrial radio access (E-UTRA); multiplexing channel coding, ETSI, Route des Lucioles, France
10. Hanzo LL, Liew TH, Yeap BL, Tee RYS, Ng SX (2010) Turbo coding, turbo equalisation and space-time coding: EXIT-chart-aided near-capacity designs for wireless channels. Wiley, New York, NY, USA
11. Iliev T, Lokshina I, Radev D (2009) Use of extrinsic information transfer chart to predict behavior of turbo codes. In: Proceedings of IEEE wireless telecommunications symposium (WTS) 2009, Prague, Czech Republic, pp 1–4
12. Iliev T (2007) Analysis and design of combined interleaver for turbo codes. In: Proceedings of the MOCM—13, vol 1, Bacau, Romania, pp 148–153
13. Kryvinska N, Auer L, Zinterhof P, Strauss C (2008) Architectural model of enterprise multiservice network maintaining mobility. IEEE, pp 1–22
14. Kryvinska N, Strauss C, Collini-Nocker B, Zinterhof P (2011) Enterprise network maintaining mobility—architectural model of services delivery. Int J Pervasive Comput Commun 7:114–131. https://doi.org/10.1108/17427371111146419
15. Lokshina I (2011) Application of Reed-Solomon codes in mobile digital video broadcasting systems. Int J Mobile Network Des Innov 3(3):204–208
16. Lokshina I (2011) Application of extrinsic information transfer charts to anticipate turbo code behavior. Int J Interdisc Telecommun Netw 3(2):31–37
17. Lokshina I, Zhong H (2017) Analysis of turbo code behavior with extrinsic information transfer charts in high-speed wireless data services. Int J Interdisc Telecommun Netw 9(4):26–36
18. Maunder RG (2015) A fully-parallel turbo decoding algorithm. IEEE Trans Commun 63 (8):2762–2775
19. Maunder RG, Wang J, Ng SX, Yang L-L, Hanzo L (2008) On the performance and complexity of irregular variable length codes for near capacity joint source and channel coding. IEEE Trans Wireless Commun 7(4):1338–1347
20. Ngo HA, Maunder RG, Hanzo L (2015) Extrinsic information transfer charts for characterizing the iterative decoding convergence of fully parallel turbo decoders. IEEE Access 3:2100–2110
21. Richardson T (2000) The geometry of turbo decoding dynamics. IEEE Trans Inf Theory 46:9–23
22. Shannon CE (1948) A mathematical theory of communication. Bell Labs Tech J 27(3):379–423
23. Sklar B (2002) Digital communications, Prentice Hall PTR, New Jersey, 2002
24. Truong TK, Jeng JH, Reed IS (2001) Fast algorithm for computing the roots of error locator polynomials up to degree 11 in Reed-Solomon decoders. IEEE Trans Commun 49:779–783
25. Wicker SB (1995) Error control systems for digital communication and storage. Prentice-Hall, Englewood Cliffs, NJ, USA

Izabella Lokshina is Professor of MIS and chair of Management, Marketing and Information Systems Department at SUNY Oneonta, USA. Her main research interests are Intelligent Information Systems, Communication Networks, Complex System Modeling and Simulation.

Hua Zhong is Associate Professor of Operations Management in Management, Marketing and Information Systems Department at SUNY Oneonta, USA. His main research interests are Supply Chain Modeling, Scheduling, Queuing Systems and Simulation.

A Qualitative Evaluation of IoT-Driven eHealth: Knowledge Management, Business Models and Opportunities, Deployment and Evolution

Izabella Lokshina and Cees Lanting

Abstract eHealth has a major potential, and its adoption may be considered necessary to achieve increased ambulant and remote medical care, increased quality, reduced personnel needs, and reduced costs potential in healthcare. In this chapter, the authors try to give a reasonable, qualitative evaluation of IoT-driven eHealth from theoretical and practical viewpoints. They look at associated knowledge management issues and contributions of IoT to eHealth, along with requirements, benefits, limitations and entry barriers. Important attention is given to security and privacy issues. Finally, the conditions for business plans and accompanying value chains are realistically analyzed. The resulting implementation issues and required commitments are also discussed. The authors confirm that IoT-driven eHealth can happen and will happen; however, much more needs to be addressed to bring it back in sync with medical and general technological developments in an industrial state-of-the-art perspective and to recognize and get timely the benefits.

Keywords IoT-driven eHealth · Knowledge management · Business models
Potential opportunities · Requirements · Limitations · Entry barriers
Security · Privacy · Deployment · Evolution

1 Introduction

There are high expectations for eHealth as a major tool to achieve the following improvements in healthcare:

- A further shift from clinical to ambulant treatment.

I. Lokshina (✉)
Management, Marketing and Information Systems,
SUNY Oneonta, Oneonta, NY 13820, USA
e-mail: Izabella.Lokshina@oneonta.edu

C. Lanting
Consulting Services, DATSA Belgium, VBR, 3010 Leuven, Belgium
e-mail: Cees.Lanting@datsaconsulting.com

© Springer International Publishing AG, part of Springer Nature 2019
N. Kryvinska and M. Greguš (eds.), *Data-Centric Business and Applications*,
Lecture Notes on Data Engineering and Communications Technologies 20,
https://doi.org/10.1007/978-3-319-94117-2_2

- Reductions in the per user/patient workload of medical and care staff.
- Improvements in the quality of medical and care services for users/patients.
- And finally, significant reductions in the medical treatment and care cost per user/patient.

The attention, and hype, around the Internet of Things (IoT), and, in particular, IoT-driven eHealth, has further increased the visibility and expectation of eHealth [26–28].

In this chapter, the authors make an effort to give a reasonable, qualitative evaluation of what can be expected of IoT in eHealth and IoT-driven eHealth itself. They look at the possible contributions of IoT to eHealth, the requirements that need to be met, the benefits and limitations of eHealth, and the entry barriers [19, 25]. Important attention is given to security and privacy, representing an important set of issues [13, 47].

However, the authors conclude that these are not the first issues to be addressed: first there needs to be a joint understanding between the users/patients and health and care providers that there are benefits for both the users/patients and health and care providers in applying eHealth [39]. The conditions for business plans and accompanying value chains are realistically analyzed, and the resulting implementation issues and commitments are discussed [35, 43]. As a result, the chapter contributes to the literature by reviewing, innovatively, business models, strategic implications and opportunities for IoT-driven eHealth, as well as its deployment and evolution.

This chapter is comprised of ten sections and is organized as follows. Section two provides a theoretical view on the IoT-driven eHealth in the context of knowledge management and considers evolution of knowledge-based management practices in healthcare. Section three focuses on contributions of IoT to eHealth, considering IoT as enabler and discussing IoT-based medical-relevant eHealth systems. Section four provides an analysis of requirements for IoT-driven eHealth. Section five considers the limitations of eHealth. Section six defines the entry barriers. Section seven outlines security and privacy issues; however, it confirms these issues are not the first topics to be addressed, but instead, the benefits of applying eHealth. Section eight analyzes the conditions for business plans and accompanying value chains and calls attention to the associated implementation issues and commitments. Section nine contains summary and conclusions, following by references.

2 Theoretical View on IoT-Driven eHealth in a Context of Knowledge Management

2.1 Views on eHealth

Everybody talks about eHealth these days, but few people have come up with a clear definition of this term. The term was apparently first used by industry leaders and marketing people rather than academics, and they used this term in line with other "e"-words such as eCommerce, eBusiness, eTrade and so on.

So, how can the authors define eHealth in the academic environment? It seems quite clear that eHealth encompasses more than a technological development.

The authors can define the term and the notion as follows: eHealth is an emerging field in the intersection of medical informatics, public health and business, referring to health services and information delivered or enhanced through the communication technology, i.e., the Internet, and related technologies. In a broader sense, the term characterizes not only a technical development, but also a state-of-mind, a way of thinking, an attitude, and a commitment for networked, global thinking, to improve health care locally, regionally, and worldwide by using information and communication technology. As such, the "e" in eHealth does not only stand for "electronic", but implies a number of other "e's,", which together, perhaps, best describe what eHealth is all about, or what it should be [11].

2.2 Views on IoT

IoT is a system that relies on autonomous communication of groups of physical objects. IoT, in context of the digital revolution, is an emerging global communications/Internet-based information architecture facilitating the exchange of knowledge, services and goods [5, 7, 8, 30]. The authors expect that main domains of IoT will be transportation and logistics; healthcare; smart environment (home, office and plant, integrated in the environment); and personal and social area [10, 31, 38, 46].

In Table 1 the authors consider realms of ubiquitous society. This entity is called the multiversity. Table 1 suggests that leaders, managers and planners must understand the fundamental nature of three elements of reality: time, space and matter.

Table 1 Realms in the ubiquitous society and in the multiverse

Variables			Realm
1. Time	Space	Matter	Reality
2. Time	Space	No-matter	Augmented reality
3. Time	No-space	Matter	Physical reality
4. Time	No-space	No-matter	Mirrored reality
5. No-time	Space	Matter	Warped reality
6. No-time	Space	No-matter	Alternative reality
7. No-time	No-space	Matter	Augmented virtuality
8. No-time	No-space	No-matter	Virtuality

The new service designs, architectures and business models are needed in the multiverse, not only in the universe. What is obvious is that managers must work in order to manage these critical eight realms of the ubiquitous society. The applications of IoT are numerous, basically meaning smart things and smart systems such as smart homes, smart cities, smart industrial automation and smart services. IoT systems provide better productivity, efficiency and better quality to numerous service providers and industries.

IoT is based on social, cultural and economic trust and associated trust management skills, which broadly speaking mean developed security services and antifragility operations. Critical issues of the IoT security field are trusted platforms, low-complexity, encryption, access control, secure data, provenance, data confidentiality, authentication, identity management, and privacy-respecting security technologies. Security of IoT requires data confidentiality, privacy and trust. These security issues are managed by distributed intelligence, distributed systems, smart computing and communication identification systems.

Finally, key systems of global economy are markets, networks and crowds. IoT can be found among these key systems of global economy. Probably, there is a lot of potential for smartness between these key systems. Data, information and knowledge about communication and interaction of these systems are vital issues for the future of management.

Especially the Internet of Intelligent Things (IoIT), defined by experts as smart Machine-to-Machine (M2M) communication, provides much potential for crowdsourcing of markets and networks. IoIT provides also much potential for smart networking (between markets and networks and between various networks).

The authors expect that one obvious consequence of IoIT will be a broader scope of deliberate democracy. Additionally, the legal framework of IoT/IoIT is still considered rather vague, or absent in a certain sense. Such issues like standardization, service design architecture, service design models, data privacy and data security create management and governance problems, which are not, or at least not completely solved inside current service architectures. IoT has also become subject to power politics because of risks of cyber war, cyber terror and cyber criminality.

Finally, the authors can see that IoT will be central for the collection of raw Big Data, captured from the environment, human beings and robots and AI applications.

2.3 Views on IoT and Big Data in a Context of Knowledge Management

The Data-Information-Knowledge-Wisdom (DIKW) model is an often-used method, with roots in knowledge management, to explain the ways to move from data to information, knowledge and wisdom with a component of actions and decisions. Simply put, it is a model to look at various ways of extracting insights and value from all sorts of data, big, small, smart, fast and slow. It is often depicted as a hierarchical model in the shape of a pyramid and known as the data-information-knowledge-wisdom hierarchy, among others [2, 9, 16, 40].

Ackoff had originally defined the traditional DIKW model as the following [2]:

- Data is the result of a relatively accurate observation, and it may or may not be inspired by a problem to be solved. Data comprises objective facts, signs and numbers, and it does not need relationships with other elements to exist, but if to take each data individually, it does not communicate anything and does not contain any meaning. Data is something perceived by the senses (or sensors) but it has no intrinsic value until it is put in a context. Data becomes information only when it is placed in context, through contextualization (in fact), categorization, processing, correction and synthesis.
- Information, deduced from the data, includes all data, giving them meaning and gaining added value compared to the data. Information is the choice to put some data in a context, fixing some as premises, and making a series of inferences, then drawing conclusions. These conclusions are called information but do not become knowledge if they are not related to the knowledge and experience of a specific person.
- Knowledge is the combination of data and information, to which is added the opinion of expert persons, competence and experience, to build a valuable asset that can be used to aid decision-making. Knowledge cannot be lost in the same way in which one can lose data and information. In the domain of competence, the more to move from data to knowledge, the greater is the dependence on the context. Davenport and Prusak had offered the following definition [9]: "Knowledge is a fluid mix of framed experience, values, contextual information, and expert insight that provides a framework for evaluating and incorporating new experiences and information. It originates and is applied in the minds of knowers. In organizations, it often becomes embedded not only in documents or repositories but also in organizational routines, processes, practices, and norm". Knowledge is always individual and cannot be transmitted because it is generated from the individual's previous experience and knowledge; what one can transmit is only the narration of the experience.
- Wisdom is immaterial, intangible. Wisdom is the judgement, the ability to add value and is unique and personal. Wisdom is something that goes beyond the concepts of information and knowledge and embraces both, assimilating and transforming these into individual experience. Wisdom accompanies knowledge and allows to make the best choices.

The traditional DIKW model is an attempt to categorize and simplify the key concepts involved in cognitive processes, especially when there is a need to manage large amounts of data. This theoretical model provides a hierarchy, consisting of a very large base of raw data, which, going towards the top of the pyramid, is subject to an aggregation–contextualization process, i.e., information, and application testing, i.e., knowledge. On top of the pyramid, as shown in Fig. 1, is confined wisdom, which assumes a level of knowledge that is beyond the scope of a specific

I. Lokshina and C. Lanting

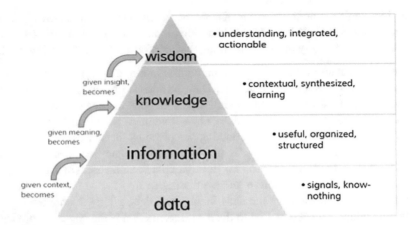

Fig. 1 The traditional DIKW pyramid

application. These cognitive states are then connected in a hierarchical manner, assuming that between them there can be a smooth transition from the bottom to the top.

Besides its presentation as a pyramid, there is an effective representation of the traditional DIKW model on a Cartesian plane, as shown in Fig. 2. Ackoff had originally indicated only one axis, the understanding one, but Rowley later showed that the size of the context or connection is also important [2, 40].

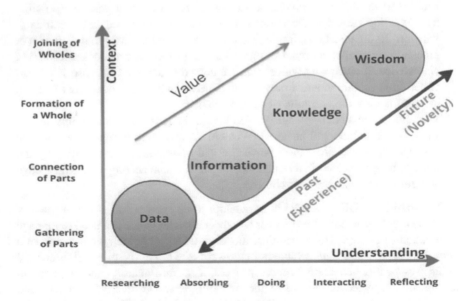

Fig. 2 The traditional DIKW model on a Cartesian plane

Figure 2 highlights the rising value, from data to wisdom. In Fig. 2, the first three categories refer to the past; they deal with what has been or what was known. Only the fourth category, wisdom, deals with the future because it incorporates vision and design. With wisdom, one can imagine the future and not just grasp the present and the past. But achieving wisdom is not easy; one must go through all the other steps/categories [40].

As in the case with all models, the traditional DIKW model has its limits. The authors suggest the model is quite linear and expresses a logical consequence of steps and stages with information being a contextualized "progression" of data as it gets more meaning. Reality is often a bit different. Knowledge, for instance, is much more than just a next stage of information. Nevertheless, the traditional DIKW model is still used in many forms and shapes to look at the extraction of value and meaning of data and information.

One of the main criticisms of the traditional DIKW model is that it is hierarchical and misses several crucial aspects of knowledge and the new data and information reality in this age of IoT, Big Data, APIs and ever more unstructured data and ways to capture them and turn them into decisions and actions, sometimes bypassing the steps in the DIKW model, as in, for instance, self-learning systems [13]. The data must be of a certain type to really add value to an organization [29]. Big Data does not necessarily mean more information: the belief, rather widespread, that more data = more information does not always correspond to reality [42]. Among Big Data, there are obviously interpretable data and data that cannot be interpreted, sometimes because of lacking metadata or place/time references [32, 44]. Among the interpretable data, there are relevant data, i.e., the signal, and irrelevant data, i.e. noise, for our aims. Relevance is a characteristic of data, not only subjective, i.e., what for one is the signal could be noise to another; but also, contextual, i.e., what may be relevant depends on the context to be analyzed.

So, a criterion to decide whether it makes sense to think of an analysis based on Big Data would be to think about the interpretability, relevance and whether the process could extract really new information from the mass of data. However, the essence still stays the same: looking at what to do with data lakes and turning data through Big Data analytics into decisions and actions, as shown in Fig. 3.

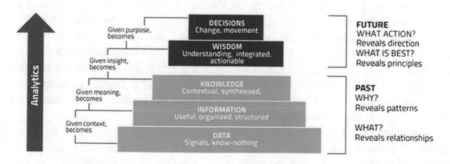

Fig. 3 What matters: actions and decisions in the DIKW model

The traditional DIKW model, as all models or ways of looking at things in a more or less structured way, has been discussed and looked upon from various angles with some suggesting to omit wisdom, others debating the exact definitions and the relationships between them and a few telling to add a dimension of truth and moral sense to it, with the addition of something even higher than wisdom: "enlightenment". The authors suggest the traditional DIKW model as one of several ways to define, illustrate and explain the various forms of data, information, etc. in a business, transformation and customer/stakeholder perspective. They have nothing against enlightenment as a step beyond wisdom, usually defined as "evaluated understanding" or "knowing why", which they would then call truly understanding the purpose of information in a context of what people need and want, beyond the more factual knowledge. The enlightened business? Who knows [26].

The traditional DIKW model is also mapped to different types of management information systems. For instance, data is related with transaction processing systems; information with information management systems; knowledge with decision support systems; and wisdom with expert systems.

On the other hand, what the authors are most interested in, is the decision and action part, because without decisions and actions there is little sense in gathering, capturing, understanding, leveraging, storing and even talking about data, information and knowledge. The authors mean the decisions and actions as in business and customer outcomes, creating value in an informed way. However, in the bigger picture, the authors state that the decisions and actions can simply be learning, evaluating, computing or anything else [26].

2.4   Effects of IoT and Big Data to Knowledge-Based Management Practices

Organizations use information and knowledge both for improving the quality of decisions and for legitimizing decisions including also decisions made by poor knowledge [4]. The authors consider that organizations often fail to use information in an effective way in decision-making because of the oversupply of information, caused by biased organizations incentives for information in result of tendency to underestimate the costs of information gathering relative to its benefits. Typically, decisions about information are made in a different part of an organization than where the actual information gathering is conducted. This separation of using and gathering information enable managers to initiate information gathering process that may have value for them, but from the organizational perspective create more costs than benefits. This kind of behavior is rational for managers as it creates an illusion of managing uncertainty [9].

Rationality of information oversupply relates also to strategic value of information. This can be seen in cases where information is not, in the first place, used for doing sound decisions, but for persuading someone to do something.

Despite of increasing academic, as well as practical efforts, there is a difference in views on knowledge in decision-making either seen as a static asset owned by an organization or as a social construction emerged from interaction. Static view on knowledge implies the manageability of knowledge, where as social view emphasizes that knowledge cannot be managed, only enabled. Static view treats knowledge as object that can be identified and handled in information systems, when social view deems the role of IT as useful but not critical because it emphasizes assessing, changing and improving human individual skills and behavior. Related to differences in the role of IT, including IoT and Big Data, the two views on knowledge have also contributed two different knowledge management strategies [30–32, 35].

The authors evaluate possibilities that come along with the emergence of IoT and Big Data. Do IoT and Big Data lay down a basis for more smart, intelligent and even wise decision-making? Do IoT and Big Data bring knowledge-based decision-making into higher level? In order to reflect on these questions, the authors have had to analyze the functions of knowledge and information in decision-making.

One possible useful approach to analyzing decision-making is defining it as a moment which divides time into two eras, before and after decision. It is important to recognize that while decisions fulfill expectations they simultaneously produce insecurity in the sense that it becomes obvious that a different decision could have been reached. To manage uncertainty-related decision-making, organizations need information and knowledge to convince internal and external stakeholders that choices are made rationally. Although, conflicting interests and problems of gathering the all relevant information means that rationality in decision-making is only bounded. The authors suggest that by information and knowledge it is possible to create an impression of rational and reasoned behavior, which, in turn, contributes to internal trust and to preserved external legitimacy. This means that sound knowledge before decision also helps the implementation of decisions. It is also good to understand that the problem of bounded rationality is key motivation for organizational foresight activities.

The discussion shows that information is gathered, and knowledge used both for improving the quality of decisions and for mitigating potential decision consequences. Occasionally organization's knowledge behavior is based on rationalistic ideal, whereas sometimes it is highly symbolic. Adopting the conventional view of IoT and Big Data, the authors suggest that the true value of IoT and Big Data in decision-making lies on their ability to simultaneously promote bounded rational behavior, i.e., provide the best possible information and to limit symbolic use of information, i.e., oversupply of information that have no value in improving decision's quality. More generally, the authors assume that IoT and Big Data predict a new start of knowledge management and the revision of the traditional DIKW model [26].

Perhaps, the division of knowledge management strategies into codification and personalization strategies should also be reconsidered [23, 45]. For instance, Jennex had stated that society and organizations manage by planning [16]. Resources are limited, time is limited, and planning applies thought before action. The output of planning is a plan or strategy, a statement of how something will be done. Society and organizations need to have a strategy for managing the layers and technologies, including IoT and Big Data, in the revised DIKW model.

The authors suggested the basic components of a knowledge management strategy can be generalized and used to manage decisions and actions in the revised DIKW model, including the following [26]:

- Identification of users of the knowledge pyramid layers and transformation processes.
- Identification of actionable intelligence needed to support organizational/societal decision-making.
- Identification of sources of the Big Data, data, information, and knowledge.
- Identification of Big Data, data, information, and knowledge to be captured.
- Identification of how captured Big Data, data, information, and knowledge is to be stored and represented.
- Identification of technologies, including the IoT, to be used to support capturing and processing Big Data, data, information, and knowledge.
- Generation of top management support.
- Establishment of metrics for Big Data, data, information, and knowledge use.
- Establishment of feedback and adjustment process on the effectiveness of actionable intelligence use.

Additionally, the authors define the following organizational dimensions as possible drivers and functions that enhance the use of IoT and Big Data at organizational level [26]:

- Interpretation of operating environment: open system.
- Agency: network (i.e., organizations as information flows).
- Accountability: horizontal + vertical.
- Organizational copying mechanism: foresight-based resilience.
- Leadership: business intelligence.
- Information flows: intra-organizational.
- Innovation philosophy: open.
- Production logic: service-based logic (i.e., customers first).
- Change philosophy: immanent, emergent, cyclical.

Certainly, there are some organizational drivers that enhance Big Data utilization in data-driven ecosystems. As organizations operate in open system as networks, the role of information becomes a valuable commodity. Knowledge, based on information from intra-organizational information flows and incorporated to organizational life through the mechanisms of foresight and planning, is the foundation of business intelligence. This requires a new understanding on the organization's accountability function, with an emphasis on measuring and analyzing accountability both vertically, i.e., reporting the outputs and outcomes of an organization from the bottom up; and horizontally, i.e., reporting to constituents including customers, citizens and the media.

This new requirement of understanding concerns the innovation and change philosophy held by organizations [26, 41]. The innovation paradigm opens because of the availability of information—tomorrow's strategies and innovations are developed together rather than in organizational silos. Big Data also strengthens the transformation from mass-production logic towards more customized and personalized-production logic. To achieve advantage in the increasing competition, more focus should be put on both the products and services that organizations deliver.

Jennex concluded the goal is a top-down strategy approach based on decisions and actions [16]. The authors also note the digital revolution in management process, i.e., integrated knowledge management, by developing and utilizing smart solutions like utilization of IoT and Big Data, impact strategies based on decisions and actions as in business and customer outcomes, creating value in an enlightened way [26].

2.5 Effects of IoT and Big Data to Knowledge-Based Management Practices: Knowledge Integration and Sharing in Healthcare

Knowledge in healthcare is a multifaceted concept with multi-layered meanings. Due to this nature, it has become important to manage knowledge in order to drive performance by ensuring that relevant knowledge is delivered "to the relevant person in the right place in a timely fashion" (or simply put right knowledge is delivered to the right person at the right place at the right time) [6].

Apart from existing in the human mind, i.e., tacit knowledge, knowledge can exist in physical records, i.e., explicit knowledge, such as patient records and medical notes, which needs to be accessed more readily in paper or in computerized form. The major focus of knowledge management in healthcare is to create environments for knowledge workers to develop, leverage and share knowledge. For this to happen effectively knowledge management requires deep-rooted strategic and behavioral change. From this point of view knowledge management represents an evolution towards greater personal and intellectual freedom empowering individuals to engage more actively in their work by sharing ideas, thoughts and experiences. Once knowledge has been discovered, storing it, reusing it and generating new knowledge from it, is important to adding value to data to create shared knowledge.

Continued progress in technology makes sharing knowledge easier, and the Internet with collective portals makes knowledge accessible to a wider range of audience. The rise of networked computers has made it easier and cheaper to codify, store and share knowledge. There is no shortage of technologies to aid in managing knowledge in a healthcare environment; rather the prevalence of such technologies can create confusion.

The goal of knowledge management in healthcare is to enhance the performance by providing efficient access to knowledge and communities of healthcare professionals. It aims to combine the information from different sources (tacit, implicit and explicit), bridge and provide it on a platform which allows applications to be built on it. It aims to prioritize, share, consolidate and provide consistent and accurate information and performance indicators to help with efficient decision-making processes.

As workers in a knowledge-intensive environment, healthcare professionals inevitably hold a considerable amount of experiential knowledge, which can be used to solve day to day problems. It is important that knowledge used to solve such problems is captured, shared and reused in order to prevent the lack of nourishment, i.e., update and replenishment, of that knowledge and to improve knowledge of context. The knowledge process in a healthcare environment can be used to increase collaboration among clinicians, nursing staff with social service agencies for purposes of innovation and process improvement. Updating of knowledge assets cultivates the collective knowledge in healthcare, enriching effective management, smoothening the flow of knowledge to enable better problem-solving and increase knowledge potential.

Knowledge management in the healthcare industry when applied effectively can result in increased efficiency, responsiveness, competency and innovation, which results in superior performance in potentially critical applications. The challenge is, therefore, to create a knowledge management system that can acquire, conserve, organize, retrieve, display and distribute what is available in a manner that informs, educates and facilitates the discovery of new knowledge to contribute to the benefit of the organizations.

Knowledge management in the healthcare industry can, therefore, be viewed as an integrator that offers a framework for balancing the technologies and approaches to provide valuable decision-making actions. It integrates into a seamless whole by aligning organizational information and practices with the organization's objectives. This fits into an employee's daily work activities, manages content effectively, and encourages the potential opportunities of knowledge sharing with external agents.

The healthcare industry has been called "data rich while knowledge poor" as its functions hold large amounts of data (e.g., patient records, outcomes of surgery and medical procedures, clinical trial data, etc.) and still the knowledge potential of many actions is yet to be fully exploited because much of the data is not translated into knowledge, i.e., there is low added value, in order to provide a wider context, a deeper understanding and to help with strategic decision-making. Knowledge appears to be underutilized at the point of care and need. This limits the ability of experienced personnel to harvest knowledge and provide a clearer understanding of the involved process and factors by providing "a window on the internal dynamics of the healthcare enterprise".

Multidisciplinary healthcare teams harvest personal expertise essential for patient safety, learn from it, adapt it to local situations and individual patients, then distribute it via reliable networks to care givers to improve care quality. The healthcare application with a high value form of information that allows sharing of the lessons learned from past experiences improves the context of knowledge potential processes in future. A fundamental challenge faced by clinical practitioners and healthcare institutions is in the ability to interpret clinical information to make potentially lifesaving decisions while dealing with large amounts of data.

Clinical practice is quantitative and very much qualitative too. The tacit knowledge acquired by clinicians and nurses over the years, mainly through experience, represents a valuable form of clinical knowledge. Knowledge management in the healthcare industry involves understanding diseases, hospital systems and, most importantly, patients. The authors note that clinical methods exist for understanding diseases and illnesses, but clinical methods or models are not so readily available for understanding patients. When quantitative and qualitative methods complement each other, and when various modalities of knowledge are used, a holistic view of a situation is best obtained, consequently leading to efficient decision-making.

Knowledge management strategies in the healthcare industry can be broadly classified into codification, where knowledge is identified, captured, indexed and made available, and personalization, where tacit knowledge is shared by means of discussion, effective communication through a multidisciplinary approach, allowing for case by case specific knowledge to be created, which in turn allow creative problem-solving. Therefore, the use of both strategies of understanding diseases and patients is according to the different scenarios. When dealing with routine cases, the codification strategy can be applied and when dealing with a situation where a more creative solution is required, the personalization strategy can be applied. This approach, however, usually only works when the required knowledge is shared, i.e., processed, successfully.

Furthermore, the authors consider IoT as a tool contributing to a solution to some knowledge management problems in the healthcare industry. As noted previously, IoT is an abstract concept in which physical objects are identified and connected to each other and have the ability to communicate with and to interact with their surroundings, providing some intelligence in their functionality. IoT is focused on the network of smart devices that communicate with each other and with cloud-based applications in order to provide a better quality of life. The term and the notion refer to implementing networked RFID technologies with a view to providing a better quality of life and increasing knowledge sharing. The implementation of IoT may drive advancements in all areas of human life by extracting knowledge from the raw data provided, with the possibilities of solving many socio-technical interaction problems [26].

Therefore, the aim of IoT in healthcare is to create a better quality of life and increasing knowledge sharing by connecting things, which can be people (e.g., patients and medical staff), objects (e.g., medical equipment) and systems (e.g., patient monitoring systems), and integrating them seamlessly into network with the internet technology and connectivity. The integration, evolution and adaptation of emerging biomedical technologies will also provide a foundation for the IoT [26, 37].

In 2008, the number of things connected to the Internet already exceeded the number of people on earth, and it is possible that 50 billion things will be inter-connected by the year 2020 [27, 28]. A technological evolution of telemedicine is taking place, in which healthcare professionals can now monitor patients' vital signs remotely and continuously.

In addition, the technology supports per-signalization, in which patients with chronic conditions can live independently in their own homes or secure housing, i.e., in a non-hospital setting, using IoT technology to support their life and life-styles. Such devices include Lifeline Home Units, Personal Pendants, Wandering Client Alarms, PIR Movement Detectors, Fall Detectors, Bed Occupancy Sensors, Temperature Extreme Sensors, Automatic Medicine/Pill Reminders and Dispensers, Talking Color Detectors for blind and partially-sighted people and assisted GPS/GSM technologies, which recognize when the user/patient goes outside the chosen safety zone [26, 33]. The number of such devices and their applications is increasing daily.

3   Practical View on IoT-Driven eHealth

3.1   Analysis of a Case Study on Improving the Patient Discharge Planning Process Through Knowledge Management by Using IoT

3.1.1   Background

The UK National Health Service (NHS), a publicly funded organization, provides healthcare for all UK citizens (currently more than 62 million people). The NHS is faced with problems of managing patient discharge and the problems associated with it, such as frequent readmissions, delayed discharge, long waiting lists, bed blocking and other consequences [1]. The problem is exacerbated by the growth in size, complexity and the number of chronic diseases under the NHS. In addition, there is an increase in demand for high quality care, processes and planning. Effective Discharge Planning (DP) requires practitioners to have appropriate, patient personalized and updated knowledge to be able to make informed and holistic decisions about a patients' discharge.

The NHS case study [18] examines the role of knowledge management in both sharing knowledge and using tacit knowledge to create appropriate patient dis-charge pathways. It details the factors resulting in inadequate DP and demonstrates the use of IoT and Big Data as technologies and possible solutions that can help reduce the problem. The use of devices that a patient can take home and devices that are perused in the hospital generate information that can serve useful when presented to the right person at the right time, accordingly harvesting knowledge. The knowledge when fed back can support practitioners in making holistic deci-sions with regards to a patients' discharge.

3.1.2 Discharge Planning Dilemma in the NHS

Discharge is defined as when an in-patient leaves an acute hospital to return home or is transferred to a rehabilitation facility or an after-care nursing center. DP should commence as early as possible in order to facilitate a smooth discharge process. Discharge guidelines have been prescribed by the UK Department of Health (DH) and different trusts implement discharge pathways or process maps following these guidelines. Several DP improvement attempts have been made and reasonable improvements have been noticed. Several methods by which DP takes place have been identified in two UK hospital trusts, including DP commences on admission: patient and care giver are involved in the decision-making process; a clinical management plan where an expected date of discharge is predicted based on actual performance in the ward or, on benchmarking information from past cases; multidisciplinary teams make a decision based on experience during their meetings. A bed management system stores information on beds occupied and weekly meetings are held to decide the discharge date for patients.

All of these methods involve knowledge management [26]. It is seen that, a rough DP is currently drafted for patients upon entry to hospital according to their diagnosis, and a tentative discharge date is provided in line with recommendations. Changes are made over the course of the patient's stay and records are manually updated by nurses, upon instruction by the doctors. This sometimes results in confusion and even disagreement on discharge dates by different doctors, i.e., when treating the patient for different symptoms; and nurses, i.e., when a change of shift occurs. This case study proposes that patient DP requires viewing the whole system and not as isolated units. In the discharge plan the patient and care giver involvement needs to be considered, however very little indication has been provided on these. To date, clear guidelines are not present on what information needs to be collected, stored and reused on patients.

3.1.3 Analysis by the Authors

The UK NHS is facing problems of managing patient discharges while having to meet waiting time, treatment time and bed usage targets. Patient discharge is currently being driven by quantitative measures such as targets (e.g. to reduce "bed-blocking") and problems resulting from this situation has received a great deal of popular press attention recently and political capital has been made from this. Targets are prioritized while compromising patient's after-care quality. Being target-driven (rather than knowledge driven) implies that the healthcare system fails to consider the factors that affect the effective recovery of a patient after treatment and discharge. Hospitals focus on accomplishing and achieving internal targets, resulting in compromised patient safety and well-being after discharge.

The exact situation with regard to patient discharge and readmissions is not really well established, as there are variations in discharge methods between trusts. However, it is reported in the popular press that doctors have to make quick decisions about patients just to "get the clock to stop ticking" resulting in deteriorating trust between doctors and patients. More precisely, doctors find themselves torn between meeting targets and providing their sick patients with the best treatment. These claims in the assorted news media have been reaffirmed by Andrew Lansley, the Secretary of State for Health in the UK Government [18]. "The NHS is full of processes and targets, of performance-management and tariffs, originally, all designed to deliver better patient care, but somewhere along the line, they gained a momentum of their own, increasingly divorced from the patients who should have been at their center." [18, 26].

Several factors result in the current inadequate DP. These factors are internal and external to the NHS along with psychosocial factors of patient and family. It is important to understand the factors behind inadequate DP to be able to analyze and identify the factors causing the problem systematically. A comparison can then be made between the factors along with the results obtained from the case study, followed by a catalogue of possible solutions underpinned by knowledge management. This will then lead to making a diagnosis, i.e., the proposed knowledge management model [26]. A Root Cause Analysis (RCA) highlights the factors contributing to inadequate DP as shown in Fig. 4, and demonstrates the patient discharge as a complex process, with various interrelated factors.

A carefully designed DP supported by KM can ensure more efficient utilization of hospital resources and will encourage better inter-department communication to ensure that tacit knowledge makes better informed decisions about patient

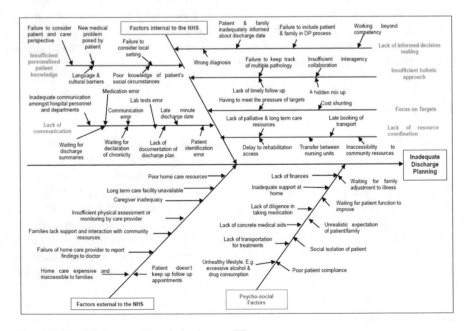

Fig. 4 RCA of factors resulting in inadequate DP

discharge. It is believed that this in turn will allow for better coordination of the external factors and will give hospital personnel more time to inform patients and their families, accordingly addressing the psychosocial factors. At discharge, preventable and undetected errors can occur. These can be reduced by knowledge sharing among hospital staff and having patient centric discharge pathway leading to improved DP. Patient participation and understanding in DP will help reduce potential readmissions and delayed discharge. Patient participation in the discharge process is a legally stated right in the UK and therefore more active participation of patients is encouraged. The failure to assess a patient's care needs correctly can result in a disproportionate delay in patients being discharged [26].

The problems caused by inadequate DP have been identified in [26] and summarized succinctly in Fig. 5. The number of patients readmitted to hospitals through Accident and Emergency (A&E) departments within 28 days of being discharged has risen steadily from 359,719 in 1998 to 546,354 in 2008, while in 2010 more than 660,000 patients were readmitted to hospital within 28 days of discharge [18, 26]. According to statistics provided by the Department of Health, in England in 2010–2011 the total number of patients who were readmitted was 561,291 [18, 26]. According to the statistics, readmission rates in England have been rising since 2001–2002 to 2010–2011 [18, 26]. Figure 6 follows the increasing trend of the percentage of patients readmitted for treatment to UK acute hospitals within 30 days of discharge and a "line of best fit" shows the regularity (and therefore the predictability) of the rise.

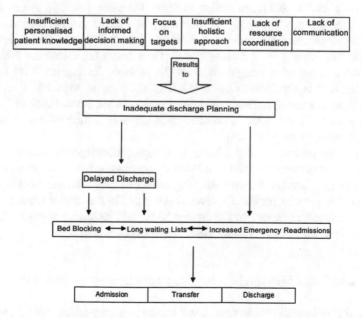

Fig. 5 Problems resulting from inadequate DP

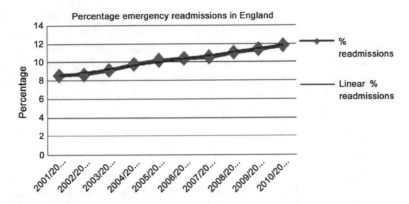

Fig. 6 Emergency readmissions in england as percentage of admissions

The problem of inadequate DP does not just concern readmissions, however. "Bed-blocking" due to delayed discharge has equivalent negative implications. It is reported by the NHS confederation that one in four patients are occupying beds when they could be recovering at home, which results in longer waiting lists, loss of confidence in the NHS and escalating expenditure. The average number of patients and days of delayed discharge per month in England for the year 2012 according to the Department of Health was 3997 patients and 114,386 days respectively [18, 26].

Approximately £250 m was spent on "delayed discharges" between August 2010 and the end of 2011, amounting to £550,000 a day [18, 26]. Apart from the financial implications the delay in discharge is clearly disadvantageous to the well-being of patients, the morale of their relatives and wastes valuable hospital resources. The King's Fund reports that if it was better organized the NHS could reduce the number of overnight stays by 2.3 million, freeing up 7000 beds and saving the NHS nearly £500 m a year [18, 26]. Mike Farrar, the Chief Executive of the NHS Confederation, indicated that these problems are the result of an "outdated hospital model of care" while a breakdown in communication may also be a possible contributory cause [18, 26].

Many older patients face the brunt of delayed discharge as due to a lack of communication between the NHS and social care homes, they are forced to stay in hospital, causing longer waiting lists for other patients who are seeking urgent treatment. The reasons for the dilemma as described in the case study are clearly a result of inadequate support for DP among NHS staff, including physicians, nurses, social workers, and possibly other health professionals [26].

3.1.4 Knowledge Management for Successful Discharge Planning

A hospital is a dynamic environment, with changes taking place rapidly as patients move from one ward to another and treatments are carried out over time. Similarly, DP involves changes from a stable temporal state to another with an element of unpredictability of what is going to happen next. In this context, the past experiential knowledge of doctors and nurses is useful in assessing situations and deciding on plans. This enables making critical decisions, as their knowledge can be reconfigured and extended to fit the new situation and provide a personalized approach in assessing patients' journey along codified guidelines.

Knowledge management may have the potential to remove bottlenecks to improve the DP process mapping and identify possible improvement opportunities. Understanding the relevant knowledge for a given situational decision is crucial to this process and a decision can never be completely separated from the context in which it is made. This implies that in a hospital setting when looking at DP the interrelated factors need to be considered in the context of knowledge management process.

Clearly, monitoring and understanding a patient's condition after discharge is a key part of successful DP. This requires the support of appropriate sensing and monitoring technologies with IoT and Big Data, i.e., medical equipment, patient monitoring systems, and smart devices supporting per-signalization; so that patients with chronic conditions are able to live independently in a non-hospital setting, i.e., in their own homes or secure housing [26].

3.2 IoT in eHealth

Although the authors prefer to use the term IoT for integrating so far not communication-able devices into a digital, communicating infrastructure, often based on the internet infrastructure and services, they hereafter include, in general, communicating sensor and actuator devices, aimed at measuring and, where applicable, controlling health-relevant parameters.

3.3 IoT as Enabler

The technological development of direct and indirect sensor systems, as well as miniaturization, are making available more and more IoT sensor systems that seem to make practical use in eHealth possible, and, thereby, eHealth feasible and accessible.

3.4 Gadgets and Medical Relevance

Most of these sensors require positioning and sophisticated and medical knowledge-based algorithms to make them medical-relevant. In absence thereof, they, unfortunately, stay at the level of gadgets with an indicative value for healthy living and exercising. Moreover, smart applications and algorithms, using the facilities of the current generation smart phones, in particular, accelerometers and cameras, have created another wealth of healthy living and exercising APPs, with even more limited medical relevance [3, 36].

3.5   Dynamic HER and Dynamic EPHR

The grand vision of Electronic Health Record (EHR) infrastructures is the interconnection and reusability of all recorded health information, regardless of where it is stored, so that all relevant health information can electronically flow to wherever it is needed. Nothing will become of this vision, however, unless critical privacy and security problems are overcome. IoT devices, if designed and used to support medical applications, may become part of a Dynamic Electronic Health Record (EHR) or a Dynamic Electronic Personal Health Record (EPHR), where IoT may be used for the following:

- To provide the on-line, dynamic, very recent past complement to the static EHR and EPHR stored information.
- As a tool in support of security mechanisms.

3.6   System Approach Versus "Whatever" Approach

In order for IoT to make an important and necessary contribution to eHealth, a system approach needs to be followed, not a "whatever" approach, as is too often the case with today's wearables. In a number of the companies and research organizations in the world, there is the infrastructure and multi-disciplinary competence, necessary to develop IoT-based medical-relevant eHealth systems, as is shown by the laboratory prototypes, such as continuous, real-time blood pressure monitoring systems; and even by pre-production prototypes, such as diabetes insulin measurement and control systems.

In the next sections, the authors provide a qualitative evaluation of the technical and business-related aspects of relevance for the system-based approach as such.

4   eHealth Requirements

Eysenbach, when explaining the term eHealth, also had given a set of requirements, such as the ten plus "e's" in eHealth [11]. The "e" in eHealth does not only stand for "electronic", but implies many other "e's," which together perhaps best characterize what eHealth is all about or what it should be, including the following:

- Efficiency—one of the promises of eHealth is to increase efficiency in health care, thereby decreasing costs: e.g., avoiding duplicative or unnecessary diagnostic or therapeutic interventions.
- Enhancing quality of care—increasing efficiency involves not only reducing costs, but at the same time improving quality: it may enhance the quality of health care, e.g., by allowing comparisons.
- Evidence based—eHealth interventions should be evidence-based in a sense that their effectiveness and efficiency should not be assumed but proven by rigorous scientific evaluation: to be done.
- Empowerment of consumers and patients—by making the knowledge bases of medicine and personal electronic records accessible to consumers, eHealth opens for patient-centered medicine.
- Encouragement of a new relationship between the patient and health professional, towards a true partnership, where decisions are made in a shared manner.
- Education of physicians through online sources (continuing medical education) and consumers (health education, tailored preventive information for consumers).
- Enabling data and information exchange and communication in a standardized way between health care establishments.
- Extending the scope of health care beyond its conventional boundaries, in both a geographical sense as well as in a conceptual sense, eHealth enables access to services online from global providers.
- Ethics—eHealth involves new forms of patient-physician interaction, posing new challenges and threats to ethical issues as online professional practice, informed consent, privacy and equity issues.
- Equity—to make health care more equitable is one of the promises of eHealth, but at the same time there is a considerable threat that eHealth may deepen the gap between the "haves" and "have-nots".

In addition to these ten-essential e's, eHealth should also be easy-to-use, entertaining (pleasant), exciting, and… it should exist! Refining this top-down, but less detailed view gives the following requirements for eHealth [26]:

- Medical and/or care relevant and usable systems, including:
  - Collection of medical relevant data with:
    direct and indirect practical measurement, compromise between user/patient comfort and data collection quality and reliability;
    suitable sensors used in a way matching the capabilities and limitations of the sensors.
  - Data pre-processing:
    data reduction to avoid data overflow;
    generation of reliable warnings (alarms) to make use of data manageable and beneficial.

- Data interchange and exploitation with:

  - Combination with other IoT and non-IoT data, e.g., location information;
  - Security and privacy;
  - Trust and reliability;
  - Anonymization of data where possible;
  - On-line and off-line data post-processing with medical relevant objectives.

- System approach versus "whatever" approach, including:

  - The users/patients: must be active committed stakeholders/beneficiaries;
  - The medical and care providers: must be committed stakeholders/ beneficiaries;
  - The infrastructure and service providers (installation, operations, maintenance and repair);
  - The IT infrastructure (middleware, cloud storage, cloud processing, applications);
  - The near/on-user/patient systems, smart systems;
  - The compromise between patient benefits versus black-box/post-mortem benefits; and
  - Hybrid/dialogue development approach with:

    the top-down requirements;
    the bottom-up possibilities; and
    cost and benefit-driven.

5   eHealth Limitations

For the foreseeable future, eHealth will not replace doctors, medical experts and care providers. Instead, it must be a joint tool used together between users/patients and eHealth professionals for the benefit of both, and this has to be fully taken into account in the development and deployment. In addition, the following limitations have to be considered [26]:

- The patient benefit versus black-box/post-mortem approach, as it simplifies:

  - Recording effects of a disease or condition than preventing or curing it; and
  - Applying negative evidence gathering, e.g. non-compliance with the prescribed diet and medication than directly contributing to overcoming an illness or condition.

- Generating warnings and alarms that are essential for the usefulness of eHealth, without risking eHealth to become the black box of Health.
- On the other hand, generating warnings and alarms is as good as the quality of the data collection and the applied algorithms; therefore, applying AI and Big

Data techniques may be helpful post-processing options. But, the absence of warnings and alarms can never be taken as guarantee for the absence of risks and conditions.

The unjustified cost-saving expectations, meaning the cost of installation, maintenance, technical and medical healthcare operation should be taken fully into consideration, already in the system design and planning phase. Additionally, it may be easier to achieve better quality health care than achieving real cost reductions.

6 eHealth Entry Barriers

Before eHealth becomes widely implemented and adopted, a number of barriers will have to be overcome. The main barriers that we see are the following [26]:

- Functionality.
  - Medical relevant data and information;
  - Time needed to accept and develop procedures and algorithms and AI to handle the reduce data, obtain information and generate reliably warnings and alarms.

- Trust.
- Security and privacy.
  - Security and privacy concerns are major impediments to eHealth.
  - If they are not properly addressed, health care seekers will not feel comfortable in participating, and health care professionals will face huge liability risks.

- Usability and "companionship" for both users/patients and health and care providers.
- Market development and the required stability in value chains and business plans.

7 Security and Privacy

7.1 Security Safeguards

Although the authors prefer the more general terms, such as Data Ownership and Access Control [24], in this section they mainly use the more familiar terms Security and Privacy [12, 26].

Developing and implementing security and privacy functions in eHealth is a prerequisite for adoption by both users/patients and health and care providers. It concerns, however, a more complex ecosystem than environments currently addressed, requiring new and more sophisticated privacy and security systems, that in turn may find their application also in other more demanding applications, e.g., in Industry4.0, energy, social networks.

In particular, the requirements include: individual privacy, temporary and permanent sharing of subsets of private information, user-controlled access between providers, transferring ownership from a provider to the user/patient or another provider, role-based access, etc., and, a controlled and regulated "break-glass" function for emergency situations.

Note that while security is related to privacy, the two concepts are quite different. The Health Insurance Portability and Accountability Act of the United States (HIPAA) and the Organization for Economic Co-operation and Development (OECD) and other influential sources on the subject of electronic health information clearly distinguish between security and privacy. The eight Fair Information Principles codified in 1980 by the OECD are: openness; collection limitation; purpose specification; use limitation; data quality; individual participation; security safeguards; accountability.

Note that "security safeguards" constitute only one of the eight principles: security safeguards are necessary to achieve privacy, but not sufficient. In fact, most real-life threats come from "secondary use" by insiders with authorized access [12, 26].

7.2 Security Issues

Security is generally defined as the extent to which personal information can be stored and transmitted in such a manner that access to the information is limited to authorized parties. The Health Insurance Portability and Accountability Act of the United States (HIPAA) requires organizations that deal with health care information in electronic form to maintain proper "security safeguards."

It requires mechanisms for identification, authentication, authorization, access controls, audit trails, accountability, encryption, digital signatures, physical security, disaster recovery, protection of remote access points, protection of electronic communications, software discipline, system assessment for vulnerabilities, and integrity of data. Ironically, many of today's commercial security technologies even have a highly adverse impact on privacy principles, which deal with the ability of data subjects to limit the collection of identifiable information [12, 26].

7.3 Privacy Issues

Privacy is defined as the claim of individuals to determine for themselves when, how, and to what extent information about them is communicated to others. This implies that the control over the release of identifiable personal information should rest with the data subject. Legislation that places this control in the hands of third parties does not contribute to privacy, nor do trusted third party solutions that are unilaterally imposed upon data subjects. Secondary use concerns information disclosed to one party for a particular purpose subsequently used for other purposes. The five threats to privacy in eHealth systems that the U.S. National Research Council identified refer primarily to insider attacks: insiders who cause accidental disclosures, who abuse their record access privileges, who knowingly access information for spite or for profit; vengeful employees [12].

Privacy is also sought by medical practitioners: many do not like the idea of central parties (e.g. health insurances) being able to monitor all their actions, since this negatively impacts their autonomy. In many situations, they prefer to be able to access information on the basis of their role rather than their identity. Role-based access is also preferred by most researchers for accessing medical information and databases. Ironically, the very technologies that are currently being considered to implement important security safeguards may make it impossible for patients and health care service providers alike to escape systemic identification throughout the HER [12, 26].

7.4 Privacy-Respecting Security Technologies

If at the technical level everything is systemically identifiable, privacy legislation becomes virtually meaningless. The problem can be solved through use of security technologies that do not violate basic privacy principles. This implies that privacy-respecting security technologies must, at the very least, allow individuals to de-identify their own personal information before disclosing it.

It is important to recognize that privacy-respecting security technologies are not all about anonymity or pseudonymity towards parties that subjects are voluntarily interacting with: they are about controlling which parties can learn what, as personal information flows through the system from one party to the next. For instance, role-based signing can be securely implemented through the use of privacy-preserving technologies, in such manner that digital signatures cannot be traced to an identifiable person but only to the role they assumed when signing; at the same time, through the magic of cryptography, in case of a dispute, error, or other mal-event, the signer cannot repudiate his action [12, 26].

7.5 Ownership of Health Records

To ensure that patients have control over their health privacy, they should have control over the access to the information. The question is: to what extent? Electronic Personal Health Record (EPHR) can include patient identification and contact information, and in theory, the aggregation and unification of all medical information about him/her. The EPHR can be stored on the individual's home computer, on a portable device (such as a laptop, smart card, or smartphone), or on a secure server on the Internet (possibly distributed across multiple trusted parties). With each new consultation with a health professional, the physician is given access to the relevant health information and may update entries in accordance with new findings [14, 26].

The idea of giving patients electronic control over their medical information is not new. Gaunt had listed over 50 internet-based EPHR systems [14]. More in line with the idea of EPHRs are smartcards storing emergency and other health data. According to Waegemann, during the mid-1980s, the vision of patients being in charge of their health information became a leading force [47]. However, from the view of the health care professional, it is one thing for a patient to view his own health information, it is quite another thing to add, delete, modify, or prevent updating of arbitrary data in the HER [14, 26].

Policies on health data ownership differ substantially between delivery networks, states and globally. Larkin had noted that whether patients or physicians provide information for the record is a subject of often intense debate. Ricci had described a health care future where security and privacy issues have been resolved [39]. The question of medical record ownership has totally shifted as consumers have embraced ownership of their own Personal Health Records (PHRs) through secured Web sites [14, 26, 34].

8 Value Chains, Business Models and Strategies, Deployment and Evolution

While eHealth has a major potential and it adoption may even be considered necessary to achieve increased ambulant and remote medical care, increased quality of care, reduced personnel needs, and reduced or reduced increase in costs, the market is not developing as hoped and expected [26].

Predominantly vertical markets have developed explosively for fitness, sports and healthy living. Their contribution to eHealth is limited, however, and the value chains less suitable for an eHealth market development. It is, in particular, the unsettled configuration of the value chains that create an uncertainty in the eHealth market, or better markets, as the parameters may be different between countries or even regions therein [26]:

- The separation and/or overlap between private and public health services provision.
- The separation and/or overlap between private and public health services insurances.
- The role of telecom and communications services providers.
- The role of equipment manufacturers.
- The role of equipment and communications services installation and services companies.

"Asymmetries" in the value chain create a separation between costs and benefits and overlapping and/or crossed responsibilities, potentially putting investments needed and benefits at different entities in the value chain, such as:

- Investments made near the user/patients would contribute to cost savings in a hospital.
- Investments made in a hospital would contribute to cost savings in the public social sector.

The unsettled configuration of the value chains results in uncertainty for the scope and hence of business plans [26]. And this uncertainty in value chains and business plans do not favor the commitment and market development, in turn leading to low interest from industry, hesitant telecom service providers and manufacturers in joint research and development and standardization, essential to arrive at coexistent and interoperable infrastructure and support for common generic and specific applications [20, 26].

Whereas telecom providers tend to try to offer "premium services" for eHealth services, it could be observed that few eHealth applications require high bandwidth, low delay, low Bit Error Rate (BER) services. Instead, eHealth requires rather a reasonable high availability including a short time to repair, 24/7. And, while eHealth, and in fact, our whole society becomes more and more dependent on access to the internet and the services it supports, the availability of networks and Quality of Services (QoS) is not improving, but rather degrading [26]. This may lead to the development of communications service providers that guarantee a service covering support for eHealth equipment and high availability telecom services to address this gap [15, 17, 21, 22, 34].

The time necessary for organizations to arrive, alone or together with partners in the value chain, to decisions to invest and deploy eHealth systems at a large scale is often not sufficiently considered or even ignored. As deployment takes a significant amount of time, and technological development keeps it pace, it is foreseeable that organizations applying eHealth systems will be working in parallel with several generations of equipment [15, 20, 26, 34], using several generations of the telecom infrastructure (second-, third-, fourth-, fifth-generation WAN, Lora, satellite, etc.).

Regarding the functionality, it may be expected that eHealth equipment will develop into fully or partially implanted systems, with an increasingly feedback and control functions.

9 Conclusions

In summary, the authors make the following concluding points:

- There are theoretical and practical views on IoT-driven eHealth.
- Theoretical view on IoT-driven eHealth concerns associated knowledge management issues.

- Practical view on IoT-driven eHealth concerns possible contributions of IoT to eHealth, the requirements, benefits, limitations and entry barriers, as well deployment and evolution.
- The eHealth has a major potential, and its adoption may even be considered necessary based on:

  - Increased ambulant and remote medical care;
  - Increased quality;
  - Reduced personnel needs; and
  - Reduced costs potential.

- However, it is subject to the following challenges and issues to be addressed:

  - While the fitness and healthy living market are developing explosively, the real eHealth market development stays behind;
  - The eHealth reduces personnel needs, but also shifts the need to higher skilled jobs;
  - Achievable savings are real, but cost estimates seem exaggerated, and do not take into account installing, operating, maintaining equipment and reliability services, nor medical services 24/7;
  - Large scale deployment of eHealth takes time, comparable to the deployment of telecom networks;
  - Unsettled value chain and business models do not favor commitment and market development;
  - Lack of commitment results in technical issues, such as lacking quality objectives, data analysis and alarms, standardization, reliability of devices and services, that are not sufficiently addressed; and
  - Security and privacy need to be addressed; however, as shown, the resolution of these issues may be more complex and require more time than many have currently recognized.

- In addition, since eHealth touches on many areas of public policy, including quality of life, public health, employment, industry, research, one could expect more active and effective government policies in favor of eHealth.

In conclusion, the authors state IoT-driven eHealth can happen and will happen; however, much more needs must be addressed in order to bring it back in sync with medical and general technological developments in an industrial state-of-the-art perspective, and to recognize and get timely the benefits.

References

1. About the NHS (2012) The UK National Health Service; NHS
2. Ackoff RL (1989) From data to wisdom. J Applies Syst Anal 16(1):3–9
3. Architecture; Analysis of user service models, technologies and applications supporting eHealth (eHealth) (2009) Technical Report ETSI TR 102 764 V1.1.1, ETSI

4. Blair DC (2002) Knowledge management: hype, hope, or help? J Am Soc Inform Sci Technol 53(12):1019–1028
5. Bucherer E, Uckelmann D (2011) Business models for the internet of things. In: Uckelmann D, Harrison M, Michahelles F (eds) Architecting the internet of things. Springer, Berlin Heidelberg
6. Cavaliere V, Lombardi S, Giustiniano L (2015) Knowledge sharing in knowledge-intensive manufacturing firms. An empirical study of its enablers. J Knowl Manag 19(6):1124–1145
7. Chan HC (2015) Internet of things business models. J Serv Sci Manag 8(4):5–52
8. Dijkman RM, Sprenkels B, Peeters T, Janssen A (2015) Business models for the Internet of Things. Int J Inf Manage 35(6):672–678
9. Davenport TH, Prusak L (1998) Working knowledge. Harvard Business School Press, Boston, MA
10. Evans D (2012) The internet of things: how the next evolution of the internet is changing everything
11. Eysenbach G (2001) What is eHealth? J Med Internet Res, preface
12. For the record: protecting electronic health information. Committee on maintaining privacy and security in health care applications of the national information infrastructure. (1997) NRC
13. Fricke M (2009) The knowledge pyramid: a critique of the DIKW hierarchy. J Inf Sci 35 (2):131–142
14. Gaunt N (2001) Initial evaluation of patient interaction with the electronic health record. South & West Devon Health Community ERDIP Project
15. Gregus M, Kryvinska N (2015) Service orientation of enterprises—Aspects, dimensions, technologies. Comenius University, Bratislava. ISBN 9788022339780
16. Jennex ME (2017) Big data, the internet of things, and a revised knowledge pyramid. Data Base Adv Inf Syst 48(4):69–79
17. Kaczor S, Kryvinska N (2013) It is all about Services—Fundamentals, drivers, and business models, The society of service science. J Serv Sci Res, Springer 5(2):125–154
18. Kamalanathan NA, Eardlay A, Chibelushi C, Collins T (2013) Improving the patient discharge planning process through knowledge management by using the internet of things. Adv Internet Things 3(1):16–26
19. Kiel D, Arnold C, Collisi M, Voigt K-I (2016) The impact of the industrial internet of things on established business models. The 25th international association for management of technology (IAMOT) Conference
20. Kryvinska N, Lepaja S, Nguyen HM (2003) Service and personal mobility in next generation networks. In: The fifth IEEE international conference on mobile and wireless communications networks (MWCN), pp 116–119
21. Kryvinska N (2012) Building consistent formal specification for the service enterprise agility foundation, The society of service science. J Serv Sci Res, Springer 4(2):235–269
22. Kryvinska N, Gregus M (2014) SOA and its business value in requirements, features, practices and methodologies. Comenius University, Bratislava. ISBN 9788022337649
23. Kumar JA, Ganesh LS (2011) Balancing knowledge strategy: codification and personalization during product development. J Knowl Manag 15(1):118–135
24. Larkin H (1999) Allowing patients to post their own medical records on the internet is becoming big business, AMNews
25. Liu L, Jia W (2010) Business model for drug supply chain based on the internet of things. In: The international conference on network infrastructure and digital content. IEEE Press, pp 982–986
26. Lokshina IV, Lanting CJM (2018) A qualitative evaluation of IoT-driven eHealth: knowledge management, business models and opportunities, deployment and evolution. In: Hawaii international conference on system sciences (HICSS-51), pp 4123–4132
27. Lokshina IV, Durkin BJ, Lanting CJM (2017) Data analysis services related to the IoT and big data: potential business opportunities for third parties. In: Hawaii international conference on system sciences (HICSS-50), pp 4402–4411

28. Lokshina IV, Durkin BJ, Lanting CJM (2017) Data analysis services related to the IoT and big data: strategic implications and business opportunities for third parties. Int J Interdisc Telecommun Netw 9(2):37–56
29. Lokshina I, Thomas W (2013) Analysis of design requirements for electronic tags from a business perspective. Int J Mobile Network Des Innov 5(2):119–128
30. Manyika J, Chui M, Brown B, Bughin J, Dobbs R, Roxburgh C, Byers AH (2011) Big data: the next frontier for innovation, competition, and productivity. McKinsey Global Institute Reports
31. Marr B (2015) Big data: using SMART big data, analytics and metrics to make better decisions and improve performance. Wiley, Chichester
32. Mayer-Schonberger V, Cukier K (2013) Big data: a revolution that will transform how we live, work, and think. Houghton Mifflin Harcourt, Boston
33. McGee-Lennon MR, Gray PD (2007) Including stakeholders in the design of home care systems: identification and categorization of complex user requirements. In: INCLUDE Conference
34. Molnár E, Molnár R, Kryvinska N, Greguš M (2014) Web intelligence in practice. The society of service science. J Serv Sci Res, Springer 6(1):149–172
35. Osterwalder A, Pigneur Y (2010) Business model generation: a handbook for visionaries, game changers, and challengers. Wiley, Hoboken
36. Personalization of eHealth systems by using eHealth user profiles (eHealth). (2010). Standard ETSI ES 202 642 V1.1.1, ETSI
37. Pfister C (2011) Getting started with the internet of things. O'Reilly Media Inc, Sebastapool
38. Porter ME, Heppelmann JE (2014) How smart, connected products are transforming competition. Harvard Bus Rev 92(11):64–88
39. Ricci RJ (2012) Future of healthcare: 2012. IBM Healthcare Industry
40. Rowley J (2007) The wisdom hierarchy: representations of the DIKW hierarchy. J Inf Sci 33 (2):163–180
41. Scuotto V, Santoro G, Bresciani S, Del Giudice M (2017) Shifting intra-and inter-organizational innovation processes towards digital business: an empirical analysis of SMEs. Creativity and Innovation Manage 26(3):247–255
42. Silver N (2012) The signal and the noise: Why so many predictions fail-but some don't. Penguin Press, New York
43. Sun Y, Yan H, Lu C, Bie R, Thomas P (2010) A holistic approach to visualizing business models for the internet of things. Commun Mob Comput 1(1):1–7
44. Van Rijmenam M (2014) Think bigger: developing a successful big data strategy for your business. AMACOM Div American Mgmt Assn, New York
45. Venkitachalam K, Willmott H (2016) Determining strategic shifts between codification and personalization in operational environments. J Strategy Manage 9(1):2–14
46. Vlacheas P, Giaffreda R, Stavroulaki V, Kelaidonis D, Foteinos V, Poulios G, Moessner K (2013) Enabling smart cities through a cognitive management framework for the internet of things. Commun Mag 51(6):102–111
47. Waegemann CP (2002) Electronic health records. Health IT Advisory Report

Izabella Lokshina Ph.D. is Professor of MIS and chair of Management, Marketing and Information Systems Department at SUNY Oneonta, USA. Her main research interests are intelligent information systems and communication networks.

Cees J. M. Lanting Ph.D. is Senior Consultant at DATSA Belgium in Leuven, Belgium. His main research interests are smart communications and IoT.

WLAN Planning and Performance Evaluation for Commercial Applications

Salem Lepaja, Arianit Maraj and Shpat Berzati

Abstract In the past, WLAN planning was coverage driven to ensure that the signal quality is satisfied in all targeted areas. As the number of users with more high-throughput applications, supported by sophisticated mobile computing devices, continues to grow rapidly on the one hand and on the other hand the new high-speed IEEE 802.11 based WLANs continue to have a significant wireless communication market spread, it has been recognized that capacity is equally important as coverage for the WLAN planning. In this paper, WLANs capacity planning and performance evaluation focusing on the impact of the applications throughput requirements, high-speed IEEE 802.11 WLAN standards and user terminal capabilities are discussed. To investigate the impact of the high-speed WLANs and user terminal capabilities on the network throughput, performance evaluation is carried out by means of measurements, using IEEE 802.11ac and IEEE 802.11n standards.

Keywords Wireless local area networks · WLAN planning · Frequency bands
Channel planning · User applications · User terminals capabilities

1 Introduction

Wireless Local Area Networks (WLANs) planning in the past, aimed to ensure that S/N (Signal to Noise and Interference ratio) and RSSI (Received Signal Strength Indicator) are satisfied in all covered areas. However, as the number of users with

S. Lepaja · A. Maraj (✉) · S. Berzati
Faculty of Computer Science, AAB College,
Pristina 10000, Republic of Kosovo
e-mail: arianit.maraj@universitetiaab.com

S. Lepaja
e-mail: salem.lepaja@universitetiaab.com

S. Berzati
e-mail: shpat.berzati@universitetiaab.com

© Springer International Publishing AG, part of Springer Nature 2019
N. Kryvinska and M. Greguš (eds.), *Data-Centric Business and Applications*,
Lecture Notes on Data Engineering and Communications Technologies 20,
https://doi.org/10.1007/978-3-319-94117-2_3

more bandwidth hungry applications, supported by sophisticated mobile computing devices such as smartphones, tablets, PDAs, laptops, continues to grow rapidly on the one hand and on the other hand high-speed IEEE 802.11n and IEEE 802.11ac based WLANs continue to mark a huge wireless communication market spread, it has been recognized that capacity is equally important as coverage for WLAN planning, or the existing and future planned applications. Hence, coverage and capacity planning must be done simultaneously in order to satisfy user requirements in terms of the data rate provisioning and the signal quality within the coverage area [1–5].

In this paper WLANs capacity planning and performance evaluation focusing on the impact of the applications throughput requirements, high-speed IEEE 802.11ac standard with 40 and 80 MHz channels, IEEE 802.11n standard with 40 MHz channel, and user terminal capabilities are presented.

The basic steps of the WLAN Planning are presented in the next section. In Sect. 3, we describe the channel planning in the 2.4 GHz ISM (Industrial Scientific Medical) and in the 5 GHz UNII (Unlicensed National Information Infrastructure) frequency bands. Calculation of the capacity planning, taking into consideration different user device capabilities and user applications throughput are given in Sect. 4. In Sect. 5 we present the measurement environment and the performance evaluation. Conclusions are drawn in the last section.

2   Related Work

So far, a lot of work has been done regarding capacity planning and performance evaluation of WLAN networks.

Kelkar in [2] compared the MAC of 802.11n and 802.11ac standards, using numerical analysis and simulations. In particular, they considered the fairness behavior between competing users between the 802.11ac and 802.11n. He analyzes various problems such as lower throughput, higher delay, and large congestion at the network and came to the conclusion that the higher the frequency (5.0 vs. 2.4 GHz), the greater the bandwidth which allows more data carrying capacity. Also, he found out that the attenuation is the reduction of signal strength during transmission.

In [12] authors propose two techniques implemented in Mraki APs to improve network capacity and performance perceived by end users. Here, authors have used a dynamic channel assignment algorithm, TurboCA, and (ii) a novel approach called FastACK, that improves the end-to-end performance of TCP. Finally, they evaluated TurboCA with metrics taken from a variety of real-world networks and evaluated TCP performance of FastACK. They did such evaluations by performing measurements in a testbed platform. Based on the data that they have collected from thousands of real-words operational WLANs they concluded that throughput is an insufficient and difficult metric to measure. Authors proposed a different metrics, such as achieved bit rates and latency.

In [13] authors describe results of wireless optimization in an overlapping environment. They have involved twelve classrooms in these analyses. As main technique used in this paper was site survey, which was applied to define optimal configuration of these twelve AP's. In APs they considered the direction of the antennas, channel combination and the transmit power. The optimization result has improved the quality the quality of WLAN in the tested environment.

Our approach is different. We have designed an experimental network with one 802.11n, ac AP and several end user devices. Our methodology is based on carrying out intensive measurements in our experimental network, in order to investigate the impact of the high speed IEEE 802.11n and ac WLAN standards and user terminal capabilities.

3 Basic Steps for WLAN Planning

In general WLAN planning is carried out in several steps as shown in Fig. 1. The starting step is the survey on WLAN communication requirements, which includes users (categories and applications) and communication environment survey for the targeted area. The next step is the channel planning development. For channel planning we have to take into consideration available WLAN frequency bands and channels, IEEE 802.11 standards, and communication environment. Having provided data for WLAN and channel planning, the next step is the WLAN dimensioning. Dimensioning of the WLANs encounters determining the needed APs to ensure that capacity, S/N (Signal to Noise and Interference ratio) and RSSI (Received Signal Strength Indicator) are satisfied, taking into consideration applications throughput, number and type of user devices. In addition, the vendor's

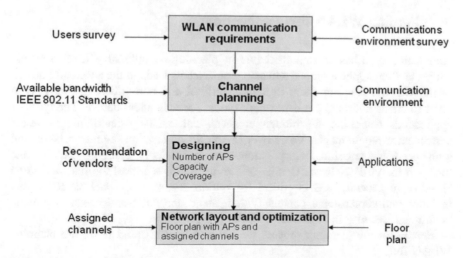

Fig. 1 Basic steps for WLAN planning

recommendation and planner's experience should be considered. In the network layout and optimization phase, the floor plan with APs location and assigned channels should be specified.

3.1 User Survey

User survey can be considered as the most crucial stage of WLAN implementation. The user survey consists of collecting data such as: number of users and their category, user density in each targeted area, user device capabilities, type of the demanded applications as well as their throughput requirements. Once the customer equipment and their capabilities are identified, it is a need to define the applications that will be used by these devices as well as their performance requirements (throughput and quality if service).

3.2 Communication Environment Survey

In addition to the user survey, for successful WLAN planning, data about the size of the coverage area, type of the communication environment (closed, open, mobile), buildings architecture (one or more floors, offices spread out in two buildings) also are crucial. Furthermore, it is important to have some inputs about existing communication infrastructure (LAN network and connection to the Internet) in the covered area.

4 Planning WLAN for Applications

Application type has an important role in planning an optimal WLAN network. Understanding application performance requirements leads to the successful design of the WLAN. It is important to identify critical and non-critical applications in order to create a throughput benchmark model for each client device. The types of applications determine the throughput needs and can help determining wireless performance requirements. Voice over IP (VoIP) application requires reliable and high-speed transmissions. Corporations are increasingly using WLAN as the transport for VoIP. Network Quality-of-Service (QoS) is critical for this application [14–29]. In general, QoS is critical for multimedia services, such as IPTV that includes multidimensional content (video, audio and data) and needs to provide quality and security in IP based networks [30–32].

Hence, for the planning process we have to bear in mind if we are planning WLAN for:

- Data applications
- Real time applications, voice or video, or
- Location based applications.

However, planning a WLAN for an application doesn't mean it will not support other applications. Depending how the particular WLAN is planned it will support all the applications or part of them. If it is planned for real-time applications, probably it will support data applications as well. If it is planned for location-based applications, it will support location and probably real time and data. If it is planned for data there might be some part of targeted area where real time applications will not be supported because of packet drops. Also, there will be area where location will not be accurate.

Most often WLAN is planned for data traffic and further on in most cases data traffic used TCP. As TCP, with sliding window flow control, is very reactive to losses, dropped packets will be resented but delay will be introduced.

Voice and other real-time applications are different from data traffic. Data traffic is sensitive to packet drops while tolerant to delays. Whereas, voice and real-time traffic in general, are sensitive to delay whereas, due to redundancy, tolerant to packet drops (up to 2%). For real-time WLAN planning one has to take care of delay and jitter. Low delay and constant jitter are very important factors for real-time applications. However, voice uses a codec to transform voice to binary value. A given codec generates a certain packet size of a certain rate, e.x. 75 packets/s each 120 bytes, hence voice is a very consistent flow.

Video is quite different from voice. It is much more bandwidth intensive than voice and also bandwidth consumption changes over the time. Hence designers of video applications try to save on bandwidth.

Real-time video applications are mainly grouped into:

- Interactive (video conferencing e.x.), where bandwidth issue is not a problem because users are interested to hearing what is being said than seeing faces.
- Streaming video, which is a one way communication. If it is a real time, live streaming, not a You tube. Users are interested on listening and watching as well.
- Elastic and non-elastic video. Elastic video means that bandwidth between receiver and sender can be detected dynamically and codec can be adopted to bandwidth consumption, i.e. on network conditions. Whereas, non-elastic video will not adopt to changing conditions on network.

So the way the WLAN is planned and designed for real-time video it will depend are we planning for interactive, streaming, elastic or inelastic codec.

In general, for data and real-time WLAN planning, planners first have to know for what type of data or real-time application are planning WLAN for. Secondly, planners have to know (or to measure) how much throughput (bandwidth) the targeted application needs.

Location service application is slightly different from data and real-time because in addition to the data rate it is also crucial to know where the access points are.

Table 1 Commercial and non-commercial applications [7]

Application class	Required throughput	QoS Class (Layer 2/Layer 3)
Web-browsing/email	500 Kbps–1 Mbps	WMM 0 (BE)/DSCP 0
Video conferencing	384 Kbps–1 Mbps	WMM 5 (VI)/DSCP AF41
SD video streaming	1–1.5 Mbps	WMM 4 (VI)/DSCP CS4
HD video streaming	2–5 Mbps	WMM 4 (VI)/DSCP CS4
Apple TV streaming	2.5–8 Mbps	WMM 4 (VI)/DSCP CS4
Apple FaceTime10	900 Kbps	WMM 5 (VI)/DSCP AF41
YouTube video streaming	500 Kbps	WMM 0 (BE)/DSCP 0
Printing	1 Mbps	WMM 0 (BE)/DSCP 0
File sharing	5 Mbps	WMM 0 (BE)/DSCP 0
E-learning and online testing	2–4 Mbps	WMM 4 (VI)/DSCP CS4
Thin-client	85–150 Kbps	WMM 4 (VI)/DSCP CS4
Thin-client (with video or printing)	600–1800 Kbps	WMM 4 (VI)/DSCP CS4
Thin-apps	20 Kbps (per App)	WMM 4 (VI)/DSCP CS4
VoIP call signaling	5 Kbps	WMM 3 (BE)/DSCP CS3
VoIP call stream	27–93 Kbps	WMM 6 (VO)/DSCP EF

DSCP Differentiated services code point
WMM Wi-Fi Multimedia

If there is only 1 AP and a user device or collection of user devices, the task is to find where these devices are located. The single AP will get only the signal level from these devices. The signal level may be translated into a distance of the devices from AP. However, this is not enough to locate devices, hence with one AP devices cannot be located. Hence, more APs are needed to accomplish device location task. Furthermore, planning should be done not only with enough AP where all the devices can be heard by APs but planner have also to put AP at appropriate places (at the corner of the targeted area as well) around area to be covered. So whenever the WLAN is planned for location services most of the time it will be planned for real-time applications, for certain user density and certain kind of small cells.

In Table 1, there are shown the most common application classes and their bandwidth requirements. In this table are also given the recommended QoS classes [7].

5 Channel Planning in 2.4 GHz ISM and 5 GHz UNII Frequency Bands

In this section, we will explore the channel planning in the 2.4 GHz and in the 5 GHz band. In IEEE 802.11 standards, there are defined fourteen channels in the 2.4 GHz band, each 22 MHz. The availability of these channels is governed from local authorities of different Countries. In Europe, there are 13 channels allowed for

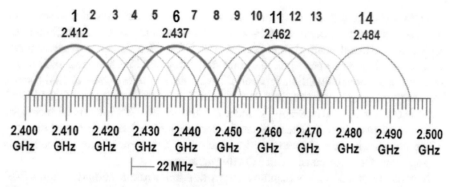

Fig. 2 WLAN channels in 2.4 GHz band [6]

use, whereas in the United States 11 channels are allowed. Furthermore, in the 2.4 GHz band, there are only three non-overlapping channels 1, 6 and 11, as shown in Fig. 2. It is clearly seen that most of the channels in the 2.4 GHz band overlap. Limited available bandwidth and channel overlapping results in the low number of the available channels in 2.4 GHz. Hence, implementation of highspeed IEEE 802.11 standards in the 2.4 GHz band is not suitable particularly in high density areas [6].

Different from 2.4 GHz band, IEEE 802.11 standard defines 23 channels in the 5 GHz band, each 20 MHz (Fig. 3). As seen from the Fig. 3, channels are spaced 20 MHz apart and separated in three UNII bands.

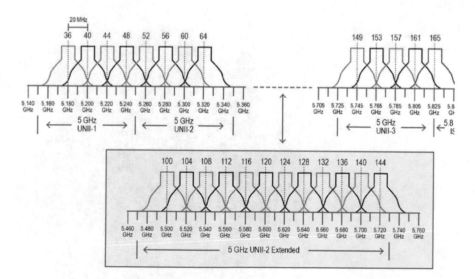

Fig. 3 WLAN channels in 5 GHz band [6]

The most critical issue in channel planning is interference avoidance. There are two types of channel interferences in WLAN environment: Co-Channel Contention (CCC) and adjacent channel interference. The CCC (also called co-channel interference) is when APs operate on the same frequency and interfere with each other. Actually, it is about contention between all users for a wireless channel on the same frequency (Carrier Sense Multiple Access with Collision Avoidance—CSMA/CA) and it is not like other RF (Radio Frequency) interference. CCI is a crucial factor for the capacity planning. Adjacent channel interference is when APs operate on the different channels and still interfere with each other. This type of interference is also an important factor to consider in WLAN planning.

In capacity based WLAN planning, network administrators should design Wi-Fi cells (Fig. 4) with associated range (adjacent channel interference) between −65 and −67 dBm. At the same time they should develop a channel plan that ensures that CCC interference remains below −85 dBm, considering the noise floor to be < −85 dBm, whereas S/N: 25–30 dB. For coverage based planning received signal strength is usually limited to −72 dBm [7].

6 Capacity Planning

The WLAN planning in the past was mainly coverage driven to ensure that signal quality (S/N and RSSI) is satisfied in all targeted areas. However, as users demands for high-throughput applications supported by mobile computing devices such as smartphones, tablets, PDAs, laptops, continue to grow rapidly, it has been

Fig. 4 Contention and association range [7]

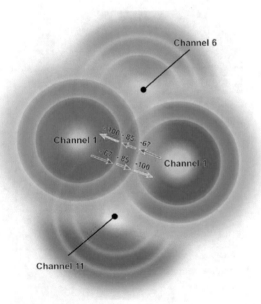

recognized that providing coverage does not mean enough capacity, particularly in high density areas. Hence, capacity is equally important as coverage for the WLAN planning to meet the user requirements of the existing and predicted IP services in the future. Capacity planning in terms of the needed number of the APs is based on airtime utilization of the shared wireless link. By airtime utilization we understand the time an application occupies on wireless link, which depends on throughput of the application and on the user device capabilities. For example, a 2 Mbps application generated by a laptop with 100 Mbps capability needs 2% airtime utilization of the wireless channel. In the following paragraphs, we will present two examples to show how the airtime utilization is calculated i.e. number of needed APs for the given application throughput and the given device capability.

Example 1 We will assume that we have 50 Laptops with $2 \times 2{:}2$ MIMO, 144 Mbps raw data rate (Table 2), and TCP throughput of 70 Mbps. Whereas E-Learning applications with 2 Mbps throughput [7]. The airtime utilization is calculated as follows: $(2/70) \times 100 = 2.85\%$, for a single laptop. For 50 laptops airtime utilization would be: $50 \times 2.85\% = 143\%$. Furthermore, taking into consideration practical case of the AP channel saturation at 80%, $(143/80) = 1.8$, i.e. 2 APs are needed to satisfy capacity requirements.

Example 2 This example is an extension of the example 1 with 30 more tablets $1 \times 1{:}1$, with 65 Mbps raw data rate (Table 2), and 30 Mbps TCP throughput. Airtime utilization for added tablets would be: $(2/30) \times 100 = 6.66\%$ for one tablet, whereas for 30 tablets, $30 \times 6.66\% = 200\%$. In total airtime utilization will be 343%. Assuming again practical case of the AP channel saturation at 80%, we will come up with $(343/80) = 4.3$, i.e. 5 APs needed to satisfy capacity requirements.

Based on the above examples, Fig. 5 illustrates the impact of the applications throughput and the user devices capabilities on the number of the needed APs. Note the difference between raw data rate and TCP throughput due to WLAN overhead, which is usually between 40 and 60%.

Table 2 Selected raw data rates in Mbps for IEEE 802.11 standards [8, 9]

MCS	Modulation & Rate	20 MHz 1 × S	20 MHz 2 × S	40 MHz 1 × S	40 MHz 2 × S	80 MHz 1 × S	80 MHz 2 × S
0	BPSK 1/2	7.2	14.4	15	30	32.5	65
1	QPSK 1/2	14.4	28.9	30	60	65	130
2	QPSK 3/4	21.7	43.3	45	90	97.5	195
3	16-QAM 1/2	28.9	57.8	60	120	130	260
4	16-QAM 3/4	43.3	86.7	90	180	195	390
5	64-QAM 2/3	57.8	115.6	120	240	260	520
6	64-QAM 3/4	65	130	135	270	292.5	585
7	64-QAM 5/6	72.2	144.4	150	300	325	650
8	256 QAM 3/4	86.7	173.3	180	360	390	780
9	256-QAM 5/6	–	–	200	400	433.3	866.7

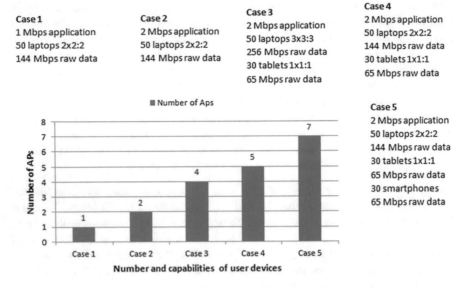

Case 1
1 Mbps application
50 laptops 2x2:2
144 Mbps raw data

Case 2
2 Mbps application
50 laptops 2x2:2
144 Mbps raw data

Case 3
2 Mbps application
50 laptops 3x3:3
256 Mbps raw data
30 tablets 1x1:1
65 Mbps raw data

Case 4
2 Mbps application
50 laptops 2x2:2
144 Mbps raw data
30 tablets 1x1:1
65 Mbps raw data

Case 5
2 Mbps application
50 laptops 2x2:2
144 Mbps raw data
30 tablets 1x1:1
65 Mbps raw data
30 smartphones
65 Mbps raw data

Fig. 5 Number of needed APs [8]

From Fig. 5 it can be seen that:

- applications with higher throughput require more APs (case 1, 2)
- devices with better capabilities need less APs (case 3, 4)
- increasing the number of the different devices more APs are needed (case 4, 5).

7 Measurement Environment and Performance Evaluation

To investigate the impact of the high-speed WLANs and user terminal capabilities on the network throughput, we have carried out thoroughly measurements using sophisticated tools. We considered only TCP throughput on Uplink.

7.1 Measurement Environment

- **Software tools**

  - iperf 3, inSSIder, Aruba OS 6.4

  - **Devices**

  See (Table 3).

Table 3 User devices capabilities

User device type	WiFi radio type	MIMO	Channel eidth
iPhone 7	802.11a/b/g/n/ac	2 × 2	40/80 MHz
Samsung S7	802.11a/b/g/n/ac	2 × 2	40/80 MHz
Xiaomi Note 3	802.11a/b/g/n/ac	1 × 1	40/80 MHz
Lenovo X1 Carbon with Intel® Dual Band Wireless-AC 7260	802.11a/b/g/n/ac	2 × 2	40/80 MHz
Lenovo L540 with Netgear 6210AC	802.11a/b/g/n/ac	2 × 2	40/80 MHz
Sony xPeria	802.11a/b/g/n/ac	1 × 1	40/80 MHz
Lenovo Idea pad, Intel Dual Band Wireless-AC(1 × 1) 3165	802.11a/b/g/n/ac	1 × 1	40/80 MHz
Server device type			
HP Probook 4530s with Gigabit LAN Card			

7.2 Network Infrastructure

To model a real WLAN environment is not an easy task because of different IEEE 802.11 WLAN types and user devices encountered. For our goal of the investigation, we have set up a possible network model with one AP and several different user devices, as shown in Fig. 6.

Measurements have been finalized using a single Access Point (AP) in the same room with the handheld devices. The distance from each device towards the access points were almost the same.

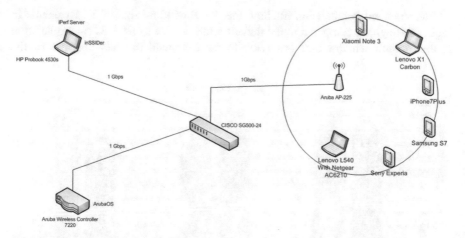

Fig. 6 Network infrastructure

7.3   Performance Evaluation

Case 1

- Throughput with one client device in the Dual-Band (2.4–5 GHz) and Single-Band (5 GHz) mode with 40 MHz channel width.

    - **Used devices**:

    Server: HP Probook 4530s with Gigabit LAN Card
    Client: iPhone 7 Plus, Samsung Galaxy S7, Lenovo L540 with Netgear 6210AC adapter
    WiFi AP Mode: Aruba AP-225 n.

As it can be seen from Fig. 7, dual-band mode operation of the AP has no impact on performances, in terms of the network throughput, of the user devices operating only in the 5 GHz band. Whereas, Lenovo L540 with Netgear 6210 AC adapter performs better (results in higher network throughput), when operating in the 5 GHz band compared to 2.4 GHz band. This outcome is due to the higher channel overlapping and Co-Channel Contention interference in the 2.4 GHz band.

Case 2

- Throughput with one client device with default and 1 Mb window size.

    - **Used devices**:

    Server: HP Probook 4530s with Gigabit LAN Card
    Clients: iPhone 7 Plus & Samsung Galaxy S7
    WiFi AP Mode: Aruba AP-225 n/ac.

Measurement results (Fig. 8) show that TCP window size of 1 MB results in higher throughput compared to the default window size of 64 KB. An explanation is that default window size is obviously far too small to make full use of the

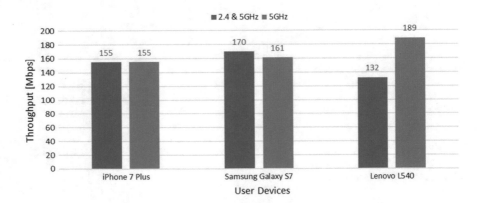

Fig. 7 Throughput with one client device

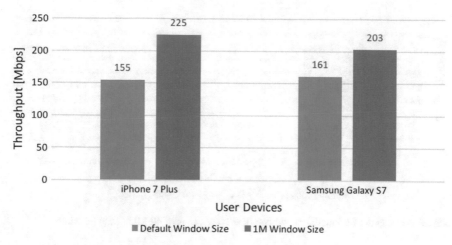

Fig. 8 Throughput with one client device with default and 1M Window Size

end-to-end wireless link capacity. Having send 64 KB of data, the sending process must wait for a window update from the receiver (sliding window flow control principle) before putting any more data on the wireless link.

Case 3

- Throughput for individual and combined devices with 40 MHz channel width.

  - **Used devices**:

    Server: HP Probook 4530s with Gigabit LAN Card
    Clients: iPhone 7 Plus, Samsung Galaxy S7 and Lenovo X1 Carbon with Intel®
    Dual Band Wireless-AC 7260
    WiFi AP Mode: Aruba AP-225 n/ac.

From the measurement results shown in Fig. 9, we notice that the overall throughput with two clients (180, 8 Mbps) is higher than in the case of the single client (155, 161 Mbps), the overall throughput with three clients (197, 7 Mbps) is higher than the overall throughput with two clients. An explanation is that in the high-bandwidth, high-latency environments (bandwidth delay product), TCP communications are often limited by the amount of data on the way at any given time. Using two or more user devices simultaneously, i.e. two or more TCP connections in our case, overall link (network) throughput is increased because the bandwidth delay product applies to each user device connection individually.

Case 4

- Throughput for individual and combined devices with 40 and 80 MHz channel width.

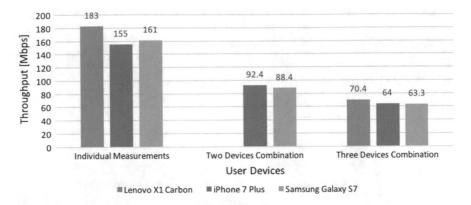

Fig. 9 Throughput for individual and combined devices with 40 MHz channel width

- **Used devices**:

    Server: HP Probook 4530s with Gigabit LAN Card
    Clients: iPhone 7 Plus and Samsung Galaxy S7
    WiFi AP Mode: Aruba AP-225 n/ac.

Figure 10 shows that operating with wider channels throughput increases significantly. However, doubling channel width does not lead to a doubling of throughput as it applies for raw data rate. Comparing raw data rates of 40 with 80 MHz channels (Table 2), we notice that for devices with the same capabilities, the raw data rate of 80 MHz channels is twice higher than with 40 MHz channels [10]. Whereas, TCP throughput, in the case of using higher modulation such as the 256-QAM (used in IEEE 802.11ac with 80 MHz channel), increases by 33% [11].

From our measurement results, shown in Fig. 10, it is noticeable a slightly throughput increase compared to the 33% in [11]. In the case of the single device measurements, the throughput results are nearly 40% higher with 80 MHz channel

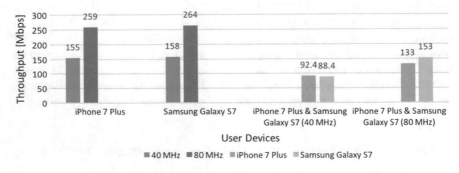

Fig. 10 Throughput for individual and combined devices with 40 and 80 MHz channels

compared to 40 MHz channel. Whereas, in the case of the two devices sharing the wireless link we have about 36% throughput increase with 80 MHz channel compared to 40 MHz channel.

8 Conclusion

We have investigated WLAN capacity planning and performances evaluation focusing on the impact of the applications throughput requirements, high-speed IEEE 8 02.11 standards, and user terminal capabilities. In the past WLAN planning was mainly coverage driven, ensuring that S/N ratio and RSSI are satisfied in all covered area. Today with the massive increase of the wideband applications and high capability user devices, capacity is a crucial issue in WLAN planning to meet users' requirements.

For WLAN planning driven by capacity, airtime utilization of the shared wireless link is the main parameter in determining the needed number of the APs. Presented calculations show that applications with higher throughput require more APs, whereas devices with better capabilities need less APs (Fig. 5).

Measurement results show that dual-band mode operation of the AP has no impact on performances, in terms of network throughput, of the user devices operating only in the 5 GHz band. Whereas, devices that can operate in both modes perform better in the 5 GHz band (Fig. 7). Furthermore, the TCP window size of 1 MB results in higher throughput compared to default window size of 64 KB (Fig. 8), because the default window size is obviously too small to make full use of the end-to-end link capacity. Measurements also show (Fig. 9) that the overall throughput with two client devices is higher than with a single client device, overall throughput with three clients is higher than overall throughput with two clients. An explanation is that in the high-bandwidth links TCP communications are often limited by the amount of data on the way. Using two or more user devices simultaneously i.e. two or more TCP connections, overall wireless link throughput is increased because the bandwidth delay product applies to each user device connection individually. Operating with wider channels, throughput increases sig-nificantly. However, doubling channel width does not lead to a doubling of throughput as it applies for raw data rate. From our measurement results (Fig. 10), it is noticeable a slightly throughput increase compared to the 33% in [11]. In the case of the single device measurements, the throughput results are nearly 40% higher with 80 MHz channel compared to 40 MHz channel. Whereas, in the case of the two devices sharing the wireless link we have about 36% throughput increase with 80 MHz channel compared to 40 MHz channel.

References

1. Vanhatupa T (2013) Wi-Fi capacity analysis for 802.11 ac and 802.11 n: theory & practice. Ekahau Inc
2. Kelkar NS (2015). A survey and performance analysis of IEEE 802.11ac WiFi networking. ISSN 2348-1196 (print) Int J Comput Sci Info Technol Res ISSN 2348-120X (online) 3 (2):808–814. (Month: April–June 2015)
3. Ravindranath NS, Singh I, Prasad A, Rao VS (2016) Performance Evaluation of IEEE 802.11 ac and 802.11 n using NS3. Indian J Sci Technol 9(26):0001. https://doi.org/10.17485/ijst/2016/v9i26/93565 (July 2016, ISSN (Print): 0974–6846, ISSN (Online): 0974–5645)
4. Maraj D, Sefa R, Maraj A (2015) QoS evaluation for different WLAN standards. In: 2015 23rd international conference on software, telecommunications and computer networks (SoftCOM), Split 2015, pp 190–194. https://doi.org/10.1109/softcom.2015.7314115
5. Maraj D, Maraj A (2015) Performance analysis of WLAN 802.11 g/n standards using OPNET (Riverbed) application. In: 2015, 57th international symposium ELMAR (ELMAR), Zadar, 2015, pp 129–132. https://doi.org/10.1109/elmar.2015.7334513
6. WiFi basics and best practices—channel planning best practices. Article ID: 2012, CISCO
7. High-density Wi-Fi design principles, white paper, Copyright, 2012, Aerohive Networks, Inc
8. Lepaja S (2014) Wireless local area networks planning. IX annual international meeting of Alb_Science Institute. Prishtina, Kosovo
9. Aruba 802.11ac Networks, Validated Reference Design, Copyright Information, © 2015 Aruba Networks
10. Matthew S (2013) Gast, 802.11ac: a survival guide—Wi-Fi at gigabit and beyond. By Publisher: O'Reilly Media, Final Release Date: July 2013, Pages: 154
11. Christian JE (2015) Evolutions in 802.11 design high-density, capacity planning and survey methodologies
12. Bhartia A, Chen B, Wang F, Pallas D, Musaloiu-ER, Lai TTT, Ma H (2017 November) Measurement-based, practical techniques to improve 802.11 ac performance. In: Proceedings of the 2017 internet measurement conference
13. Witono T, Dicky Y (2017) Optimization of WLAN deployment on classrooms environment using site survey. In: 2017 11th international conference on information & communication technology and system (ICTS), Surabaya, 2017, pp 165–168. https://doi.org/10.1109/icts.2017.8265664
14. Kryvinska N, Lepaja S, Nguyen HM, (2003) Service and personal mobility in next generation networks. In: Proceedings fifth IEEE international conference on mobile and wireless communications networks (MWCN 2003), 27–29 October, Singapore, 2003, pp 116–119
15. Bashah NSK, Kryvinska N (2012) Quality-driven service discovery techniques for open mobile environments and their business applications. J Ser Sci Res 4(1):71–96
16. Lepaja S, Lila A, Kryvinska N, Nguyen HM (2003) A framework for end-to-end QoS provisioning in mobile internet environment. In: The fifth IEEE international conference MWCN-03, 27–29 October 2003
17. Zimmerman Tim (2012) Best practices for WLAN site surveys that save money. Gartner, Inc
18. Auer L, Kryvinska N, Strauss C (2009) Service-oriented mobility architecture provides highly-configurable wireless services. IEEE, pp 1–1
19. Bashah NSK, Kryvinska N, van Thanh D (2012) Quality-driven service discovery techniques for open mobile environments and their business applications. J Ser Sci Res 4:71–96. https://doi.org/10.1007/s12927-012-0003-4
20. Shatri B, Imeri I, Maraj A (2008) Broadband wireless access (BWA) implementation in NGN network. Communication theory, reliability, and quality of service, 2008, CTRQ'08
21. Kamal Bashah NS, Kryvinska N, van Thanh D (2012) Novel service discovery techniques for open mobile environments and their business applications. In: Snene M (ed) Exploring services science. Springer, Berlin Heidelberg, pp 186–200

22. Kryvinska N, Strauss C, Auer L (2010a) Next generation applications mobility management with SOA—a scenario-based analysis. IEEE, pp 415–420
23. Kryvinska N, Strauss C, Collini-Nocker B, Zinterhof P (2010c) A scenario of service-oriented principles adaptation to the telecom providers service delivery platform. IEEE, pp 265–271
24. Kryvinska N, Strauss C, Collini-Nocker B, Zinterhof P (2008f) A scenario of voice services delivery over enterprise W/LAN networked platform. ACM Press, p 332
25. Maraj A, Imeri I (2009) WiMAX integration in NGN network, architecture, protocols and Services. WSEAS Trans Commun 8(7):708–717
26. Kryvinska N, Strauss C, Collini Nocker B, Zinterhof P (2011) Enterprise network maintaining mobility—architectural model of services delivery. Int J Pervasive Comput Commun 7:114–131. https://doi.org/10.1108/17427371111146419
27. Kryvinska N, Strauss C, Thanh DV, Zinterhof P (2011) Managing global calls placing in mash-upped enterprise. Int J Space-Based Situated Comput 1:244. https://doi.org/10.1504/IJSSC.2011.043505
28. Kryvinska N, Strauss C, Zinterhof P (2009b) Mobility in a multi-location enterprise network, case study: global voice calls placing. IEEE, pp 1–7
29. S. Lepaja (2005) Mobility and quality-of-service in global broadband communication networks. Ph.D. dissertation, institute for broadband communication networks, Vienna University of Technology
30. Shehu A, Maraj A, Mitrushi R (2010) Studying of different parameters that affect QoS in IPTV systems. In: Proceedings international conference on telecommunications and information (WSEAS), 2010
31. Shehu A, Maraj A, Mitrushi RM (2010) Analysis of QoS requirements for delivering IPTV over WiMAX technology. Int Conference on Soft Telecommunications Comput Netw (SoftCOM), 2(1):380–385
32. Shih CH, Xu YY, Wang YT (2012) Secure and reliable IPTV multimedia transmission using forward error correction. Int J Digital Multimedia Broadcast 2012:8. https://doi.org/10.1155/2012/720791 (Article ID 720791. Research Article)

Traffic Fluctuations Optimization for Telecommunication SDP Segment Based on Forecasting Using Ateb-Functions

Ivan Demydov, Ivanna Dronyuk, Olga Fedevych
and Vasyl Romanchuk

Abstract This work proposes a network traffic prediction method based on a theory of differential equations with them solutions can be written as Ateb-functions. The proposed method first uses the hypothesis about cyclic nature of network traffic pulsations. Then, description of the traffic intensity fluctuations into the computer network builds a prediction model, as nonlinear oscillating system with single degree-of-freedom under the conditions of small-scale disturbances. From the simulation, the proposed prediction method in distinction from existing methods gives more qualitative solution. Thereafter, the obtained values of the predicted traffic intensity are used as one of the key indicators in solving the problem of optimal resource allocation in heterogeneous network platforms by correcting the metrics or priorities of traffic flows routing. Decisions taking on a basis the procedures for evaluating alternatives by mutually related and mutually controversial criteria using fuzzy logic approach with triangular fuzzy numbers aggregation.

Keywords Traffic fluctuations · Forecasting · Ateb-function · Optimization

I. Demydov (✉) · I. Dronyuk · O. Fedevych · V. Romanchuk
Lviv Polytechnic National University, S. Bandera Street, 12, Lviv, Ukraine
e-mail: ivan.demydov@gmail.com

I. Dronyuk
e-mail: ivanna.m.droniuk@lpnu.ua

O. Fedevych
e-mail: olha.fedevych@gmail.com

V. Romanchuk
e-mail: romanchuk@lp.edu.ua

© Springer International Publishing AG, part of Springer Nature 2019
N. Kryvinska and M. Greguš (eds.), *Data-Centric Business and Applications*,
Lecture Notes on Data Engineering and Communications Technologies 20,
https://doi.org/10.1007/978-3-319-94117-2_4

1   Introduction

This research work contributes to the methods of forecasting and modeling of traffic flows in computer networks, which are widely used and developed to date, being in a state of constant elaboration, and are continuously implemented in the latest concepts such as IoE, XaaS cloud solutions for effective processing significant volumes of custom content and associated metadata, BigData, and more. The purpose of the study is to create information technology for monitoring and adaptive control of pulsating (non-stationary) traffic flows in computer networks by predicting load intensity and redistribution of node capacity in a given segment of a network service delivery platform.

In our understanding, which coincides with the classical definitions for packet switching technologies, the term "traffic" denotes the network load or amount of information, in other words, the number of packets transmitted over a specified time unit [1]. In turn, bandwidth is the limitation of traffic volumes, which exists depending on the software and hardware on a specific area of the telecommunication network [2].

Scientific research of heterogeneous network traffic shows that it's nature is self-similar or fractal in fact [3]. We can postulate, that the methods used previously for the calculation and modeling of network systems and based on the use of Poisson arrivals theory do not provide a complete and accurate picture of what processes occur in the computer network of service delivery platform [4].

In addition, self-similar traffic has a special structure that is persisted when using multiple scaling. In implementing this approach, as a rule, there is a certain amount of intensity surges even at relatively low amounts of traffic. This phenomenon worsens the characteristics (increases latency, packets' jitter) when passing such traffic through the nodes of computer network equipment. In practice, this is manifested in the fact that packets in the conditions of high speed of their transmission through a computer network, arrive at the node not separately, but rather large in volume groups, which can lead to their losses due to the limited volume of buffer memory of commutation equipment that was pre-calculated using classical techniques.

The described features of network traffic caused a large increase in the number of publications and scientific research on the methods of analysis, modeling, and prediction of self-similar traffic [1–4].

Traffic of a Gigabit Ethernet channel exhibits self-similar properties with a likely high value of the Hurst index, as shown in the research in [5]. Attention deserves time intervals during which the network is impacted by anthropogenic nature, because at that time there is a high probability of buffers overflow in the network nodes, which can lead to queues in the system and, consequently, a sharp deterioration in the quality of service for the whole range of existing services provided by the communication provider. It is also shown that the self-similar properties of traffic allow with a sufficient degree of credibility to predict the appearance of time intervals in the segment of the network platform where there is an overload due to the productivity of equipment and communication lines, which in turn makes it possible to construct a system with dynamic control of bandwidth reserves for specified traffic types [6]. So, work [5] confirms our view that traffic forecasting plays a significant role in the development of traffic management algorithms aimed at improving the quality of service. It should be noted that with the growth of volumes of info communication services, there is a shortage of network resources, throughput capacity of data transmission channels, which, as already mentioned above, affects the quality of services provided to users.

In the works [5, 7], the authors developed the computational schemes for forecasting traffic using three methods: the Boxes-Jenkins method, the derivative method and the modified Boxes-Jenkins method using splines. In [8], based on this approach, a method for forecasting self-similar traffic, which can be used for modeling data flows in packet networks, is proposed. The developed algorithm is based on the Boxes-Jenkins procedure. The results of the forecast obtained by this algorithm are based on a large number of data, the values of which more influence the result of the forecast, the closer this data in time to the moment of forecasting.

Based on the aggregated models using wavelet transformations [9], the forecasting of traffic has been realized considering statistical characteristics, as well as the properties of scaling invariance.

There is interesting and quite novel the method for traffic forecasting, presented in [10], based on a deep learning architecture and the Spatiotemporal Compressive Sensing method to extract the low-pass component of network traffic by adopted discrete wavelet transform, and then high-pass component, and obtain a predictor of network traffic.

Therefore, research on a certain topic has shown that in general methods for modeling and predicting the transmission of traffic in modern computer networks have not been sufficiently studied. This indicates the need to develop new models and methods for forecasting traffic to increase the efficiency of using computer network resources.

In addition, it is important that the task of controlling the workload of SDP is not only multicriteria, but often criteria that are chosen to solve its optimization component are mutually controversial, and the solutions themselves are subject to numerous limitations. Examples of such parameters are transmission delay, jitter, allocated bandwidth, current or predicted traffic intensity, economic performance, etc. Difficulty also occurs when an attempt is made to make an unambiguous decision in the context of additional technological constraints, for example when sensor networks with a limited energy budget are considered [11] or wireless networks with limited radio frequency and energy resources, ad hoc, etc. [12]. To effectively overcome such situations, several methods have been developed based on the use of the theory of fuzzy logic. In papers [13–17] aspects of load management in telecommunication networks are considered, in [18] the approach of fuzzy logic application in Random Early Detection is shown to equalize peak loads in systems of users' load balancing and distribution. To a lesser extent, this direction also applies improvement of the technical efficiency indexes of sensor nets, that discussed in [11], and increasing the efficiency of routing in wireless network systems, delay tolerant nets [12, 19, 20, 21]. Despite this variety, all these works have not considering a simple and effective fuzzy-logic approach with TFN (Triangle Fuzzy Numbers) aggregation of parametric estimates based on a set of scales, which, for example, is used in [22, 23].

In the next subchapters of this work we discuss the principles of mathematical modeling to create a mechanism for predicting traffic fluctuations based on the theory of differential equations, whose solutions have the form of the Ateb-function. After that, the received values of the predicted intensity of traffic are used as one of the key indicators in solving the problem of optimal resource allocation in heterogeneous network platforms by correcting the metrics or priorities of traffic flows' routing and making decisions based on the procedures for evaluating alternatives according to the given criteria using fuzzy logic on the basis of TFN.

2   Mathematical Model and Forecasting of Network Traffic Fluctuations

As a rule, the dynamics of changes in the network system's workload is described, and, consequently, its forecasting is performed considering a certain volume of the known, or rather, fixed statistical information on the state of specific areas of the telecommunications platform. There are two structural approaches to predicting the traffic intensity. The first involves the distributed operation of some network data collection software, such as the Wireshark type. Second—centralized—involves the use of SDN-controller. Frequency of calculations of the traffic intensity is determined by the dynamics of load fluctuations since the nature of these fluctuations impacts on the accuracy of the forecasting. For public networks, these indicators are not very important, although they have a significant impact on service quality metrics. In networks with guaranteed quality of traffic flows servicing, load fluctuations have a detrimental effect on the stability of their operation (packet losses and delays). Centralized flow management allows us to control completely the load balancing process at the specified segment of the network platform, to form a complete matrix of predictive intensities of data channels utilization by their traffic-handling capacity.

Let us consider the fluctuations of traffic in a computer network as the function of time, as a nonlinear oscillating system with single degree-of-freedom in conditions of small disturbances. The simulation of the traffic intensity behaviour in the computer network represented by the function $x(t)$ is performed by a simple differential equation with a small parameter ε in the form

$$\ddot{x} + a^2 x^n = \varepsilon f(x, \dot{x}, t),\tag{1}$$

where $x(t)$—is the number of packets in the network at time t; α—is the constant that determines the period of traffic fluctuation, $f(x, \dot{x}, t)$—is an arbitrary analytic function used to simulate small deviations of traffic from the main component of oscillations, n—is the number that defines the degree of nonlinearity of the equation that affects the period of the main component of oscillations.

Under the following conditions on α and n $\alpha \neq 0$, $n = \frac{2k_1 + 1}{2k_2 + 1}$, $k_1, k_2 = 0, 1, 2\dots$, is was proved [24], that the analytic solution of Eq. (1) is represented in the form of Ateb-functions. For tasks of traffic flows' forecasting in the information network or telecommunication platform, it is important to choose the type of function f, in such a manner the peculiarities of the operation of given network system are taken into consideration. An approach that considers small disturbances in the form of periodic functions is appropriate for modeling a network with a smooth change in the intensity of traffic flow. However, the authors [25] decided to consider perturbation as a convolution of weighted delta functions. Such a description is more in line with the network, which has sharp changes in the intensity of traffic. Let's consider the function f in the form

$$f(x, \dot{x}, t) = \sum_{i=1}^{N} a_i \delta(t_i),\qquad(2)$$

where N is the number of disturbances per time interval $[0, T]$, a_i—disturbance range $-A \leq a_i \leq A$, A—the maximum disturbance range (is generated randomly), δ—Dirac delta function, t_i—the moment of time in which there is an i-th disturbance that is generated randomly.

To construct a solution, we first consider Eq. (1) without perturbation function

$$\ddot{x} + a^2 x^n = 0.\qquad(3)$$

If we write the differential equation of the second order (3) in the form of a system of differential equations of the first order, considering the replacement of the variables $y = \dot{x}$, then the differential equation of the second order (1) is transformed into the following system of differential equations of the first order:

$$\begin{cases} \frac{dx}{dt} - y = 0 \\ \frac{dy}{dt} + \alpha^2 x^n = 0 \end{cases},\qquad(4)$$

and the solution (4) is represented by the periodic Ateb-functions in the following form

$$\begin{cases} x = aCa(n, 1, \varphi) \\ y = a^{\frac{1+n}{2}} hSa(1, n, \varphi) \end{cases},\qquad(5)$$

where a—an amplitude of intensity oscillations, $Ca(n, 1, \varphi)$, $Sa(n, 1, \varphi)$—Ateb-cosine and Ateb-sine, respectively, $h^2 = \frac{2a^2}{1+n}$. The variable φ is corresponding to the time t by equation of the form

$$\varphi = \frac{a^{\frac{n-1}{2}}}{L} t + \varphi_0, \tag{6}$$

where L is some constant, φ_0—the initial phase of oscillations, which are determined from the initial conditions for the Eq. (3).

The following system of equations expresses periodic conditions:

$$\begin{cases} Ca(n, 1, \varphi + 2\Pi) = Ca(n, 1, \varphi) \\ Sa(1, n, \varphi + 2\Pi) = Sa(1, n, \varphi) \end{cases}, \tag{7}$$

where Π is half-period of Ateb-function. When we put the expressions (5) and (6) in Eq. (4) and consider condition (7), we obtain the following expression for calculating the constant L in formula (6):

$$L = \frac{2B\left(0.5, \frac{1}{1+n}\right)}{\Pi(1+n)h}. \tag{8}$$

In expression (8) $B(x, y)$ defines the full Beta-function with arguments $x = 0.5$, $y = 1/1 + n$. Considering formula (6) and expression $Ca(n, 1, \varphi)^{m+1} + Sa(1, n, \varphi)^2 = 1$, we receiving resulting formula to calculate the half-period of the Ateb-function

$$\Pi(n, 1) = B\left(0.5, \frac{1}{1+n}\right). \tag{9}$$

Let us consider the initial conditions for the system of differential Eqs. (4). In [26], these initial conditions for the differential Eq. (1) are considered as follows

$$x(0) = 0, \quad \dot{x}(0) = const, \tag{10}$$

when $n = \frac{1}{2k+1}$ and $k \to \infty$. Such initial conditions cannot be used to simulate the intensity of traffic flows. We assume that the values of traffic flows and changes in such traffic values are constantly determined at the time $t = 0$. According to these assumptions, we determine the initial conditions as

$$x(0) = c_1, \quad \dot{x}(0) = c_2, \tag{11}$$

where c_1 determines the initial traffic and $c_1 \neq 0$, c_2 determines the change of initial traffic and can be zero at the initial time moment $t = 0$. Considering, that $Ca(n, 1, 0) = 1$ and $Sa(1, n, 0) = 0$ and Eq. (5) we obtain from the initial conditions (11) that $c_1 = a$ i $c_2 = 0$.

We use the asymptotic method for constructing a solution of Eq. (1) based on (5). The asymptotic method builds a solution in the form of a numerical series with a small parameter ε

$$x(t) = \sum_{i=1}^{\infty} \varepsilon^i x_i(t). \tag{12}$$

For numerical simulations, we must discard terms of series that contain ε by in order of magnitude more than M. Therefore, we obtain a solution with the accuracy of the order ε^{M+1}. We will find solutions in the form of series for a small parameter ε, as follows

$$x(t) = \sum_{i=1}^{M} \varepsilon^i x_i(t). \tag{13}$$

We substitute a numeric series (13) into the left part of (1), and after that expressions with the same order of small parameter ε were equated with each other. Now consider the Eq. (1), which is transformed into a system of differential equations with variables (just as done in (3) in (4)) of the form

$$\begin{cases} \frac{dx}{dt} - y = 0 \\ \frac{dy}{dt} + \alpha^2 x^n = \varepsilon f(t, x, y) \end{cases}. \tag{14}$$

The degrees of a small parameter ε represent the asymptotic approximation. To create the first approximation in (13) according to the first order of ε ($M = 1$ in the formula (13)), we make a replacement of the variables by type

$$x = \xi + \varepsilon f(t, \xi, \zeta), \ y = \zeta = \dot{x}. \tag{15}$$

and equate the coefficients at the same degrees of ε, and, we discard the terms of the higher order relative to ε. As shown in [27], the variables ξ, ζ are calculated with respect to the solution without disturbance (5), can be taken as an improved first approximation of solutions (14) and (15).

On the basis of the developed mathematical model, a new method was proposed for simulating the intensity of traffic flows in info communication network systems. The basic hypothesis is that the nature of network traffic has a periodic nature. In this case, the simulation of the daily cycles of periodicity was carried out.

The first step was to simulate the main traffic trend. However, this approach does not consider the sharp fluctuations in traffic. Therefore, it was proposed to use the equation to simulate and predict the behavior of the intensity of traffic flows in a network platform, namely, use differential equations describing nonlinear oscillating systems with single degree-of-freedom and with small disturbances. It is assumed that the main periodic component simulates the daily volatility of traffic in the network, while the deviation of traffic from the main trend is modeled using a

78   I. Demydov et al.

delta-function with a small parameter. For the solution of differential equations with small parameter, the asymptotic method of Bogolyubov-Mitropolskiy was used [28].

Figure 1 presents the results of application of the developed software complex, within the framework of the implemented new method of monitoring and fore-casting the intensity of traffic flows. On their basis, it is suggested to carry out a redistribution of the load on a computer network segment, for example, using a modified Dijkstra algorithm applying heuristics or by estimating and prioritizing routes in the process of multi-path routing based on the theory of fuzzy logic.

On Figs. 1 and 2 are shown samples of network traffic intensity profiles collected at the Department of Automated Control Systems of the Lviv Polytechnic National University, obtained from 20 working computers roughly enrolled to BigData processing in the local computing cluster from 8:30 to 17:30, as well as 4 computers loaded from 8:30 to 21:00 and 32 workstations enabled and working from 8:30 to 16:00.

Figure 1 shows a sample of traffic profile (red line), and a predicted forecast (black). To calculate the predicted traffic values, the Ateb-function was selected with parameters m = 1/7, n = 3. Figure 2 shows the results of the normal distri-bution of the delta function application for the Ateb-function with the parameters m = 1/5, n = 1 to improve the process of traffic intensity values forecasting according to the developed method.

To improve the characteristics of the forecast, the mechanism of embedding of delta functions was additionally introduced. The graph is redrawn automatically. The results of the program operation in this case are presented on Fig. 2.

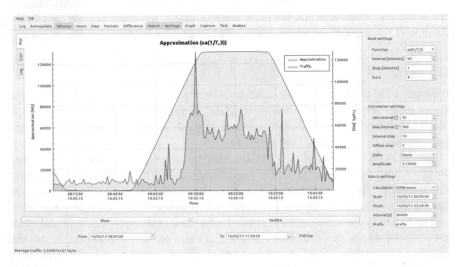

Fig. 1 The graphical comparison of traffic intensity and scalable predictive values based on the Ateb-function with parameters m = 1/7, n = 3

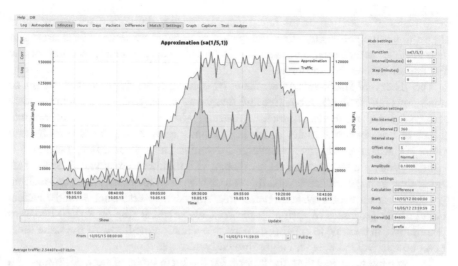

Fig. 2 The graphical comparison of the traffic intensity in prediction after the application of the normal distribution of the delta function for the Ateb-function with parameters m = 1/5, n = 1

From Figs. 1 and 2 it should be concluded that it is possible to achieve a rather qualitative forecasting of the profile form of the traffic intensity, particularly the nature of emissions, even with small volumes of data on its previous values. The prediction accuracy, as well as the methods for raising it, are part of future authors' research plans. At present, in [26] authors found that it is possible to achieve correlation between predicted and real (exemplary) traffic profiles at the level of 97–99% when performing short-term forecasting (for a period of 5–10 min), with medium-term forecasting—85 to 88% (per the term up to 1 day), with the worst recorded result not less than 75%. The results contributed show an improvement with respect to the derivative method up to 30% [7].

3 Approach to Correction of Routing Metric and Priorities Based on the Use of Fuzzy Logic Theory with Triangle Fuzzy Numbers

Optimization of the processes of redistribution (load balancing) between different routes in network platforms can be based on a certain set of criteria describing the quality of service (QoS): network latency, jitter, packet loss indexes (PER), network throughput (which is obtained on the basis of load intensity forecasting); QoE indicators; the priority of user servicing, the speed of their movement, the type of service, and for the case of mobile radio access systems—the signal level of the stations of each particular access network [29]. The combination of these criteria in the process of operation of the heterogeneous service delivery platform by some algorithms can significantly affect the quality of network services for end users and their level of satisfaction, and also the technical and economic efficiency of the distribution of network resources.

The theory of fuzzy sets is a means of solving tasks for the aggregation of ambiguous (ambivalent), subjective and fuzzy evaluation judgments about the state of some partial parameter or metric (metric), which directly affects the optimal choice or priority of the route, which is especially relevant for the case multi-path routing, that is, the existence of several alternative paths of data transmission.

Fuzzy sets, unlike the classical set theory, can divide the results on a scale corresponding to linguistic terms, and this can also be used for further decision-making by experts or an expert system. Using this approach to construct a mathematical model of the processes of assigning network resources, it (the approach) becomes quantitative, in contrast to existing subjective evaluation methods. At the same time, it can be relatively simple and algorithmically automated.

The scheme for calculating the criteria for the main types of network traffic of a heterogeneous telecommunications platform is presented on Fig. 3. Thus, it is proposed to divide all parametric criteria into two groups: QoS-dependent and those that depend directly on the properties of the radio interface (for the case of the wireless network access system analysis performing). The first group traditionally relates delay, jitter, packet loss ratio (PLR), network platform bandwidth, and, accordingly, to the second—predicted load intensity, relative cost of network

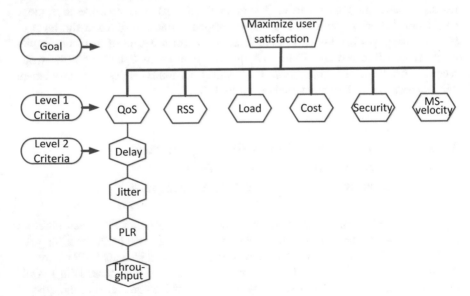

Fig. 3 A set of criteria for deciding on the initialization of the process of resource redistribution/prioritization of routes

resources usage, security level, for wireless telecommunications access platform—users' mobility level [km/h] (velocity), signal strength (RSS, [dBm]) (see Fig. 3; Table 1).

After preparing the parameters for computing to decide on the priorities for the routes (their areas selected on the basis of the next hop), we introduce the scale of the values of each criterion [for their presentation in the format of triangular fuzzy numbers (TFN)] regarding the initiation of such resource redistribution, and put it in line with the Likert scale [1—"cannot be applied" (0,0,0.25), 2—"can be used" (0.0.25.0.75), 3—"average", (0.25, 0.5, 0.75), 4—"highly likely to be used" (0.5.0.75.1), 5—"to be applied" (0.75.0.75.1)], taking into account the estimates of Table 1.

Accordingly, in the process of fuzzification, the calculation of triangular fuzzy numbers (Triangle Fuzzy Number) $\tilde{Q}_{mn}$ for the aggregated estimation of selected sections (alternatives) of the route to some platform network node. The latter is performed in accordance with the values of the criteria of the service quality group from Table 1 which are additionally multiplied with normalized weighting factors, that is, according to the estimates of the values of the criteria parameters for making a decision on the redistribution of network resources on the scale given in Table 1. As a matter of fact, the re-routing procedure itself (or changing the priorities of alternative route paths) may indeed lead to changes in QoS indices for the served user.

$$\tilde{Q}_{mn} = (q_1, q_2, q_3)_{mn} = \sum_{i=1}^{4} \left(W_{imn} \times \tilde{L}_{imn} \right) \qquad (16)$$

Table 1 Estimates of criteria parameters of network systems for initiating the process of resource redistribution/giving priority to routes on the Likert scale

Scale	1	2	3	4	5	Min/max values
Predicted workload (%)	70–100	50–70	40–50	30–40	0–30	0–100
Cost	1–2	2–3	3–4	4–8	8–10	1–10
Security	1–2	2–4	4–6	6–8	8–10	1–10
Velocity (km/h)	<120	80–120	60–80	40–60	0–40	0–160
Delay (ms)	<300	200–300	100–200	50–100	10–50	10–500
Jitter (ms)	<30	20–30	10–20	5–10	1–5	10–30
Packet loss ratio (PLR) (%)	<8	6–8	4–6	3–4	1–3	1–8
Network platform band-width (Mbps)	<0/1	0.1–10	1–50	50–100	100–200	0/1–200
RSS (dBm)	< −110	−100 to −110	−90 to −100	−75 to −90	−55 to −75	−110 to −55

$$q_{j_{mn}} = \sum_{i=1}^{4} \left(W_{imn} \times l_{ijmn} \right), \quad (j=1,2,3; \; n=1,2), \tag{17}$$

where q_1, q_2, q_3 is the lower level of the generalized estimate $\tilde{Q}$, its main value and the upper level, respectively, $\tilde{L}_{imn} = (l_{i1}, l_{i2}, l_{i3})_{mn}$ is a triangular fuzzy number that characterizes the parameter of the network node according to the i-th criterion, and the node itself belongs to a heterogeneous network platform with serial number m and uses n-th access technology [22]. Here l_{i1}, l_{i2}, l_{i3} are the lower level of the linguistic variable, its main value, and the upper level, according to the format of triangular fuzzy numbers (Triangular Fuzzy Number) [23].

At the same time and in the same way (16–17), a generalized assessment of priority for a selected (alternative) route section to an access node of network platform is calculated $\tilde{P}_{mn}$, which is calculated in accordance with the technology which was applied in this section and partial estimates of the group of technical (technological group) parameters that based on the interpretation of input data (see Table 1), in particular the predicted load intensity for defined part of the network platform. This score, together with the value of expression (16), will qualitatively characterize a particular route or its sections, which for example, was laid down to a specific node of the network access platform.

The aggregated qualitative assessment of the priority of a route to a given network node can be defined as follows

$$\tilde{R}_{mn} = (r_{1mn}, r_{2mn}, r_{3mn}) = 1/2 \times \left(\tilde{Q}_{mn} + \tilde{P}_{mn} \right)$$
$$= 1/2 \times (q_{1mn} + p_{1mn}, q_{2mn} + p_{2mn}, q_{3mn} + p_{2mn}). \tag{18}$$

Finally, we will make a defuzzification of the received fuzzy (triangular) number (18) in accordance with the method proposed in [23] for obtaining an aggregated quantitative estimate:

$$R_{mn} = 1/3 \times \sum_{t=1}^{3} r_{tmn}. \tag{19}$$

In [22], an example of the implementation of fuzzification and defuzzification procedures into some network platform is presented to optimize the distribution of its resources and the execution of vertical handover through mathematical and probabilistic statistical simulation (see Fig. 4).

For greater clarity, we assume that we have 6 alternative route sections (Section 1–Section 6) under conditions of the need to route a path from a user of a heterogeneous network access platform, when using GSM and LTE technologies to

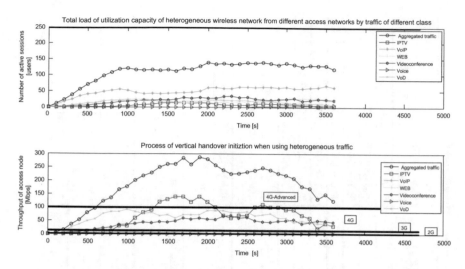

Fig. 4 The load intensities on a heterogeneous network platform from various network access systems and the threshold for initiating an intelligent vertical handover

organize its radio interfaces. We need to choose a better alternative (in fact—to select the "last mile" route) under the current operating conditions, that is, when the system has some set of parameters, see Table 2. The choice will be made in accordance with Fig. 3 and scales of Table 1.

The most optimal result for establishing the highest priority of the route segment for a particular user and, accordingly, implementing the redistribution of the network resources of the heterogeneous network platform was obtained for section 5 of the route, that is, the access node BS2 (network subsystem of LTE technology) (shown in Fig. 5, according to Table 2). The result for the route section 4, that is, the access node BS1 (also the network subsystem of the LTE technology), is lower, although, according to the membership function, belongs to the same linguistic term "High" (see Fig. 5). Thus, the theory of fuzzy sets demonstrates the quantitative difference between the qualitatively homogeneous solutions of the optimization problem of the converged resources distribution at network-dependent levels of heterogeneous distributed network platforms, which is useful, in particular, for solving the problems of multi-path routing.

Table 2 Estimation results based on fuzzy sets (triangular fuzzy numbers) for network access nodes and their defuzzification ($\tilde{R}_{mn}$)

	Section 1 BS1 (GSM)1/2 × ($\tilde{Q}_{mn} + \tilde{P}_{mn}$) ($\tilde{R}_{mn}$)	Section 2 BS2 (GSM) 1/2 × ($\tilde{Q}_{mn} + \tilde{P}_{mn}$) ($\tilde{R}_{mn}$)	Section 3 Section 3 BS3 (GSM) 1/2 × ($\tilde{Q}_{mn} + \tilde{P}_{mn}$) ($\tilde{R}_{mn}$)
Jitter	(0,2; 0,005; 0,005)	(0,2; 0,005; 0,005)	(0,105; 0,005; 0,005)
PLR	(0,04; 0,053; 0,03)	(0,04; 0,053; 0,03)	(0,008; 0,053; 0,003)
Bandwidth	(0,15; 0,095; 0,075)	(0,1; 0,0095; 0,06)	(0,01075;0,0057; 0,1750)
Delay	(0,105; 0,04)	(0,05; 0,105; 0,04)	(0,1; 0,105; 0,0055)
Cost	(0,0025; 0,0125; 0,01)	(0,0025; 0,0125; 0,01)	(0,006; 0,011; 0,001)
Security level	(0,0225; 0,0525; 0,15)	(0,0225; 0,0525; 0,07)	(0,00375;0,05375; 0,15375)
User velocity	(0,032; 0,017; 0,12)	(0,032; 0,017; 0,02)	(0,0052; 0,012; 0,12)
RSS	(0,0125; 0,0875; 0,125)	(0,0625; 0,0375; 0,052)	(0,125;0,00475; 0,0475)
Predicted intensity of workload	(0,00525; 0,00525; 0,00525)	(0,00525; 0,00525; 0,00525)	(0,00525; 0,00525;0,00525)
Fuzzified estimate ($\tilde{R}_{mn}$)	(0,51475; 0,43275; 0,56025)	(0,51475; 0,29725; 0,29275)	(0,36895; 0,2555; 0,516)
Result of de-fuzzification R_{mn}	0,502583333	0,36825	0,38015
	Section 4 BS1 (LTE) 1/2 × ($\tilde{Q}_{mn} + \tilde{P}_{mn}$) ($\tilde{R}_{mn}$)	Section 5 BS2 (LTE) (conditionally best alternative route) 1/2 × ($\tilde{Q}_{mn} + \tilde{P}_{mn}$) ($\tilde{R}_{mn}$)	Section 6 BS3 (LTE) 1/2 × ($\tilde{Q}_{mn} + \tilde{P}_{mn}$) ($\tilde{R}_{mn}$)
Jitter	(0,15; 0,005; 0,005)	(0,2; 0,005; 0,005)	(0,105; 0,005; 0,005)
PLR	(0,04; 0,053; 0,023)	(0,04; 0,053; 0,03)	(0,04; 0,053; 0,008)
Bandwidth	(0,25; 0,0095; 0,175)	(0,25; 0,195; 0,175)	(0,01; 0,005; 0,175)
Delay	(0,1; 0,105; 0,0055)	(0,1; 0,105; 0,04)	(0,1; 0,105; 0,0055)
Cost	(0,006; 0,0225; 0,001)	(0,0051; 0,0225; 0,01)	(0,006; 0,006; 0,001)

(continued)

Table 2 (continued)

	Section 4 BS1 (LTE) $1/2 \times (\tilde{Q}_{mn} + \tilde{P}_{mn})$ $(\tilde{R}_{mn})$	Section 5 BS2 (LTE) (conditionally best alternative route) $1/2 \times (\tilde{Q}_{mn} + \tilde{P}_{mn})$ $(\tilde{R}_{mn})$	Section 6 BS3 (LTE) $1/2 \times (\tilde{Q}_{mn} + \tilde{P}_{mn})$ $(\tilde{R}_{mn})$
Security level	(0,0325; 0,075; 0,15)	(0,0325; 0,0525; 0,25)	(0,0375; 0,0525; 0,15)
User velocity	(0,032; 0,012; 0,12)	(0,032; 0,017; 0,12)	(0,032; 0,012; 0,12)
RSS	(0,1025; 0,2375; 0,225)	(0,125; 0,2375; 0,225)	(0,125; 0,2375; 0,225)
Predicted intensity of workload	(0,00525; 0,00525; 0,00525)	(0,00525; 0,00525; 0,00525)	(0,00525; 0,00525; 0,00525)
Fuzzified estimate $(\tilde{R}_{mn})$	(0,71825; 0,52475; 0,70975)	(0,78985; 0,69275; 0,86025)	(0,46075; 0,48125; 0,69475)
Result of de-fuzzification R_{mn}	0,6509166	0,78095 (conditionally best alternative route)	0,545583333

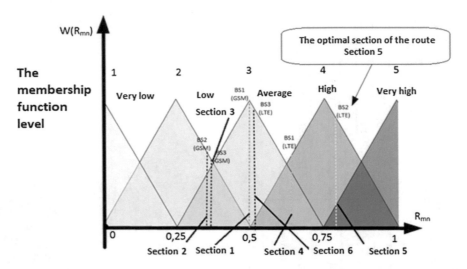

Fig. 5 Determination of the optimal alternative when selecting the route section from the user of the heterogeneous network access system using the Likert scale for the set of criteria values (Table 1)

4 Conclusion

The work proposes a combination of several approaches to increase the effectiveness of traffic management and the adoption of optimal solutions. Namely, the introduction of the Ateb-prognostication method for constructing a traffic matrix in the segment of the network system and further aggregated estimation of the chosen sections (alternative variants) of the route to some network node of the platform based on the theory of fuzzy sets is proposed. The result is an improvement in the method of routing the traffic of data flows in the service delivery platform segment, which consists in predicting the intensity of the pulsations of this traffic on the set of existing routes, providing the correction of the data transmission route and possible reducing the load of the buffers of the network equipment. And, accordingly, the delays that occur when the traffic ripples. In this way, the quality of the computer network platform operation increases, in particular, when the user content is transmitted according to the criteria of the QoS group.

To solve the problem of accepting the optimal (according to the criteria of the QoS group and other technological criteria) solution to the routing procedure while servicing users of network service delivery platforms, a centralized method for managing the priority of routes or their sections was proposed (could also be applied for "next hop" routing type). The proposed approach avoids ambiguity in the interpretation of ambiguous, fuzzy-evaluated and subjective judgments in the process of multi-criteria optimization of the network resources utilization, especially in conditions of significant load intensity in the network platform using the SDN controller. To study the processes of real heterogeneous network systems operation in the conditions of users' high velocity authors have been developed a software simulation model, which, in turn, implements the contributed mathematical model of the decision-making process for mentioned procedures based on the use of fuzzy logic methods. This allows us to configure a large number of simulation parameters using Ateb-prognostication, auxiliary mathematical models, for example—to describe and predict the movement (mobility) of users on heterogeneous network platforms. Obviously, the choice of the optimal route or its alternative sections at such platforms based on the solving of the multicriteria optimization task of making a decision on the redistribution of network resources is non-trivial, since it is affected by several dynamic and mutually interrelated (often contradictory) factors. Thus, their aggregation is effective in accordance with the approaches proposed in this work.

Acknowledgements The authors acknowledge the support from the Ministry for Education and Science of Ukraine and from the joint Ukrainian-Austrian R&D project titled "Traffic and telecommunication networks modelling" (0117U001612).

References

1. Feldmann A, Gilbert AC, Willinger W (1998) Data networks as cascades: investigating the multifractal nature of internet WAN traffic. In: Proceedings of the ACM SIGCOM, pp 42–55
2. Leland W, Taqqu M, Willinger W, Wilson D (1994) On the self–similar nature of ethernet traffic. IEEE/ACM Trans Netw 2(1):1–15
3. Taqqu M, Willinger W, Sherman R (1997) Proof of a fundamental result in self–similar traffic modeling. Comput Commun Rev 27(2):5–23
4. Paxon V, Floyd S (1995) Wide-area traffic: the failure of poisson modeling. IEEE/ACM Trans Netw (3):226–244
5. Kryukov YA, Chernyagin DV (2009) Investigation of the traffic self-similarity in the high-speed packet data transmission channel [Yssledovanye samopodobyia trafika vysokoskorostnoho kanala peredatchy paketnykh dannykh]. In: 3rd international conference "modern problems of informatization in the simulation, programming systems and telecommunications". Available at: http://econf.rae.ru/article/4819
6. Kryvinska N (2010) Converged network service architecture: a platform for integrated services delivery and interworking. In: Book, electronic business series, vol 2. International Academic Publishers, Peter Lang Publishing Group; ISBN-13:978-3631595251; ISSN:1868-646X
7. Muranov OS, Kochergin YA, Chupryn VM (2009) Investigation of the influence of the traffic forecasting mechanism on the quality of the system of adaptive control of the switch [Doslidzhennya vplyvu mexanizmu prognozuvannya trafiku na yakist` systemy adaptyvnogo keruvannya komutatorom]. In: Problems of informatization and management: proceedings of National Aviation University, vol 1, issue 25, pp 137–143
8. Muranov OS (2010) Enhancing the quality of adaptive control techniques in packet telecommunication networks: author's thesis [Pidvyshhennya yakosti texnologiyi adaptyvnogo upravlinnya trafikom paketnyx telekomunikacijnyx merezh: avtoref. dis. ... kand. techn. nauk], Kyiv, 30 p
9. Aljomar M (2014) Prediction of self-similar traffic in packet networks [Prognozuvannya samopodibnogo trafika u paketnyx merezhax]. In: Improving the development of complex systems, isssue 20, no 12, pp 102–109

10. Kuchuk, GA, Mozhaev OO, Vorobyov OV (2006) The method of fractal traffic forecasting [Metod prognozuvannya fraktalnogo trafika]. In: Radio electronic and computer systems, no 6, pp 181–188. Available at: http://nbuv.gov.ua/UJRN/recs_2006_6_34
11. Jiang J, Liu Y, Song F, Du R, Huang M (2015) The routing algorithm based on fuzzy logic applied to the individual physiological monitoring wearable wireless sensor network. J Elect Comput Eng 2015:7p. Article ID 546425
12. Gupta S, Bharti PK, Choudhary V (2011) Fuzzy logic based routing algorithm for mobile ad hoc networks. High Perform Archit Grid Comput 169:574–579
13. Chrysostomou C, Pitsillides A (2009) Fuzzy logic control in communication networks. Found Comput Intell 2:197–236
14. Cibira G, Dulik M (2016) Fuzzy logic routing within international academic networks. In: Proceedings of the 11th international conference ELEKTRO'2016, Strbske Pleso, Slovakia, 16–18 May 2016, pp 36–42
15. Kryvinska N (2008) An analytical approach for the modeling of real-time services over IP network. Elsevier Trans IMACS J Math Comput Simul (MATCOM) 79:980–990. ISSN: 0378-4754, http://dx.doi.org/10.1016/j.matcom.2008.02.016
16. Kryvinska N (2004) Intelligent network analysis by closed queueing models. Kluwer/Springer J Telecommun Syst 27(1):85–98. ISSN: 1018-4864 (print version), ISSN: 1572-9451 (electronic version), Journal no. 11235 Springer US
17. Kryvinska N, Zinterhof P, van Thanh D (2007) An analytical approach to the efficient real-time events/services handling in converged network environment. In: The first international conference on network-based information systems (NBiS2007), in conjunction with the 18-th international conference on database and expert systems applications (DEXA 2007), 3–4 September, Regensburg, Germany. Springer LNCS-4658, ISBN-10 3-540-74572-6, ISBN-13 978-3-540-74572-3, pp 308–316
18. Khatari M, Samara G (2015) Congestion control approach based on effective random early detection and fuzzy logic. Magnt Res Rep 3(8):180–193
19. Sabeetha K, Kumar VAA, Wahidabanu RS, Othman WA (2015) Encounter based fuzzy logic routing in delay tolerant networks. Wireless Netw 21(1):173–185
20. Veselý P, Karoviè V, ml. VK (2016) Tools for modeling exemplary network infrastructures. Procedia Comput Sci 98:174–181. https://doi.org/10.1016/j.procs.2016.09.028
21. Tabatabaei S, Hosseini F (2016) A fuzzy logic-based fault tolerance new routing protocol in mobile ad hoc networks. Int J Fuzzy Syst 18(5):883–893
22. Klymash M, Stryhaliuk B, Demydov I, Beshley M, Seliuchenko M (2014) A novel approach of optimum multi-criteria vertical handoff algorithm for heterogeneous wireless networks. Int J Eng Innov Technol (IJEIT) 4(5):42–52
23. Zhao X, Hwang BG, ASCE AM, Low SP (2013) Developing fuzzy enterprise risk management maturity model for construction firms. J Constr Eng Manage 9:1179–1189
24. Nie L, Wang X, Wan L, Yu S, Song H, Jiang D (2018) Network traffic prediction based on deep belief network and spatiotemporal compressive sensing in wireless mesh backbone networks. Wireless Commun Mobile Comput 2018:10 p. Article ID 1260860, https://doi.org/10.1155/2018/1260860
25. Sokol BI (1997) Asymptotic approximations of a solution for one nonlinear nonautonomous equation [Asymptotycheskye pryblyzhenyya reshenyya dlya odnogo nelynejnogo neavtonomnogo uravnenyya]. Ukrainian Math J 49(11):1580–1583
26. Dronyuk I, Fedevych O (2017) Traffic flows Ateb-prediction method with fluctuation modeling using dirac functions. Commun Comput Info Sci (Springer) 718:3–13
27. Awrejcewicz J, Andrianov IV (2002) Oscillations of non-linear system with restoring force close to sign (X). J Sound Vibr 252(5):962–966
28. Bogolyubov NN, Mitropolskiy YuA (1974) Asymptotic methods in the theory of nonlinear oscillations [Asymptotycheskye metody v teorii nelynejnyh kolebanyj]. Nauka, Moscow, p 503
29. Kryvinska N (2006) M/GX/1 queuing system for the modeling of real-time services over IP network. In: Fifth international symposium on mathematical modelling, 5th MATHMOD, 7–10 February, Vienna, Austria, ISBN: 3-901608-30-3, pp DMP: 8-1–8-10

IP-Telephony—Comparative Analysis of Applications

Maria Ivanova

Abstract This paper deals with the Internet Protocol Telephony or simply IP Telephony. We explain the meaning of this technology, its conveniences and disadvantages, evolution and history. The last part of the work informs about the most popular and widely used programs, which provide IP Telephony: Skype as a pioneer in a modern IP Telephony market with its peer-to-peer system and Whatsapp, a relative newcomer with a client-server system, which has gained much on the popularity among both young and elder generations in the last few years. Finally, a comparison matrix is given, which briefly displays the main differences and similarities between other popular VoIP applications (Viber, Telegram, Facebook Messenger, Google Hangouts, including Skype and Whatsapp).

1 Introduction

In today's progressing world technologies are becoming smarter and everything is being computerized or being linked to the global Network. Due to fast developments and technological advances many ordinary things of humans' routine have already become faster, easier and cheaper. And so has telephony.

As a result, a large number of users nowadays hardly even uses normal telephone communication, preferring modern and cheaper ways to connect with each other. And the name of this new way of communication is IP Telephony.

There are many definitions of this term, but the most precise one in my point of view is the following one: "IP telephony (Internet Protocol telephony) is a general term for the technologies that use the Internet Protocol's packet-switched connections to exchange voice, fax, and other forms of information" [1, 2]. IP Telephony is—simply put—a general concept for online voice transmission and to implement this concept a special protocol is needed. A protocol used to transmit voice over the Internet is called VoIP—what stands for "Voice over Internet". Another related

M. Ivanova (✉)
Vienna University of Economics and Business (WU Wien), Vienna, Austria
e-mail: marikamaria2307@gmail.com

© Springer International Publishing AG, part of Springer Nature 2019
N. Kryvinska and M. Greguš (eds.), *Data-Centric Business and Applications*,
Lecture Notes on Data Engineering and Communications Technologies 20,
https://doi.org/10.1007/978-3-319-94117-2_5

term is SIP—Session Initiation Protocol. It is a communication protocol for management of multimedia communication sessions (voice or video call over the Internet) [3]. It basically deals with call establishment, the first and important phase of a call. After a call has been set up, various other protocols come into play, which handle data transmission (voice packets) between telephones [4].

The main difference between SIP and VoIP is that VoIP is responsible generally for making or receiving IP calls and SIP establishes, adapts and finishes this multimedia session [5]. The terms "IP Telephony" and "VoIP" are often mixed up. In this paper, we will still use them both as synonyms because they both mean the same concept in general.

Basically, IP Telephony has the same principles as cellular (also called mobile) network or traditional analogue phone system [6]. Traditional phone system is called "Plain Old Telephone Service" or POTS. Though these technologies are completely different, they all are based on "signaling, channel setup, digitization of the analogue voice signals, and encoding" [7, 8] In fact, no real difference can be observed superficially: the voice is transmitted from user to user both, be it through POTS, cellular network or VoIP technologies. The real difference is hidden behind the transmission itself: while traditional phone communication is carried out through circuit-switched connections (telephone network) in which two nodes establish one communication channel and have to stay connected for the whole period of communication, mobile network "transmits data through a global network of transmitters and receivers" [9, 10].

IP Telephony transfers information on commonly shared lines in a form of IP packets and has to deliver packets of data in a certain order [11]. The voice or other data has to be compressed and split into packets or "packetized" and then it can be transmitted through the network one-by-one (it is also said "hop-by-hop", where a "hop" means a route taken by a packet while travelling from one router to another). Each user who sends or receives data becomes his own routing information—IP address—so that data can be sent or received [12, 13]. Figure 1 demonstrates basic principles of packet-switched and circuit-switched networks.

1.1 Advantages and Disadvantages

Even though it seems that IP Telephony has a great deal of advantages, there are still some inconveniences in usage of this system. Firstly, we would like to point out two most notable advantages of VoIP technology:

- Cost saving: Internet calls reduce costs of communication because network infrastructure is used both by data and voice [15]. Cost saving is notable in relation to long distance and international calls. The only fee to charge is an Internet fee. An example of the Virgin Entertainment Group proves this idea and shows a big difference among POT and VoIP: since they switched to VoIP instead of POTS, the cost for long distance calls reduced in $700,000 per year [16].

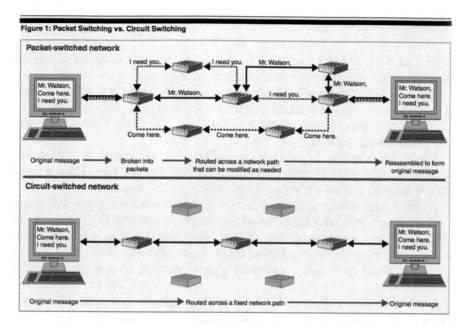

Fig. 1 Packet switching versus circuit switching [14]

- Additional features: Standard telephone line, as well as mobile communication, cannot even compete with modern VoIP software as the latter offers a huge variety of additional facilities, for instance: video conversation, messaging, number identification, data files transfer during a call, voicemail, identifier of user's availability (if a user is currently busy or can take a phone)

These two advantages are in fact significant and make IP Telephony to the most efficient method of communication in today's world. But nevertheless, there are still some problems concerning VoIP technology usage. These are:

- Security threats: VoIP telephone systems can be attacked by hackers and as a result, calls can be intercepted. Furthermore, confidential information can be stolen if some important services are integrated to the VoIP software. However, it is possible to address to VoIP-ready firewall, which control for risks of IP telephony [17].
- Dependence on the Internet: to provide constant availability, it is necessary to maintain constant Internet availability. Furthermore, the quality of calls depends on the quality of internet connection, Internal Network configurations, router [18].
- Dependence on power: VoIP phone service, as well as any mobile phone system, needs power supply to function properly.

2   Literature Review

In this part, a short literature review will be presented in which an appearance of the term "IP telephony" in various databases will be inspected.

The sources which have been taken into account were the following: Scopus, Web of Science, Google Scholar, Science Direct, ProQuest, Wiley, IEEE, ACM (Guide to computing literature), EBSCO (Library information science and technology abstracts), Springer.

Firstly, it is important to point out the reasons for choosing the named databases and libraries. The first 3 of them from the list (namely, Scopus, Web of Science and Google Scholar) are known to be major bibliometric databases.

As for Scopus, it currently covers over 69 million of records and over 1.4 billion cited references, including cited references dating back to 1970 [19]. It delivers not only papers on science, engineering, social sciences, but also gives a deep overview of interdisciplinary fields. In addition, it offers powerful tools for sorting and filtering the results by country, document type or research area to get a detailed overview.

Concerning Web of Science (WoS), its coverage goes back to 1900 (for Science Citation Index) [20]. Apart from books, patents and conference proceedings, it indexes approximately 12,000 scientific journals, containing in its' core collection literature in the art, humanities, sciences and social sciences [20]. Its platform also offers various tools for filtering the results, such as: filtering by year, document type, research area or category.

And finally, Google Scholar—a search engine that indexes scientific and scholarly literature, covering books, academic journals, conference papers, patents and other scholarly literature [21]. Google Scholar does not provide any information on the size of its database, but according to third-party researchers, it contained approximately 160 million documents as of May 2014 [22]. The remaining libraries were chosen because they offer rather rich research databases.

To start with, we analyzed the progress of the number of publications of the terms based on the results of the databases from our list. After series of rough investigations, it became clear that the most significant period which would be rich in content and would help us to make some assumptions and predictions, is the period of the last 15 years. The results are presented in the Fig. 2.

The most noticeable thing about this chart is that the number of results based on ProQuest is the highest. The number of publications starts to grow considerably and quickly in 2003, reaching the peak in 2005. Afterwards, it immediately started to slump and in 2008 the decline took on a moderate character. After reaching its lowest point in 2013, it started slightly to recover.

As for the results from "Google Scholar", there are no dramatical ups and downs. The line chart is more or less flat and the highest point (which is honestly not considerably high compared to the remaining results) was in 2008.

The results based on Scopus, Web of Science and ScienceDirect all follow the following pattern: the line started with a gradual growth reaching the highest

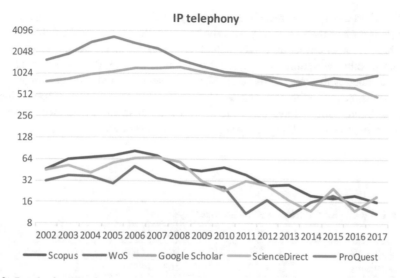

Fig. 2 Results for "IP telephony" by year

point in 2006. Afterwards, the number of results stated to decline rapidly and by 2017 the number of results is lower than ever (according to Scopus and Web of Science they keep falling; based on the results of ScienceDirect, there is a slight increase).

In a next step, we are going to inspect the document types which are the most popular with the terms. The results were acquired by filtering the results of databases by category "document type" where it was possible. It came out that 6 databases offer such an option, namely: ScienceDirect, Wiley, Scopus, Web of Science, ProQuest and Springer.

Unfortunately, not all categories of document type had same names (titles) in all named sources. For that reason, it was decided to combine similar categories as they in fact have the same content though are called differently. These combined categories can be recognized by a slash sign ("/") between document types names.

Another thing concerning merger of different document types into 1 common group, is a combined category called "Miscellaneous". In this category, we included the following document types: editorial, editorial material, review, conference review, feature, dissertation/thesis. It was done because the numbers in each category were relative low and there was no sense to keep them as separate categories.

As a result, the following categories of document types took shape: "Article/journal", "Miscellaneous", "Proceedings paper/conference paper", "Book chapter/Chapter", "Research/Market research", "Book" and "Reference work/Reference Work Entry/Encyclopedia".

Concerning the way of presenting of the results, a Table 1 as well as a pie chart (Fig. 3) were chosen for better visibility.

The chart in Fig. 3 represents document types for a term "IP telephony". At first glance, it is clear that a category with the most result is "Article/journal" with more than half of total number of results (namely, 53.4%). "Miscellaneous" is again at the second position with almost one quarter of a whole number (24%). The next on the list are document types "Research/Market research" and "Proceedings paper/ conference paper" with 9.9 and 9.1% respectively. Categories "Book chapter/ Chapter" and "Book" have the same score of 1.7% while a category "Reference work/Reference Work Entry/Encyclopedia" gained less than 1% (0.2%).

Table 1 Results for "IP telephony" by document type

	SD	Wiley	Scopus	WoS	ProQuest	Springer	Total	%
Article/journal	0	200	317	200	5866	267	6850	53.4
Miscellaneous	18	0	77	107	23	0	225	24.0
Research/market research	356	0	0	0	907	0	1263	9.9
Proceedings paper/ conference paper	0	0	413	338	0	418	1169	9.1
Book chapter/chapter	202	0	6	10	0	0	218	1.7
Book	0	211	0	2	0	0	213	1.7

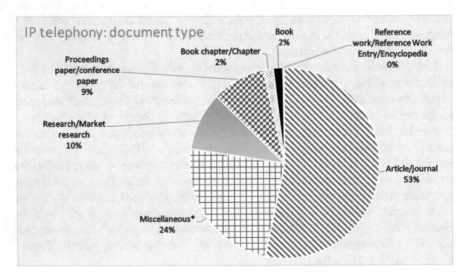

Fig. 3 Results for "IP telephony" by document type: pie chart

3   History and Evolution

According to Joe Hallock, two main technologies that made VoIP communication possible, are telephone and the Internet [12]. It means it is necessary to address to the history of these two great inventions of the humanity to observe the roots of the VoIP technology. The aim of this chapter is to demonstrate the evolution of telecommunication from the beginning until now and to prove that the advances in communication technology have changed the way people communicate in a relative short time. To begin with, as telephone was invented earlier as the Internet, we would like firstly to do a short digression into the 19th century when it all began.

3.1   Telephone

Though the dispute about telephone invention still remains unsettled, a person whose name is often mentioned in a context of a telephone invention is Alexander Graham Bell. Despite all claims and complaints, Alexander Graham Bell was the first inventor to be granted with a U.S. patent for a telephone in 1876 [23]. Originally, he intended to improve telegraph but in 1876 ended up inventing a telephone [24]. The telephone, invented by Bell, was a short-range one [25].

Originally, he wanted to invent an optimized telegraph (so-called "harmonic telegraph") [26], so that it could transmit multiple messages over a single wire at once [27].

The first versions of a telephone were telegraph-like as there was only one wire both for sending and receiving voice using a single-wire ground return path [28].

The next inconvenience was that users could not speak and listen simultaneously as there was only one opening for sound [29].

The telephone, invented by Alexander Bell, transformed sound into electrical signal with a help of a "liquid transmitter ". The sound passed through a receiver onto a stretched membrane. A cork with a needle on the other side, as shown in Fig. 4, connected with a cup with sulphuric acid and a contact, accepted sound vibrations caused on a membrane which passed between the needle and the contact.

First users could use only paired telephones in their own lines till 1889, when an American inventor Almon B. Strowger came up with an idea how to connect one line with up to 100 other lines [31].

At the beginning of the 20th century telephone users could not make long distance calls from their own telephones and had to use a special telephone booth at an arranged time [29].

During the whole 20th century a telephone design stayed unchanged. A rotary dial phone was the most wide-spread one until a new technology—namely a push-button telephone—came into play [32].

Fig. 4 Alexander Graham Bell's original telephone [30]

3.2 Internet

The next essential component of the VoIP concept is undoubtedly the Internet. The original idea was to create an independent packet switching network for data and information exchange. Its forerunner was called ARPANET (standing for Advanced Research Projects Agency Network) and was created in 1969 [33]. ARPANET was funded by the U.S. Department of Defence and was used for communication purposes within a private network relying on packet switching [34]. It was also claimed that a decentralized model of ARPANET will be robust and resistant enough to withstand in a case of a global war [12].

This network originally consisted of 4 nodes (routers):

- University of California, Los Angeles (UCLA)
- The Augmentation Research Center at Stanford Research Institute (now SRI International)
- University of California, Santa Barbara (UCSB)
- The University of Utah School of Computing [35]

In 1981 due to the Computer Science Network (CSNET) [funded by the National Science Foundation (NSF)] accessibility of the ARPANET was extended. The combination of the ARPANET and NSFNET became known as the Internet. Some years later the NSF enabled several universities interconnection with its project called National Science Foundation Network (NSFNET), which enabled network accessibility to the US supercomputer sites from academic as well as scientific establishments [36]. Figure 5 demonstrates ARPANET network in 1974.

Since the first steps were taken, a real revolution has begun. In only several years the Internet has been taken over by commercial Internet providers. Technological revolution including emerging of mobile- and smartphones as well as relative

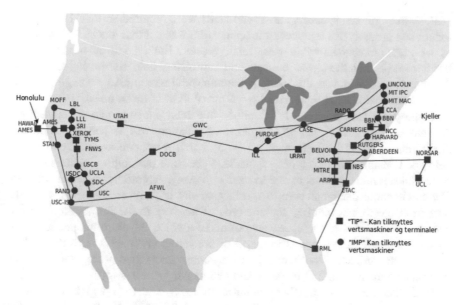

Fig. 5 ARPANET 1974 [37]

availability of PCs in combination with the Internet have made a communicational breakthrough. Nowadays the Internet is a means for education, business, entertainment as well as communication including chatting, social networks, messengers and VoIP calls.

3.3 Voice Over IP

As the background has been outlined, it is time to observe an evolution of VoIP itself.

It is hard to believe, but in fact a Voice over IP technology will celebrate its 45th birthday in 2018. Though it seems to a be modern way of communication, which is nowadays widely used in communication apps and messengers, it firstly became its shape in 1973 due to Danny Cohen [7], a computer scientist, who developed a protocol to transmit packets with voice on a real time through the network (called ARPANET) [38]. After termination of ARPANET project in 1990, it became a real matter of interest for commerce, involving such technology giants as Microsoft and Intel.

As for the first publicly available Voice over IP application, it was developed in 1995 by a small Israeli telecom provider company called VocalTec, Inc. and was called InternetPhone for voice calls over the Internet. A VoIP audio transceiver was invented and patented by VocalTec co-founders Alon Cohen and Lior Haramaty [39]

Originally, the application could carry out only PC-to-PC voice calls. 3 years later, in 1998, PC-to-phone and even phone-to-phone connections became possible

[12]. A microphone, speakers and sound cards were required so that the software functioned properly. As a pioneer, the application was not that flawless as its sound as well as connection quality were relative poor. But on the other side, it has become clear that there was a huge and promising field for investigation and development. In addition, this modern solution did not require any charge even for international calls, what represented a real revolution of those times.

Some years later not only PC-to-phone but even phone-to-phone connections became possible. The first PC-to-phone application was issued by company Net2Phone in 1996 [40]. Though a public switch telephone network and an IP network transmit different types of media, it became possible to establish voice communication between them due to media gateway. According to Search Unified Communications.techtarget.com, a media gateway is "any device...that converts data from the format required for one type of network to the format required for another" [41]. Thus, if one user wants to make a voice call from his phone to his friend's computer, the sound stream will be converted into packets and transferred over the IP network. Vocaltec, Vienna Systems and Cisco were among those companies which were first to make and use media gateways [40].

Since then the era of digital communication began. Due to new gadgets, devices and technology, which enabled sending and routing VoIP traffic (including media and VoIP gateways), a new way of communication started to expand firstly in the United States and then in other (mainly in European) countries. According to available data, by 1998 Voice over IP traffic made 1% of the whole traffic in the United States. 2 years later, in 2000, the percentage came to roughly 3% of all voice traffic [40] and in 2003 of voice calls expanded considerably and as a result reached 25% of total number of voice traffic [12].

In 2003 a new player—Skype—entered a VoIP market. The next chapter deals with its evolution and features.

4   Skype

Nowadays Skype is widely known all over the world as it connects millions of users every day. According to available data, the number of Skype users reached 660 million worldwide in 2010 [42]. What is more, Skype makes up 13% of world's total international calls in minutes [43].

Skype is compatible with the on following operating systems: Windows, mac OS, Linux, Android, iOS, Windows mobile, BlackBerry, HoloLens, Xbox One, Amazon Kindle Fire HD X, Amazon Fire HD [45, 46]. Its web interface is shown in Fig. 6.

4.1   Skype: Historical Digression

Skype was released in 2003 by a group of Estonian developers as a service for peer-to-peer free Internet calls and later low-cost PC-to-phone calls [47].

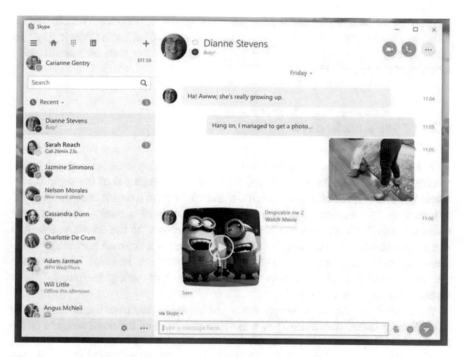

Fig. 6 Skype web interface [44]

The name "Skype" derives from original project's name "Sky peer-to-peer", which was later shorted to "Skyper" and then transformed into "Skype" [48].

By the time of its launch, Skype could only make PC-to-PC calls, which were available only for Skype users.

In July 2004, Skype launched "SkypeOut", which allowed users to call mobile or standard phones all over the world at low cost, charged per minute [49].

In April 2005 Skype extended its offer with another similar concept called "SkypeIn". It offered users to receive calls from external landlines [43].

In 2005, when eBay acquired Skype [50], videotelephony between Skype users was launched for the 1st time [51].

In 2008, when 35% of users used Skype for business purposes, Skype for SIP, a service for businesses, was launched [52]. This product allowed businesses with SIP phone systems to connect to Skype directly and make low cost international calls [43].

In May 2011, Skype was taken over by Microsoft. Shortly after that, Microsoft began to replace Skype peer-to-peer service by centralized Microsoft Azure cloud computing service. In addition, Microsoft developers adopted Skype app user interface so that instant messages became the main issue of the app, shifting from the idea, that Skype is primarily a VoIP service [53].

Though there are lots of other messengers with the same features of voice and video calls, a VoIP service of Skype has remained the most reliable solution both for private as well as business purposes. Today the application, which can be installed and used both on a PC and a mobile device, offers online messaging (for sending text, images, documents and videos), voice, video calls and additionally video conference calls [54].

4.2   Skype Architecture

At the beginning, Skype used to combine peer-to-peer system with client-server technology. In decentralized P2P architecture, each computer is at the same time a client and a server, which can establish a direct connection to other computers. Such system does not require high costs (e.g. maintenance of a central server), what enables to develop networks inexpensively. Client server, on the contrary, involves one or more central servers which connect other users and store all their data. When a user wants to connect to another PC, this connection will be established through the central server [55]. A simple visualization of the main principles of these technologies is shown in Fig. 7.

As a peer-to-peer system, Skype stores and transmits data over the Internet using devices of its users, which are known as "nodes". Some of the most suitable nodes (no firewall or operating system restriction, sufficient CPU power, memory etc.) are then selected and used by Skype as "supernodes", which handle data transmission between Skype users (normal nodes which have firewall restriction, for instance), indexing as well as serve as networking address translation (NAT) nodes [57]. According to Mark Gillett, CVP, Skype Product Engineering & Operations, calls do not pass through supernodes, they just help "normal" nodes find each other [58].

After Microsoft acquisition all supernodes were replaced by servers hosted by Microsoft [58].

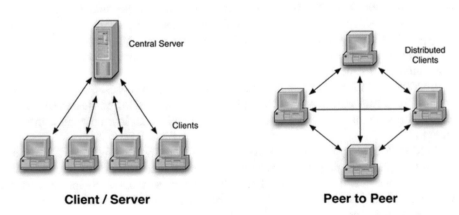

Fig. 7 Client server and Peer to peer [56]

As a client server, Skype had the only central element—a login (authentication) server that stored users' accounts and backup copies of their contact lists. The central server was required only for connection establishment. Once a successful connection through a central server has been established, devices could transmit data directly to each other or through a "supernode" [59]. This principle is shown in Fig. 8.

Looking at Fig. 7, it is possible to see that Skype represents a combination of both systems. On one side, nodes can establish connection between each other directly or through "supernodes". On the other side, there is still a central server that stores account information and contacts.

As it has already been mentioned in the historical part of this chapter, Skype changed from P2P to central cloud computing model. The main issue was model's inefficiency to handle with a huge (and continuously growing) number of users. Supernodes could not maintain appropriate connection and a diversity of emerging platforms like iOS and Android required a new and a common solution [57].

Due to a cloud infrastructure with dedicated supernodes instead of randomly chosen ordinary users' PCs, Skype can perform faster and more efficient, especially at large scales. Thanks to cloud server, it became possible to save battery on a device as well as to send messages to users who are currently offline [61]. This happens because cloud storage can hold data when a recipient is unavailable, so that a sender does not need to wait for a peer to show up online [62].

Skype protocol is proprietary. According to PCmag, a proprietary protocol is "a non-standard communications format and language owned by a single organization or individual" [63]. Moreover, protocol's specifications are openly unavailable. Skype software is closed-source as well [64], what means that software developers retain property rights for it [65]. In other words, the software cannot be reviewed by anyone who is not its developer and users have to rely on developers when it comes to security [66].

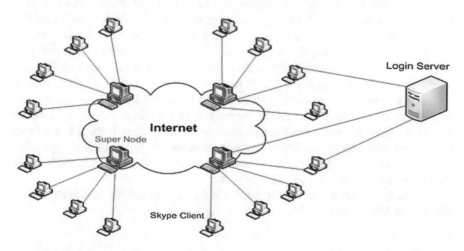

Fig. 8 Skype Network [60]

Skype's compatibility with many other VoIP systems is impossible without Skype license [67].

In 2014 Microsoft replaced former Skype protocol with MSNP24 [68].

Concerning, calls to landline and mobile phones, they are not encrypted.

Concerning security, Skype is not really a safe way to transmit sensitive information. According to the Guardian, calls over Skype network are not affected by end-to-end encryption. It allows third parties like Microsoft as well as government to intercept and wiretap calls easily [69]. When end-to-end encrypted, all sent data get encrypted on sender's system and only an addressee can decrypt it [70].

Nevertheless, Skype-to-Skype communication (voice, video, files, instant messages) is protected by client server encryption, which uses public key, which is less secure compared with end-to-end encryption [71]. But if Skype user makes calls over PSTN (to mobile or landline phones), they will not pass through encryption [72].

5 WhatsApp

Similarly, to Skype, Whatsapp is considered to be a pioneer in its branch—mobile messaging. Opposed to Skype, its main function is not VoIP calls, though it offers this service as well.

WhatsApp Messenger is a free instant messaging app for smartphones, which offers VoIP service as well. The software is adopted for multiple platforms. Besides text messages, Whatsapp sends images, GIFs, videos, document, locations and audio files/messages. To sign in, a mobile number is required. Its' mobile interface is represented in Fig. 9.

In contrast to Skype, all data, transmitted through Whatsapp, pass through end-to-end encryption. It secures, that nobody except a user and his recipient, can access and read sent information. According to Whatsapp official website, due to end-to-end encryption, each message is protected with an individual lock and it is only a recipient device which owns a special key to unlock the message [74].

Opposed to Skype proprietary protocol, Whatsapp is an open encryption protocol Signal [66]. Though Whatsapp encryption protocol is an open standard, the application itself is closed source. Moreover, an open source software can be reviewed to verify the presence of malware (malicious code). Closed source (proprietary) software, like Skype, on the contrary, does not allow anyone to get in and thus users have to trust the developers unconditionally [66] It means, the other application parts except its protocol, are not available for others and it is impossible to verify other issues which can influence the security (as encryption is not the only factor which is responsible for security) [66].

Moreover, its client software is available for Android, iOS, Windows Phone, BlackBerry OS, Symbian [75].

Fig. 9 Whatsapp mobile interface [73]

5.1 WhatsApp: History

The story of Whatsapp took its roots back in 2009, when former Yahoo employees Brian Acton and Jan Koum decided to create a messaging app, inspired by iPhone's App Store and strongly believing, that such an app would be a real breakthrough [75]

At first, the main concept of Whatsapp was different as it was not intended to develop a messenger, rather than a status indicator, which would inform all contacts about user's availability for a call. This is how Whatsapp functioned by the time of its 1st release. Shortly after that, Apple launched push notifications and this made Jan Koum rethink the whole concept of his development. The main pillars, on which new Whatsapp would rest, were: compatibility with multiple platforms (cross-platform), easiness of logging in (mobile phone number as a login) and user's phone contact folder as a base for potential Whatsapp contacts. According to Jan Koum, he wanted to develop an app which would be easy in use and would "just work" [76].

Already in the year of its release, 2009, the function of picture messaging (sending photos) was added [75].

In 2013 voice messages were introduced [77].

Further important milestone of Whatsapp history is Facebook acquisition in 2014 [78].

In July 2017, the number of Whatsapp users reached 1300 million, making Whatsapp the most popular messaging app in the world of all the time [79].

In 2015 Whatsapp Web was launched, enabling users to stay connected using a PC, syncing with a device [80].

Furthermore, in 2015 Whatsapp started offering VoIP service—a function of making voice calls since then was available for Whatsapp users [81].

In March 2016, WhatsApp introduces its document-sharing feature, initially allowing users to share PDF files with their contacts [82].

In 2016 Whatsapp stops charging annual fee in the amount of 99 cent (1st year of usage was free). Instead of charging a fee, Whatsapp owners intended to develop a service for businesses to create a revenue from this offer [83]. In February 2016, Whatsapp hits another strike with 900 million users, remaining the most popular messaging app at the time [84].

Additionally, in 2016 video calls became available to Whatsapp users [85].

5.2   WhatsApp: Encryption and Server

As it has already been mentioned, Whatsapp uses end-to-end encryption. Concerning system, opposed to Skype, Whatsapp has a client server system. The latter means, that all messages and data sent through Whatsapp have to pass through the server and are not directly sent from user to user, as Skype system does [86].

Figure 10 shows how end-to-end encryption works. Furthermore, all data pass through encryption, including plain text, images, videos and voice messages [88].

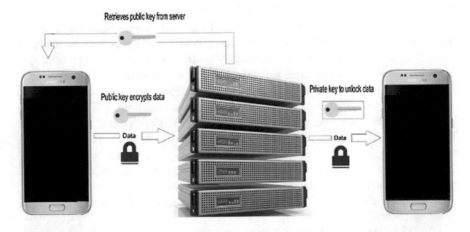

Fig. 10 End-to-end encryption [87]

Before sending a message to B, A requests a public key which applies for B by a central Whatsapp server. To encrypt his message, A uses a public key. Afterwards, the message goes from A to the server. Then the server directs the encrypted message to B. To decrypt a received message, B uses his private key, which is available only on B's device and which B receives from the server [86].

All in all, Whatsapp is a safe and secure solution for sending sensitive data: if messages are once intercepted, it still will be almost impossible to read them as the private key to encrypt them is stored on a recipient device. In addition, Whatsapp regularly changes key, so if someone steals user's keys, he will be able to encrypt only a fraction of user's messages-the whole part will still be unavailable [89].

6 Comparison of Applications

It is time, at last, to review some more other application and software to see what specific features they can offer and mark both advantages and disadvantages of each of them. After a brief review, a comparative matrix with the main characteristics for each program will be provided.

6.1 Viber

Viber is a cross-platform VoIP application with 920 million users in June 2017 [90], which was originally developed in 2010 by the Israeli company "Viber Media" and later acquired by a company "Rakuten" (Japan). It offers instant messages, but it is also possible to send multimedia messages with video, audio or images [91]. It operates on Windows, macOS, Linux, Android, iOS, Windows Phone, Win 10 Mobile, Bada, BlackBerry OS, Series 40, Symbian [92]. Its' mobile interface is shown in Fig. 11.

Concerning price, Viber is free. Additionally, it offers landline-calls or calls to mobile phones with a help of its' paid service called "Viber Out" with a price starting at 1.9 cents/minute, depending on a country. Viber users call each other for free [94].

As for its Voice over IP feature, voice calls over Viber are available only to iPhone, Android, Microsoft's Windows Phone and from a desktop [95].

Viber allows users to link one account with multiple devices.

Concerning the PC version (desktop application), it syncs all contacts and a record of calls from the phone account automatically [96].

Moreover, it offers push notifications and enables to create chats with up to 100 participants [97].

The app is convenient to sign in, as no usernames and passwords are required. Users need only a mobile number for verification [98]. In addition, communication

Fig. 11 Viber [93]

over Viber is rather secure, as end-to-end encryption is used to protect VoIP calls, video calls, text and files. Thus, Viber has the same architecture as Whatsapp: all messages and calls pass through Viber server. Before establishing a connection, user A requests public key by Viber server, sends a message to B through the server, the server directs the encrypted message to B, which decrypts it with its private key [99].

6.2 Telegram

Telegram is another messenger which was developed in Russia in 2013 for both mobile and desktop systems. Compatible operating systems are: Android, iOS, WD, Microsoft OS, Windows, Mac, Linux, macOS; there is additionally a Web version and a Chrome app [100]. Its mobile interface is represented in Fig. 12.

According to the data available in February 2016, the number of active monthly users reached 100 million [102].

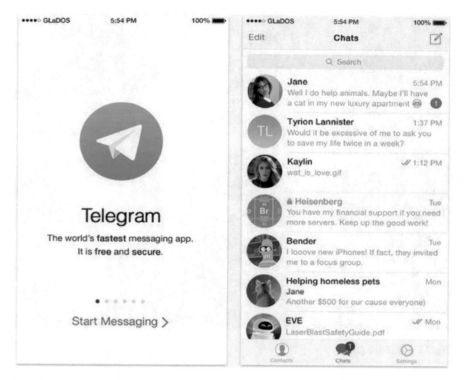

Fig. 12 Telegram [101]

Each account is linked with a phone number of a user and it is possible to link multiple devices to one account. Moreover, it is possible to change an active number any time. Like in almost every messenger, both text and multimedia messages can be sent and groups can be formed with up to 5000 participants [103].

One of the most rare and useful feature of Telegram is that it is possible to delete mistakenly sent messages on both sides within 48 h [104].

Telegram is considered to be secure as well, as it uses end-to-end encryption [105].

Concerning architecture, Telegram is a mixture of peer-to-peer and client server systems. If it is possible, peer-to-peer connection will be established. If not, Telegram servers will establish connection among users [105].

6.3 Facebook Messenger

Facebook Messenger is another popular messaging service, originally a Facebook chat which turned to a separate messenger-app in 2011 adapted for Android, iOS [106] and a big number of other operating systems including PC [107]. Its mobile interface is shown in Fig. 13.

Fig. 13 Facebook [108]

As a usual standard messenger, it offers text and voice messages, voice and video calls as well as file exchange and even playing games [109].

To log in a user needs to have a Facebook account and it is possible to use one account from multiple devices.

According to Techcrunch, Facebook messenger reached 1.2 billion users in April 2017 [110].

Moreover, Facebook Messenger offers end-to-end encryption [111, 112].

6.4 Google Hangouts

One more popular VoIP service is a platform developed by Google called Google Hangouts. Besides its calling feature, it offers chat, video calls. It is a convenient and free solution as it comes pre-installed on some Android devices and offers text, calls to landline and mobile numbers, voice and video calls as well [113]. Its mobile interface is presented in Fig. 14.

The main advantages are: the program is free and every Android user has automatically a Google account, so he can easily use Google Hangouts without any additional registrations and passwords.

It is applicable for Mac OS X, Windows, Chrome, Ubuntu and other Debian-based Linux distributions [115].

Fig. 14 Google Hangouts [114]

Google Hangout uses peer-to-peer connection. If it is not possible, it switches to client server model [116].

Google Hangouts does not use end-to-end communication, though sent data are encrypted "in transit". Thus, Google has a full access to text, photos, voice and video conversations [117]. In addition, Google Hangouts users can make calls to mobile and domestic phone numbers [118].

To sign in, user can use their Google+ account [119].

6.5 Comparison Matrix

In the last part of the chapter, a comparison matrix is given, which relies on the information, which has been given above and additional sources (Table 2).

Table 2 Comparison matrix

	Operating system	License	Cost	Price inside network	Calls to mobile/PSTN	Encryption	Features	Protocol
Google Hangouts	Mac OS X, Windows, Chrome, Ubuntu and other Debian-based Linux distributions [115]	Proprietary [92]	Free	Free	Yes; PSTN, mobile	SRTP3 (Secure real-time transition protocol)	Text, voice, video, screen sharing [71]	Open: XMPP (Extensible messaging and presence protocol) [92]
Facebook Messenger	iOS, Android 6; PC version [106]	Proprietary [71]	Free	Free	No	End-to-end	Text, voice, video [71]	Open; MQTT (Message queue telemetry transport) [112]
Telegram	Android iOS WD, Microsoft OS, Windows, Mac, Linux, macOS, web version, Chrome app [100]	Proprietary [71]	Ftee	Free	Yes; PSTN, mobile	End-to-end	Text, voice, video [71]	Proprietary [92]
Viber	Windows, macOS, Linux, Android, iOS, Windows Phone, Win 10 Mobile, Bada, BlackBerry OS, Series 40, Symbian [92]	Proprietary [92]	Free	free	Yes	End-to-end	Text, voice, video [71]	Unknown [92]
Whatsapp	Android, iOS, Windows Phone, BlackBerryOS, Symbian [75]	Proprietary [71]	Ftee	Free	No	End-to-end	Text voice, video [71]	Open:XMPP (Extensible messaging and presence protocol) [92]
Skype	Windows, macOS, Linux, Android, iOS, Windows mobile, BlackBerry, HoloLens, XboxOne Amazon Kindle Fire HD X, Amazon FireHD [45, 46]	Proprietary [92]	Free	Free	Yes; PSTN, mobile	Client saver	Text, voice, video, P2P extension (games) [71]	Proprietary P2P [92]

References

1. What is IP telephony (Internet protocol telephony)?—Definition from WhatIs.com. http://searchunifiedcommunications.techtarget.com/definition/IP-telephony
2. Kryvinska N Converged network service architecture: a platform for integrated services delivery and interworking. International Academic Publishers, Peter Lang Publishing Group
3. SIP vs. VoIP, What' the difference. IP telephony explained. SIP blog. https://www.sip.us/sip-vs-voip-whats-difference/
4. Nobody: what is SIP? https://www.networkworld.com/article/2332980/lan-wan/lan-wan-what-is-sip.html
5. Key differences between SIP & VOIP. https://www.telstraglobal.com/insights/blogs/blog/key-differences-between-sip-voip
6. Kryvinska N (2008) An analytical approach for the modeling of real-time services over IP network. Math Comput Simul (MATCOM) 79:980–990
7. Voice over IP (2017) https://en.wikipedia.org/w/index.php?title=Voice_over_IP&oldid=809315847
8. Kryvinska N, van Thanh D (2006) Enabling voice services in hybrid/cross network framework. In: Presented at the London communications symposium 2006, London, UK, 14 Sept 2006
9. How mobile networks work. Understanding cellular phone technology. InformIT. http://www.informit.com/articles/article.aspx?p=2021961
10. Kryvinska N, van Thanh D (2006) IN-managed SIP network: a case study in network convergence. In Presented at the IEEE 20th international conference on advanced information networking and applications (AINA 2006), Vienna, Austria, 18 Apr 2006
11. Kryvinska N, Zinterhof P, van Thanh D (2007) An analytical approach to the efficient real-time events/services handling in converged network environment. In: Presented at the the first international conference on network-based information systems (NBiS2007), in conjunction with the 18-th international conference on database and expert systems applications (DEXA 2007), Regensburg, Germany, 3 Sept 2007
12. Hallock J (2004) A brief history of VoIP. Evolution
13. Kryvinska N, Zinterhof P, van Thanh D (2007) New-emerging service-support model for converged multi-service network and its practical validation. In: Presented at the the IEEE first international conference on complex, intelligent and software intensive systems (CISIS-2007) in conjunction with ARES 2007, Vienna, Austria, 10 Apr 2007
14. Packet switching. http://itlaw.wikia.com/wiki/Packet_switching
15. How business can use internet technology for voice communications. Ontario.ca. https://www.ontario.ca/page/how-business-can-use-internet-technology-voice-communications
16. King R (2006) Internet telephony: coming in clear. Business Week Online 8:16–17
17. VoIP security threats and ways of mitigating the risks. http://it.toolbox.com/blogs/voip-news/voip-security-threats-and-ways-of-mitigating-the-risks-68463
18. 5 curable causes of poor voip call quality—voip-info.org. https://www.voip-info.org/wiki/view/5+Curable+Causes+of+Poor+VoIP+Call+Quality
19. Content—scopus—solutions. Elsevier. https://www.elsevier.com/solutions/scopus/content
20. Ekstrøm J, LibGuides: web of science: about web of science. http://libguides.dtu.dk/c.php?g=654844&p=4601037
21. Google scholar search tips. https://scholar.google.com/intl/us/scholar/help.html#coverage
22. Orduna-Malea E, Ayllón JM, Martín-Martín A, Delgado López-Cózar E (2015) Methods for estimating the size of Google scholar. Scientometrics 104:931–949. https://doi.org/10.1007/s11192-015-1614-6
23. Black H (1997) Canadian scientists and inventors: biographies of people who made a difference. Pembroke, Ontario
24. Alexander Graham Bell. http://www.pbs.org/transistor/album1/addlbios/bellag.html
25. Puleo S (2011) A city so grand: the rise of an American Metropolis. Beacon Press, Boston

26. Watson TA (2017) The birth and babyhood of the telephone. Library of Alexandria
27. Alexander Graham Bell. Biography, inventions, & facts. Britannica.com. https://www.britannica.com/biography/Alexander-Graham-Bell
28. Why positive earth? http://mgaguru.com/mgtech/electric/et098.htm
29. History of the telephone—wikipedia. https://en.wikipedia.org/w/index.php?title=History_of_the_telephone&oldid=808426293
30. How Bell's telephone worked. How It Works Magazine. https://www.howitworksdaily.com/how-bells-telephone-worked/
31. Bellis M, The history of the telephone is off the hook. https://www.thoughtco.com/history-of-the-telephone-alexander-graham-bell-1991380
32. AT&T: history: milestones in AT&T History—1963. https://web.archive.org/web/20061228092155/, https://www.corp.att.com/history/milestone_1963.html
33. ARPANET. Definition & history. https://www.britannica.com/topic/ARPANET
34. Evan Andrews: who invented the internet? http://www.history.com/news/ask-history/who-invented-the-internet
35. The first four nodes on the ARPANET (5 Dec 1969): HistoryofInformation.com. http://www.historyofinformation.com/expanded.php?id=3823
36. FedTech, P.G.T.P.G. is a web editor for, trends, B.B. keeping up with the latest in technology, Yankees, he is also an avid lover of the N.Y., poetry, photography, traveling, Humidity, E.: NSFNET served as a precursor to the internet, helped spur scientific research. https://fedtechmagazine.com/article/2016/11/nsfnet-served-precursor-internet-helped-spur-scientific-research
37. Yngvar: English: symbolic representation of the Arpanet as of September 1974 (2007)
38. Callari R, Pm, 5/13/2015 8:25:47: a brief history of VoIP. https://telzio.com/blog/a-brief-history-of-voip/
39. Cohen A, Haramaty L (1994) US patent no US 5825771 A. Google patents
40. VoIP Born http://www.sipnology.com/en/company/19-voip-born
41. What is media gateway?—definition from WhatIs.com. http://searchunifiedcommunications.techtarget.com/definition/media-gateway
42. Global social media ranking 2017. Statistic. https://www.statista.com/statistics/272014/global-social-networks-ranked-by-number-of-users/
43. The evolution of Skype: about the new SkypeKit SDK. https://www.shoretel.com/blog/evolution-skype
44. Tim Anderson 24 Mar 2016 at 15:00 tweet_btn(): Microsoft introduces yet another Skype for Windows 10. https://www.theregister.co.uk/2016/03/24/microsoft_introduces_yet_another_skype_for_windows_10/
45. Skype—wikipedia. https://en.wikipedia.org/w/index.php?title=Skype&oldid=810547242
46. What platforms is Skype available on?. Skype Support. https://support.skype.com/en/faq/FA12041/what-platforms-is-skype-available-on
47. A brief history of Skype—the peer to peer messaging service. https://www.dsp.co.uk/history-of-skype/
48. Skype: the name Inspector, http://www.thenameinspector.com/skype/
49. What is SkypeOut? http://www.vodburner.com/skypeout
50. eBay completes acquisition of Skype—eBay Inc. https://investors.ebayinc.com/ReleaseDetail.cfm?releaseid=176402
51. Skype: from Swedish brainchild to Microsoft target. http://www.telegraph.co.uk/finance/newsbysector/mediatechnologyandtelecoms/digital-media/8504942/Skype-from-Swedish-brainchild-to-Microsoft-target.html (2011)
52. Fowler GA (2009) Skype targets businesses to ring up new revenue. http://www.wsj.com/articles/SB123776338990608661 (2009)
53. The new Skype for desktop is here. https://blogs.skype.com/news/2017/10/30/the-new-skype-for-desktop-is-here/
54. Features, Find out what Skype can do for you. Skype. https://www.skype.com/en/features/
55. Wolf M (2002) Home networking: what type is best? Presented at the 19 Apr 2002

56. The P2P Witch Hunt. https://blog.peer5.com/the-p2p-witch-hunt/
57. Unuth N, Skype changes from P2P to client server model. https://www.lifewire.com/skype-changes-from-p2p-3426522
58. Skype replaces P2P supernodes with Linux boxes hosted by Microsoft (updated). Ars Technica. https://arstechnica.com/information-technology/2012/05/skype-replaces-p2p-supernodes-with-linux-boxes-hosted-by-microsoft/
59. Branch PA, Heyde A, Armitage GJ (2009) Rapid identification of Skype traffic flows. In: Presented at the 18th international workshop on network and operating systems support for digital audio and video, New York, June 2009
60. Network control: peer-to-peer networks versus client/server. Home networking: what type is best? InformIT. http://www.informit.com/articles/article.aspx?p=26437&seqNum=3
61. Matthew Kaufman on why Skype is dropping peer-to-peer. https://www.infoq.com/news/2013/06/Skype-Servers
62. Send files over Skype when the recipient is offline. https://mybroadband.co.za/news/software/170993-send-files-over-skype-when-the-recipient-is-offline.html
63. Proprietary protocol definition from PC magazine encyclopedia. https://www.pcmag.com/encyclopedia/term/49868/proprietary-protocol
64. Skype protocol—wikipedia. https://en.wikipedia.org/w/index.php?title=Skype_protocol&oldid=800180764
65. Proprietary software—wikipedia. https://en.wikipedia.org/w/index.php?title=Proprietary_software&oldid=807868424
66. Collective TT, Why we still recommend signal over WhatsApp …even though they both use end-to-end encryption. https://securityinabox.org
67. What is Skype protocol? Definition from WhatIs.com. http://searchunifiedcommunications.techtarget.com/definition/Skype-protocol
68. Microsoft Notification Protocol | Open Access articles | Open Access journals | Conference Proceedings | Editors | Authors | Reviewers | scientific events. http://research.omicsgroup.org/index.php/Microsoft_Notification_Protocol#MSNP24
69. Microsoft handed the NSA access to encrypted messages. US news. The Guardian. https://www.theguardian.com/world/2013/jul/11/microsoft-nsa-collaboration-user-data
70. What is end-to-end encryption (E2EE)? Definition from WhatIs.com. http://searchsecurity.techtarget.com/definition/end-to-end-encryption-E2EE
71. Comparison of instant messaging clients—wikipedia. https://en.wikipedia.org/w/index.php?title=Comparison_of_instant_messaging_clients&oldid=810728378
72. Does Skype use encryption? Skype Support. https://support.skype.com/en/faq/FA31/does-skype-use-encryption
73. WhatsApp messenger gets 3D touch support with peek and pop. http://www.iclarified.com/51964/whatsapp-messenger-gets-3d-touch-support-with-peek-and-pop
74. WhatsApp FAQ—end-to-end encryption. https://faq.whatsapp.com/en/general/28030015
75. WhatsApp (2018) https://en.wikipedia.org/w/index.php?title=WhatsApp&oldid=819163493
76. Rowan D, WhatsApp: the inside story. http://www.wired.co.uk/article/whatsapp-exclusive
77. Voice messaging comes to whatsapp. TechCrunch, https://techcrunch.com/2013/08/07/voice-messaging-comes-to-whatsapp/
78. Constine J, A year later, $19 billion for whatsapp doesn't sound so crazy. http://social.techcrunch.com/2015/02/19/crazy-like-a-facebook-fox/
79. WhatsApp: number of users 2013-2017. Statista. https://www.statista.com/statistics/260819/number-of-monthly-active-whatsapp-users/
80. WhatsApp web. https://blog.whatsapp.com/614/WhatsApp-Web
81. WhatsApp voice calling—Everything you need to know. https://gadgets.ndtv.com/apps/features/whatsapp-voice-calling-everything-you-need-to-know-672018
82. Perez S, WhatsApp adds support for document sharing, but only PDFs at launch. http://social.techcrunch.com/2016/03/02/whatsapp-adds-support-for-document-sharing-but-only-pdfs-at-launch/

83. Scott M, WhatsApp, the internet messenger, to become free. https://bits.blogs.nytimes.com/2016/01/18/whatsapp-the-internet-messenger-to-become-free/
84. Sun L, Facebook Inc.'s WhatsApp hits 900 million users: what now? https://www.fool.com/investing/general/2015/09/11/facebook-incs-whatsapp-hits-900-million-users-what.aspx
85. Nov 15, B.P. 2016-11-15T09:10:01Z, 2016: WhatsApp video calls are now available to everyone. https://www.androidpit.com/how-to-make-whatsapp-video-calls
86. Business, A.C.M.C.M.: Forget Apple vs. the FBI: WhatsApp just switched on encryption for a billion people. https://www.wired.com/2016/04/forget-apple-vs-fbi-whatsapp-just-switched-encryption-billion-people/
87. What is encryption? Security. Techworld. https://www.techworld.com/security/what-is-encryption-3659671/
88. Whatsapp Inc.: WhatsApp encryption overview technical white paper (2016)
89. Lab K, WhatsApp switches to secure end-to-end encryption. https://me-en.kaspersky.com/blog/whatsapp-encryption/5408/
90. Viber: number of registered users 2017. Statistic. https://www.statista.com/statistics/316414/viber-messenger-registered-users/
91. Make calls on Viber. https://support.viber.com/customer/portal/articles/1340905-how-do-i-make-a-free-call-with-viber-
92. Comparison of VoIP software—wikipedia. https://en.wikipedia.org/w/index.php?title=Comparison_of_VoIP_software&oldid=807816483
93. Viber app review. https://www.lifewire.com/viber-app-review-3426625
94. Five best mobile VoIP apps. https://lifehacker.com/five-best-mobile-voip-apps-1580922421
95. Supported platforms. https://support.viber.com/customer/portal/articles/1354549-supported-platforms
96. Sync your Viber chat history to Viber for desktop. https://support.viber.com/customer/en/portal/articles/2850614-sync-your-viber-chat-history-to-viber-for-desktop
97. The best VoIP apps for Apple and Android. http://www.toptenreviews.com/services/articles/best-voip-app-apple-android/
98. Battle of the mobile messaging apps—WhatsApp, LINE, WeChat, Kakao Talk, ChatON and Viber—Lowyat.NET. https://www.lowyat.net/2013/13273/battle-of-the-mobile-messaging-apps-whatsapp-line-wechat-kakao-talk-chaton-and-viber/
99. Viber encryption overview | Viber. https://www.viber.com/ru/security-overview/
100. Telegram apps. https://telegram.org/apps
101. News feeds (India and world). https://www.thenewsminute.com/news-feeds
102. 100,000,000 monthly active users. https://telegram.org/blog/100-million
103. Telegram pushes supergroup limit to 5,000 people and makes groups viewable to anyone in the public. https://venturebeat.com/2016/03/14/telegram-pushes-supergroup-limit-to-5000-people-and-makes-them-viewable-to-anyone-in-the-public/ (2016)
104. Unsend messages, network usage, and more. https://telegram.org/blog/unsend-and-usage
105. Voice calls: secure, crystal-clear, AI-powered. https://telegram.org/blog/calls
106. Hilfebereich. https://www.facebook.com/help/messenger-app/197039404112757?helpref=uf_permalink
107. Facebook Messenger—wikipedia. https://en.wikipedia.org/w/index.php?title=Facebook_Messenger&oldid=810501520
108. Messenger de facebook, alcanza el segundo lugar en EEUU. https://tecnoinnovador.com/2015/09/05/messenger-de-facebook-alcanza-el-segundo-lugar-en-eeuu/
109. Messenger—features. https://www.messenger.com/features
110. Constine J, Facebook messenger hits 1.2 billion monthly users, up from 1B in July. http://social.techcrunch.com/2017/04/12/messenger/
111. You can finally encrypt facebook messenger, so do it. https://www.wired.com/2016/10/facebook-completely-encrypted-messenger-update-now/
112. MQTT protocol—powering facebook messenger to IoT devices. http://blog.hackerearth.com/mqtt-protocol

113. Google Hangouts. https://en.wikipedia.org/w/index.php?title=Google_Hangouts&oldid= 817544722 (2017)
114. Google Hangouts. https://hangouts.google.com/
115. Get started with Hangouts—computer—Hangouts help. https://support.google.com/hangouts/answer/2944865?co=GENIE.Platform%3DDesktop&hl=en
116. Peer-to-peer calling in Hangouts—Hangouts help. https://support.google.com/hangouts/answer/6334301?hl=en
117. Smith C, Google Hangouts encryption doesn't offer you the privacy protection you thought. http://bgr.com/2015/05/12/google-hangouts-encryption-privacy-security/ (2015)
118. Make a phone call with Hangouts—Computer—Hangouts help. https://support.google.com/hangouts/answer/3187125?co=GENIE.Platform%3DDesktop&hl=en
119. Making calls from Hangouts—in Gmail and across the web, https://gmail.googleblog.com/2013/07/making-calls-from-hangouts-in-gmail-and.html

Product Usage Data Collection
and Challenges of Data Anonymization

Ümit G. Peköz

Abstract Data collection and processing through out product life-cycle can help producers to identify the usage patterns to design, redevelop their products to meet customer needs, and also to achieve higher efficiency in production process. This paper is to demonstrate the importance of the product usage data collection and processing for the future enterprise and to highlight the technical challenges in data analyzing. The Big Data processing and analyzing can bring also risks to the individuals, such as privacy threads. It is the aim of this paper to show briefly the Big Data consequences for a private person and to review existing technical privacy protection methods. It is worth to compare state-of-art data anonymization techniques with each other to show the privacy challenges in Big data. The limitation in the findings is that each technique can have its special application area and running a full comparison can be not possible.

1 Introduction

Data collection and processing throughout product life-cycle can help producers to understand the usage patterns and to design, adopt and redevelop their products to better meet the customer need, and at the same time to achieve higher efficiency in production process by decreasing resources usage [1, 2]. Developments in computing, communication, processor and storage technologies make it possible to collect vast amount of data from products even after its delivery and during its product life cycle.

Unique identifier for each device or thing makes it here possible to assign and collect attributes for each element in this environment monitored separately [3–5].

Data collection and processing with the help of sensors, actuators, (radio frequency identification) RFID tags and second layer interface in wireless sensor networks (WSN), which is altogether called internet of things (IoT), create a large

Ü. G. Peköz (✉)
Faculty of Management, Comenius University, Bratislava, Slovakia
e-mail: upekoz@hotmail.com

© Springer International Publishing AG, part of Springer Nature 2019
N. Kryvinska and M. Greguš (eds.), *Data-Centric Business and Applications*,
Lecture Notes on Data Engineering and Communications Technologies 20,
https://doi.org/10.1007/978-3-319-94117-2_6

scale data coming from the real physical world [4]. This Big data gained from the physical world has different characteristics than online collected customer data through smart phone application, web browser cookies or user profiles, while in a virtual world the Big Data has already homogeneous characteristics and is already stabilized. In IoT world, the collected data should be unified and pre-processed [4, 6–8]. This new IT phenomena called IoT let the things to communicate and collaborate with each other, as a result of this, IoT not only monitor the changes in the environment, but can also effect these physical conditions around them through actuators by activating devices in this environment connected to IoT itself, and can form smart objects with the ability to make decisions which is different than just data collection in a sense. Here things (and not only products or devices, but also plants, animals, resources and eventually people) anytime, anywhere, any device, any media are in connection through internet communication technology [7, 9–11]. IoT is formed with appliances, devices, things we have around us in production, in supply chain environments or in our daily lives. IoT has basic computing and communication capabilities to answer a query to communicate readings from embedded sensors or at least to identify itself passively with RFID tags [4].

As an example we see smart grid meter in England, where household have their own smart meter for electricity, which monitors and record energy usage during the day to analyze energy usage patterns and to help customer better control energy consumption for a better pricing by rescheduling routine activities like washing the clothes accordingly [12, 13].

We can see that product usage data can help not only producers for a better product design but also help customer themselves to adopt their habits to profit from energy costs, which help at the end for a better electricity grid management to facilitate green energy usage effectively [1, 6].

Big Data is defined: "…as being generated by everything around us at all times. Every digital process and social media exchange produces it. Systems, sensors and mobile devices transmit it. Big data is arriving from multiple sources at an alarming velocity, volume and variety. To extract meaningful value from Big data, you need optimal processing power, analytic capabilities and skills" [6].

Big data can help companies to win a competitive edge by shortening the product to market times, addressing customer needs with a better design, and by reducing production costs [2, 6]. It is inevitable to accept the importance of Big data collection and processing to understand usage patterns for product usage and to predict future trends and demands in the market place [1].

There are two challenges ahead of future manufacturing enterprise and we can see the first one is to be able to collect product usage data through sensors and communication technology embedded products and process the collected data to understand usage characteristics. Second it to protect the privacy of customer by data anonymization against misuse of personal information. Although we can see other challenges apart from these two mentioned in a report of European Commission, Directorate-General for Research and Innovation, and European Factories of the Future Research Association, Factories of the future: multi-annual roadmap for the contractual PPP under Horizon 2020 published in 2013 [1]. In this report a research need presented for developing mark-up languages to capture and decipher the usage Data easily, to analyze these data on the go [4, 5].

IoT is the backbone in this data capturing and it faces the challenge in data collection from different sensors, actuators, Big RFID and wireless sensor networks, since collected data must be processed in a second layer to be unified or in some cases also to be pre analyzed, so that it can be transferred through gateway to the server or into a cloud solution [4, 14].

Future enterprise should overcome the technical challenges in data collection and usage, which is already developing itself on the way, but another challenge is even more complex than this technical one. Privacy protection of a person is a direct responsibility for the enterprise as they are the one who captures and must own these data in an accountable way. National, European and international law defines this responsibility accordingly [15]. With the increase of digital data available around us, and actually about us, the privacy concerns are addressed globally and for a time-being law makers and civil organizations request the data anonymization to secure privacy protection [15–17].

2 Paper Organization and Methodology

In this paper I will try to demonstrate the importance of the product usage data collection and processing for the future enterprise and to highlight technical challenges in data analyzing. Apart form benefits Big Data processing and analyzing brings also risks to individuals, by privacy threads. It is also the aim of this paper to show Big Data consequences for a private person and to review existing privacy protection methods.

As a methodology I used literature research based comparison analyses for the-state-of-art anonymization techniques in practice, to identify their short comings in privacy protection and in maintaining data quality.

In Sect. 2 the large scale data collection methods and IoT will be presented, to pinpoint the challenges in discovering non-obvious and valuable information from product usage data. Later in this section the potential risks of Big data to individuals will be also presented.

In Sect. 3 the importance of privacy protection will be presented in details, together with the legal framework to control Big-data against targeted privacy attacks. In the second part of Sect. 3 the-state-of-art data anonymization will be with the help of examples presented. The main idea behind this presentation is to get the reader understand the types of anonymization techniques in practice, so that we can eventually help them in their participation decisions in a world of IoT, by keeping the privacy concerns in their minds.

The shortcomings of data anonymization techniques will be presented in Sect. 4. In this section not only the shortcomings for data quality after anonymization treatment, but also the robustness of the presented techniques against de-identification attacks will be presented. It will be worth to compare then these techniques with each other to show the privacy challenges in Big data. It is to keep in mind each technique can have its special application area and running a full comparison can be not possible.

At the end of the paper in Sect. 5, I will try to draw a conclusion to underline the main findings in this work and to identify open questions remaining for usage data processing and the privacy protection.

3 Background and Related Work

Anonymity is not profitable, as companies are using collected user data to generate revenue. This is done trough user profiling, to find targeted offerings for product and services, which are custom selected for the personal interest and persons need at a time. Thanks to such practices they can sell us more and control in a way our behaviors.

This is a starting point in data collection, and more data the others have about us, more control they have upon us, regarding our buying decisions or our personal opinions. It is not only about a shopping experience, but through mass media methodologies, media enterprises can influence public opinion and they can do this starting with an individual. Big Data phenomenon is gaining more and more importance in business decisions, and it is interesting to see data mining techniques are not a new findings as they work on machine learning analytically capabilities with ever cheaper becoming CPUs end storage space. But what we can see is the increase in data collection and the available data from individuals [6, 8, 16].

Internet, Broadband and mobile technologies made it possible for companies to collect user information easier than ever. In a classical approach it was expected to collect user information through surveys, questionnaires or observation. But today data generated through users activities in internet flows continuously into the enterprise data warehouse and this data not only for product usage but also for personal profile creation used. Each user can be unique identified through assigned IP addresses to mobile devices or PC's or through IOT to connected household product [4, 8, 15].

Even if some corporations claim and collect indeed data anonymity still some unique ID assigned to each user through hash or pseudo identifier. Once critical mass for information is arrived, it is possible to deanonymize these data sets by cross matching with other data sources [18–20]. Typical example for this deanonymization is AOL case, where user search queries publish to let researchers work with this data to find out what usage this data can find. But these anonymized data used to identify individuals through cross linking of survey data where for people zip code, sex and the date of bird was available.

By using these so called personal identifiers it was possible for attackers to identify individuals with their search queries. Another example is Netflix case, where user ID's in platform anonymized but through background-knowledge data available in other platforms again it was possible to identify individuals and their preferences for movie types. Based on movie types that people were interested was possible to generate personal profiles and again identify them uniquely [9, 20], [21, p. 08], [22].

Data is regarded as an oil for our century, and without capabilities to collect user and customer data to forecast future trends, suggesting products with user profiling, understanding usage patterns for products, future enterprise risk their existence in this new challenging online integrated market place [1, 23, 24].

Production companies need to integrate not only their supply chain, which is already a complex task, what can be a separate subject to investigate and research. But they should further integrate their data processing practices with a collection and analyzing of product usage data, after product delivery to the end user. Which means even if already established the integrated data flow systems between production and supply chain in place, they need to extend these data networking and management system further after the end of supply chain, which is a product delivery, but through out product life cycle. Which means even if already established the integrated data flow systems between production and supply chain in place, they need to extend these data networking and management system further after the end of supply chain, which is a product delivery, but through out product life cycle. already predefined [3, 4].

Product data collection brings new challenges to the enterprise. These challenges start with a data collection from end user through connected devices, which with help of sensors or data readers are able to capture usage data for the devices, products, appliances or machines during their usage or while they are in operation [4, 7, 8].

This is a new concept, since in a normal case machines or electronic products, home appliances or cars for instance has offline interface to reed the condition of components at the time of servicing or while running diagnostics, but these readings are often not recorded in the product itself or somewhere else and they are instant readings. Enterprise need to redesign products in a way that those readings can be either recorded in a memory embedded into the product itself or transferred into a central data warehouse for analyzing while products are in use and continuously [4, 7].

The aim of this paper is not to go into the technical details of new era production redesign problematic for data collection, but to name this problematic to give a broader view of the current situation in the market.

In coming section I will try to show what solutions are available to overcome usage data collection problem, and than further highlight the data analyzing challenges ahead of the future enterprise. After highlighting these challenges I will try to identify consequences of these new happenings of large volumes data analyzing to a society and to an individual through over dimensional exposure of private data in public.

3.1 Product Usage Data Patterns and Benefits to the Enterprise

Horizon 2020 show the importance of data collection and processing as a important factor in product development to achieve energy and resources savings, and at the same time to meet customers needs individually, which create competitive advantage for the firms to stay in today's market competitive [1].

With the brought diffusion and acceptance of internet technologies within the society we run part of our daily activities online and let companies monitor, record and analyze our habits, product preferences and platform usage. Online active companies are than able to understand our product interest and prognoses future trends based on our collected data which we left online [6, 23, 24].

The new challenge for future enterprise is to bring this online data collection and proved concept of Big data analyzing into a real world with the monitoring of real products usage in our daily lives [4, 7]. The idea is to collect not only functional, component readings for products, but also with a development of new sensors to read usage data to understand eventually the usage patterns [5]. When we are able to understand and to find out for instance for a washing machine, a mean average wait of clothes for users in a certain age group we might be able to redesign washing machine to satisfy very special need of users in this group only. This new design can be completely different then conventional hardware product solution, but as a new service design for senior population as an example. If we will see so extreme innovations in solutions design is a question of time but there are already existing examples, where car producers are investing heavily into new service solution generation like care2go (Mercedes-Benz) [25, p. 2], drivenow (BMW) [26] examples.

With connected cars these companies are able to collect vast amount of data about driver habits and car usage itself, so that they offer now mobility solutions instead of car manufacturing only.

Another change we see is cloud solutions where hardware and software infrastructure are offered as a utility in IT environment. IT hardware producers are interested to learn customer computing habits as smart phone and mobile devices users are monitored for their usage patterns to help in redesigning better solutions for a Mobil computing. When we can extend these data capturing and analyzing capabilities further into all products, from household appliances and machines to the all things we have around us, we can manage thanks to the data collected and processed a better product design, reduce resources usage in production and reach sustainability in supply chain [1].

3.2   Data Collection and Internet of Things (IoT)

"The explosive progress in networking, storage, and processor technologies is resulting in an unprecedented amount of digitization of information. In concert with this dramatic and escalating increase in digital data, concerns about privacy of personal information have emerged globally" [27]. IoT can be also explained in a broader context as following: 'a world-wide network of interconnected objects uniquely addressable, based on standard communication protocols" [3].

As mentioned the classical data flow process designed for products to capture the data till the delivery of the products itself to an end user. Further data captured for product reside in a claim management process by customer service. In total quality management the customer feedback for the product is integrated into the management system and there is a continuous improvement program for product design and production. This is not a new but well established practice can be found in a competitive production company. In a normal functioning these feedback is captured by text mining in customer center logs, feedback from a customer with the help of online questionnaire or blog [28, 29, 45].

The new expectation from future enterprise is to capture feedback and customer data actively on the way, while the product is in use. Which means without one time efforts like surveys or questionnaires with the customer consciousness. New concept requires continuous, s ongoing data capturing from thinks with a help of sensors, actuators, wireless connection or RFIDs [3, 4, 8, 15].The new data collection solution for the enterprise is IoT, where every device, appliance, machine, thinks are with a help of internet connected to each other and to a cloud or on-site server to feed the readings from sensors into a data processing logic. Here data captured flows into a data warehouse continuously, that conventional data bank design and BIS has difficulty to work with this data. We can call this large volume, sometime as a unstructured data a Big data coming from a real world [5, 8].

3.3 Data Processing and Non-obvious, Valuable Information

Big data defined as continuous, structured or unstructured, with multi-dimensional characteristics flowing from different sources is already a challenge for conventional management decisions models since it is difficult to capture and decipher not obvious, valuable data from large scale data and than to evolve it into a knowledge [6, 8, 24].

Coming from this Big picture we can mention here the characteristics of Big data which will highlight the difficulties in data processing. In a literature we see many Vs to show the characteristics of Big data. It is obvious this list can not be taken as a complete but it can give a sense for Big data challenges in itself.

• Volume

Data volume increase enormously also through machine generated data (like IoT, industry 4.0)

• Velocity

Online continuous data processing is needed to validate and capture flowing data

- Variety

Data can be in a structured, or unstructured in many forms and formats, like graphical, and audio like

- Validity

Since data flows into processing logic continuously, current state of data changes on a continuous base, and the data state evolve while analytic process in use.

- Value

It is possible to gain insights and new knowledge from the data set, even if these knowledge is not known for the collection purpose of the data [8, 23, 24].

Big data defines usage of analytic and data processing techniques, like machine learning, data mining, to find out new information through search, aggregate and cross-reference of data to generate knowledge and intelligence by using statistical methodologies like finding correlations to forecast future happenings [16, 24].

As mentioned before it is not possible to process the Big data for a conventional databases and business-information-systems (BIS). The main difference between BI and Big data is the structure of managing the collected and the processed data.

There are new approaches and methods to work with Big data, as:

- Map Reduce

"Map Reduce is a programming model and an associated implementation for processing and generating large data-sets that is amenable to a broad variety of real-world tasks" [30].

- NoSQL

Non-relational databases, including hierarchical, graph, and object-oriented databases.

Their primary advantage is that, unlike relational databases, they handle unstructured data such as word-processing files, e-mail, multimedia, and social media efficiently [31].

- Hadoop framework

Distributed File System (HDFS) is designed to store very large data sets reliably, and to stream those data sets at high bandwidth to user applications. In a large cluster, thousands of servers both host directly attached storage and execute user application tasks. By distributing storage and computation across many servers, the resource can grow with demand while remaining economical at every size [24, 28, 32].

It is a condition for future enterprise to manage Big data challenges in the first place, before going into a product usage data analyzing. Because of the shared characteristics, product usage data itself is a Big data.

As long as this condition is not met, future enterprise cannot design sensors, actuators, RFID, wireless readers with exact requirements, predefined data types and information for things to collect [3–5, 7].

3.4 Big Data Challenges, but not only Technical

Internet is not the reason for Big data, it just a tool to easily collect and analyze the data, data grows day by day and Corporate have interest to work with these available data, to gain competitive advantage, through providing solutions to individual needs [8]. On the one hand the enterprise should adopt its business intelligent systems to cope with Big data, and own management decisions practices to accommodate Big data in strategic, operational and design decision makings, and on the other hand should focus on customer privacy protection [1, 8, 24].

Privacy concern is expressed globally because of increase in publicly available micro-data sets about individuals, which collected, processed and owned by many independent private and public organizations. These organizations can make these data public without any coordination with each other and can expose individual persons sensitive data with the help of cross referencing method or prior knowledge problem [15, 33].

We can give a study of Sweeney [34] as an example for cross referencing problem for data privacy, where Massachusetts voter registration records, which is public data set, and medical record data from a group insurance commission with patient specific information for 135.000 for state employees and their families were cross-referenced. It was believed to be anonymous data in both data sets, since the sensitive information like name, social security numbers were removed from this health record data and and sensitive identifiers from the voter records.

In a voter registration information zip code, date of birth and gender was not removed, while they were classified as non-sensitive information. When these data sets cross-referenced with each other by combining Zip-code, date of birth and gender, Governor of Massachusetts, Willran Weld was identified as being one of 6 people lived in Cambridge with the same date of birth, and 1 of 3 men, and the only one with in this zip-code area [18, 34].

We can give a Netflix case as an example for prior knowledge problem to identify a person uniquely in a database. In a study from Narayanan and Shmatikov (2008), they demonstrated that, in a public micro data set of 500.000 Netflix video subscription service customers, where data was anonymized by removing names and personal identifiers, it was possible to identify people with a high probability by using background information available from a internet movie database, where people commented on movies and rated these. By matching rated movie combination from IMDB, and the Netflix data set, they were able to identify person uniquely in the Netflix anonymized database and uncover his/her political orientation and other potentially sensitive information [20].

These two examples can identify which thread to privacy arise because of the increase in data granularity and data volume generated and recorded with a help of internet and data mining technologies.

Privacy and cross referencing identification problem with publicly available databases is not new and the study from Sweeney [34] was just a recent demonstration of privacy exposures arising from collected data and micro data sets. There are other examples 70 ties, like presented in a study from Adam and Worthmann [35], they referred to a study by Dalenius [36], where we can read a direct citation from a work by Miller [17].

"Some deficiencies inevitably crop up even in the Census Bureau. In 1963, for example, it reportedly provided the American Medical Association with a statistical list of one hundred and eight-eight doctors residing in Illinois. The list was broken down into more than two dozen categories, and each category was further subdivided by medical specialty and area residence; as a result, identification of individual doctors was possible ..." [17, 35, 36].

Big data and machine learning with the increase in data availability makes the situation critical for privacy protection and it is identified as an important thread for the society [16, 27, 33, 34, 37]. The future enterprise should address the privacy problem at a same time as it develops the data capturing systems and analyzing capabilities [1].

It is worth to highlight here also the privacy as understood by different cultures, and how law makers, civil and government organizations address the privacy issue with different approaches with same consequences for everyone on the planet with the privacy exposure.

4 Need for a Privacy Protection and Privacy Declarations

Privacy is understood differently in different parts of world and cultures. We can expect political, cultural and religious differences can effect the understanding of privacy and this can be a separate research topic to study and it is not handled in the scope of this paper. But one think is clear, that privacy is important for everyone within a different degree and granularity and must be protected to secure personal freedom [38]. We can see similar understanding for privacy as in the declarations below.

It is stated in a UNESCO web site as:

"The right of privacy is well established in international law. The core privacy principle in modern law may be found in the Universal Declaration of Human Rights. Article 12 of the UDHR states 'No one shall be subjected to arbitrary interference with his privacy, family, home or correspondence, nor to attacks upon his honour and reputation. Everyone has the right to the protection of the law against such interference or attacks'" [38].

It is stated in a EU Declaration for Privacy as:

"Everyone has the right to the protection of personal data. Under EU law, personal data can only be gathered legally under strict conditions, for a legitimate purpose" [39].

And further documents, the privacy rooted into:

- Article 14 of the United Nations Convention on Migrant Workers;
- Article 16 of the UN Convention on the Rights of the Child;
- Article 10 of the African Charter on the Rights and Welfare of the Child;
- Article 4 of the African Union Principles on Freedom of Expression (the right of access to information);
- Article 11 of the American Convention on Human Rights;
- Article 5 of the American Declaration of the Rights and Duties of Man,
- Articles 16 and 21 of the Arab Charter on Human Rights;
- Article 21 of the ASEAN Human Rights Declaration; and
- Article 8 of the European Convention on Human Rights.

Over 130 countries have constitutional statements regarding the protection of privacy, in every region of the world [16].

Law maker defines according to these declaration the framework to protect the sensitive information of an individual. Sensitive information is defined as being data to identify an individual in a database. In practice law makers request these identifiers to be removed from the database per record to secure the privacy. Removing sensitive information and identifiers from the database is called data anonymization. In a legal framework data anonymization is regarded as a technological safeguard for privacy protection. But the increase in data availability and the developments in data mining makes this praxis of database anonymization obsolete [22]. Some examples of privacy breaches (exposures) with anonymized databases presented in coming section.

5 State-of-the-Art Data Anonymization Techniques

Because of increase in micro data and availability of data, due to the development in internet and communication technologies as well as die user acceptance and the diffusion of internet usage into our daily lives, the increase in privacy protection risk is regarded as a tread to personal freedom of future generations [22, 33].

On the one hand the benefits of Big data to our society and to prosper is not to deny. Thanks to Big data, data mining and analyzing techniques we are able to prognose, predict and find new solutions to our problems in medical, social, environmental and humanitarian areas [6, 24]. On the other hand we let companies and organizations to collect and know more about us, and sometime without knowing the purpose of this data collection, since collected data can be aggregate and enriched further by combining, cross referencing with other data sets to serve new algorithms and analytic. This new mutated knowledge can expose our private data or attackers can try to identify us uniquely with a help of this new data, which is a concern for everyone out of question [19, 34].

Even though privacy is understood at a different intensity and level in different parts of the world, its a fundamental right for a personal freedom and the basic human rights to keep private data just for us and share it as long as we are willing to and it is right to got forgotten in case of a need [16].

In this respect data anonymization is a save guard to protect our sensitive data and personal identifiers against privacy exposures. We are willing to serve the knowledge generation and participate in surveys and census, but finding out the data is used to identify us in a different records, like in a medical record, it must be frustrating for anyone if a discrimination act against us results from this new knowledge [22, 27].

Law maker and governmental organization require data anonymization as a technical solution to safeguard our privacy and as a result we can see many different methods in practice to secure our sensitive data in micro data sets. These methods try to hide personal identifiers and sensitive attributes by suppression, generalization, perturbation, randomization, obfuscation, encryption or by geometrical transformation [11, 27, 33].

The aim of data anonymization is to hide sensitive data but at the same time to maintain the data integrity and quality to let data processing, analytic algorithm or statistical calculation run without interruption to achieve same results as the data in original state, before anonymization applied would deliver [40].

A database keep and maintain the model of an object or a part of a real word, which is constructed from entities. These entities are the elements of that world or subject. Attributes are used to present the characteristics and the relationships between different entities. Entities with a same attribute represent a particular entity type [35].

As an example, a patient and treatments in a medical record are entities where name, social security number and diagnosis type are attributes. In a statistical database it is possible to calculate, query aggregate statistical values like sample mean, count, correlations and similar for a subset of entities recorded in a database. Another database type is to serve data for entities on a micro level with high granularity. A patient database in a hospital can serve both proposes, where statistical researcher can retrieve aggregate statistics and doctors can read any particular data for an entity in this database [35, 44].

Data should be collected to serve society and people, as Adam and Wortman [35] "The problem here is "the inevitable conflict between the individual's right to privacy and the society's need to know and process information"".

Because of a given direction by public and private organizations to safeguard the privacy, we see new developments in database management to have sensitive data anonymized or by simply restricting the query privileges [41].

In query restriction it prohibits queries, which can find personal sensitive data from the database, where the seize of the query results are restricted, the overlap among continuous queries controlled and audit trial for all queries are recorded and for potential compromise checked [41].

For anonymization solution we see different evolving techniques, which we can have under four groups.

(a) suppression and generalization, where sensitive data is removed for the data set to reduce granularity, or information is group into sets within a range to prevent unique identification of an entity.
(b) Perturbation, where noise is added to sensitive data to prevent its identification.
(c) Randomization (permutation), where sensitive data is swapped between entities to prevent exact identification.
(d) Obfuscation (Encryption or geometrical transformation) [11, 33, 41, 42].

Data anonymization introduce uncertainty and noise into data, where data was initially certain, which brings its own shortcomings for data quality assurance [35]. We will look into these quality issues in Sect. 4.

5.1 Suppression and Generalization (Reduction of the Data Granularity)

Suppression

In this method sensitive data and personal identifiers are removed before releasing or publishing the data set. It the most commonly used anonymization technique and law makers requests this method to get sensitive information removed before publishing any database.

Advantages

It is a simple process to anonymize data-sets by removing sensitive data without a need to know whole data collection.

Disadvantages

Reduction in granularity of data which can effect also the data quality for statistical purposes.

Privacy risks

We saw as in an example in [34], that only by removing personal identifiers it is not possible to secure a privacy protection, since it is easy to deanonymize sensitive data by simply cross referencing anonymized data with other publicly available data-set.

It is also necessary to remove so called quasi identifiers like date of bird, zip code and gender information. In a study [13], it was shown that in a U.S. Census summary data from 1990 it was possible to identify 87% of US population uniquely with a help of these 3 quasi identifiers. Since the combination of those 3 attributes shown that uniqueness.

Generalization

In this technique instead of removing sensitive data, they are grouped into data ranges or sets. As an example it is possible to group salaries between 30.000 and 40.0000 per year per a person in a group to reduce data granularity and to prevent unique identification of an entity.

Advantages

It is believed that generalization overcome the shortcomings of suppression by simply avoiding unique identification of a person by attributes.

Disadvantages

The reduction in data granularity has an effect on a data quality.

Privacy risk

In her work 2002 Sweeney showed a methodology k-anonymity [34], where for a set of k-records with unique identifier combination we need at least $k - 1$ similar records to prevent unique identification of an entity.

This method prevent cross referencing de-anonymization attempts but fails to safeguard against background knowledge attics, where attackers possess a knowledge about a person to identify this person with high probability within a database.

Suppression and generalization is most commonly used anonymization techniques but with an increase with data availability about individuals and because of the developments in data mining technologies they fail to secure privacy completely.

In the case of Netflix example [20] Movie rental company Netflix had an intention to find out what future benefits could be gained from a database, which represents its 500.000 customers and their subscription data. Netflix removed the name, address and other sensitive information to anonymize the database.

5.2   Perturbation (Noise Addition)

In Perturbation technique, a noise is added into a data set for sensitive information to prevent exposure, or noise is added into a query results to prevent overlaps or discovery of sensitive data by combining multiple query results.

Some common Perturbation practice is to replace original Database with a sample from the same distribution, adding noise to the values or to the results of a query or sampling the result of a query from the same original query [41].

Advantages

It provides better protection than suppression and generalization by completely replacing original values with a pseudo values, and practical to apply into large dimensional databases.

Disadvantages

This proposed technique can not satisfy the high quality statistical results expected from the data set, and to meet privacy protection requirements [43].

Privacy Risk

As long as we perturbate sensitive data with known data mining algorithm it is not possible to protect data against unknown, or new developing algorithms to extract sensitive data with a bad intentions. Perturbation can be regarded as a good anonymization method, but again it has short comings to fully protect the privacy, if a background knowledge about a entity is available in other databases or resources [11, 41].

5.3 Randomization (Sensitive Data Swap)

The randomization is a method to swap sensitive data between entities, to mask and hide the attribute values of records [41], more precious then generalization [11]. Another application of this method is to return a value $xi+r$ instead of xi, where r is a random value from some distribution [33].

For instance for an age we can choose a random value from a range of $(-50,50)$ and add these to the age data. But the problem arise with the data mining algorithm, when data reads 120 in a data set for an age, machine can learn, if the range of $-50,50$ is known, that real age is minimum 70 years with a 100% correctness, which can help in exploration [27].

It is traditionally used to distort data by probability distribution in surveys and census [33]. In practice it is not possible to recover original records but it is possible to explore the distribution of the original records.

Advantages

It is found to be easy to implement randomization algorithm to data sets during the collection time, since it does not require the distribution knowledge for other records. In a case of k-anonymity methodology proposed by [34], we need to know other records to justify $k - 1$ calculation to apply suppression. Randomization on the other hand doesn't require a server with all the original data collected to generate random values in a record [33].

Disadvantages

Randomization handle all records in a same way and priority, that applying anonymization to data sets with a different intensity can bring performance problems for the utility of data [33].

Privacy risks

We need to apply data swapping to all entities for all attributes but if a background knowledge for an entity is known, it is possible to identify a person in a database based on this knowledge. In the case of the randomization, if a range for a random noise value is known to a data mining algorithm, than through machine learning attackers can exacerbate individual records with a help of approximation approach [33, 41].

5.4 Obfuscation (Encryption or Geometrical Transformation)

The obfuscation method apply encryption or geometrical transformation to a original data set to safeguard privacy.

Encrypted data is unusable in this form and only a person with a decryption key is able to work with this data. We need to provide than different access control for different users for each data-mining application, for each data-sets. This is adding additional complexity into a data management problem with not much benefits for privacy protection [42].

Another approach of obfuscation is to apply geometrical transformation, where sensitive data transformed randomly and geometrically within a data set to mask and hide this information.

Advantages

Encryption provide private security by limiting access to data only for authorized person with a need to know approach. It brings additional security measures for whole data-set and provide access control management.

Disadvantages

It brings additional complexity into data management and data is unusable in encrypted form. Data collection and validation of the data is also complicated through encryption process.

It is actually a stopper in data publishing and open data for research purposes which is in conflict with data collection and database creation in the first place.

Privacy risk

Although encryption method secure the privacy, the real risk of privacy exposure stays with a person who has the decryption key. It brings the data anonymization concept into a access and data control field, which doesn't serve the anonymization purpose itself. Geometrical transformation method is found to be inefficient in privacy protection because of sensitive data exposure once geometrical transformation algorithm is known or learned by data mining logic [42].

6 Compromise Between the Privacy and the Data Quality

The data anonymization add an uncertainty and perturbate values into data sets, which was before certain to represent the part of the world, or a population. This brings data quality problematic into a data analyzing to be addressed.

There was many research in recent decades on methods to limit the disclosure in the publication of data Research in this field looked into different opponents and different models, which proposed anonymization techniques by strict safeguards against attacks. However, none of these techniques was deployed as part of a useful tool. Because of the non-interactive nature of these techniques: the only interface they provide to the data editor is a set of parameters that control the degree of protection of privacy must be executed in the data anonymous [10].

There are new approaches and statistical algorithm to overcome data-quality loose, which works with probability and proximity approach.

Every method of data anonymization can be considered for a given situation and database type. Some of the data anonymization methods, like encryption or data obfuscation cannot handle data quality and usability assurance certainly, which makes open data usage difficult.

Suppression and generalization methods are shown to be easy to deploy, like the possibility of applying anonymization during the data collection. But they fail to protect the privacy in an environment where background knowledge and cross-referencing data is available.

There are enhancements created to limit de-anonymization attacks to data-sets, with suppression and generalization, so called k-anonymity and further l-anonymity. The aim is to define minimum requested anonymization level to secure the privacy. But we saw examples where unique identification of an individual in a data-set with background knowledge was is possible. Those enhancements methods proved themselves as insufficient to secure the privacy but raised the public awareness for the privacy exposure in data collection.

Perturbation technique with a noise addition to data-sets or to query results again raise quality issues, where new data-mining methods should be developed to overcome quality loss in data.

Randomization and data-swap between entities seems to serve the privacy purpose and data quality assurance, as sample after anonymization still represents the original distribution.

Randomization methods shortly fails to secure privacy as data analyzing technologies develops themselves and thanks to machine learning capabilities, if only a fraction of the original data, or a range of the random values are known, than analyzing logic can learn from the data to exacerbate targeted entity attributes.

If we put these findings in a cross matrix table to give an overview of anonymization methods, compared under usability, data quality, robustness against privacy attacks with background knowledge and cross referencing and limitation factor in applicability, it might help to see state of art anonymization methods against each other (Table 1).

We can see that each data anonymization method has its own benefits and limitations. For each situation and data-sets type the anonymization method should be separately decided also within the same organization.

7 Identified Problems and the Outlook

It is in conflict with a data collection propose in the first place not to publish data at all, and keeping them encrypted, hiding the data in order to protect the privacy, which makes database ownership meaningless. Just hiding the data doesn't serve

Table 1 Comparison of data anoymization methods

	Usability	Data quality	Robustness		Limitations
			Cross-reference	Background-knowledge	
Suppression	High	Mid-Low	Low	Low	K-anonymity, no real time
Perturbation	Low	Low	High	Mid	Ongoing supervision
Randomization	High	High	Mid	Low	Rang to hide
Obfuscation	Low	Low	High	High	Publication not possible

Created on the base of literature research, presented in this paper

the purpose of data collection and new hybrid anonymization methods are to be developed and applied.

The observed problem is that, all of the methods are static and fail to react interactively to de-anonymization attacks. When the anonymization is applied the privacy protection is secured against the data mining process and logic known to the database owner organization, but privacy intruders develops their own algorithms and analytic logic to overcome applied techniques.

Purpose solution can be an interactive query rights to data warehouse on demand, which monitor and control queries into data-sets and provide aggregated statistical results. Interactivity means data should be kept continuously hidden and maintain. Which restricts the whole data-set publication even after data anonymization applied.

All the usage cases we reviewed in this paper represents open public data-sets, collected by public and civil organizations, which show just the part of the problem for a known cases.

Future research should focus on unknown part of this picture, namely on Big data collected by corporations for a personal profiling through social networks, search queries, internet and shopping activities, as well as with an increasing dimension the IoT. Which methods or technologies should be developed to help individual hide and protect his–her privacy in daily online and offline activities is also a research question. Data anonymization at the data resource seems to be a sustainable solution to a privacy protection problematic.

References

1. European Commission, Directorate-General for Research and Innovation, and European Factories of the Future Research Association, Factories of the future: multi-annual roadmap for the contractual PPP under Horizon 2020. 2013
2. Mladenow A, Kryvinska N, Strauss C (2012) Towards cloud-centric service environments. J Serv Sci Res 4(2):213–234

3. EU INFSO, "INFSO D.4 Networked Enterprise & RFID INFSO G.2 Micro & Nanosystems, in: Co-operation with the Working Group RFID of the ETP EPOSS, Internet of Things in 2020, Roadmap for the Future, Version 1.1, 27 May 2008.," INFSO D, vol 4, 2008

4. Gregus M, Kryvinska N (2015) Service orientation of enterprises—Aspects, Dimensions, Technologies. Comenius University in Bratislava. ISBN: 9788022339780

5. Atzori L, Iera A, Morabito G (2010) The internet of things: a survey. Comput Netw 54(15): 2787–2805

6. IBM, IBM big data—What is big data—United States. Available: http://www.ibm.com/big-data/us/en/. Accessed: 30 Dec 2015

7. Gubbi J, Buyya R, Marusic S, Palaniswami M (2013) Internet of Things (IoT): a vision, architectural elements, and future directions. Future Gener Comput Syst 29(7):1645–1660

8. Jacobs A (2009) The pathologies of big data. Queue 7(6), 10:10–10:19

9. Xiaoli X, Yunbo Z, Guoxin W (2011) Design of intelligent internet of things for equipment maintenance. In 2011 International Conference on Intelligent Computation Technology and Automation (ICICTA), 2011, vol 2. pp 509–511

10. Xiao X, Wang G, Gehrke J (2009) Interactive anonymization of sensitive data. In Proceedings of the 2009 ACM SIGMOD international conference on management of data, New York, NY, USA, 2009, pp 1051–1054

11. Cormode G, Srivastava D (2009) Anonymized data: generation, models, usage. In Presented at the Proceedings of the 2009 ACM SIGMOD international conference on management of data, 2009, pp 1015–1018

12. Kursawe K, Danezis G, Kohlweiss M (2011) Privacy-friendly aggregation for the smart-grid. In Presented at the privacy enhancing technologies. pp 175–191

13. Darby S (2010) Smart metering: what potential for householder engagement? Build Res Inf 38(5):442–457

14. Molnár E, Molnár R, Kryvinska N, Greguš M (2014) Web intelligence in practice, the society of service science. J Serv Sci Res, Springer 6(1):149–172

15. Weber RH (2010) Internet of Things-New security and privacy challenges. Comput Law Secur Rev 26(1):23–30

16. privacyinternational.org, What is Privacy | Privacy International. Available: https://www.privacyinternational.org/node/54. Accessed 02 Jan 2016

17. Miller AR (1971) The assault on privacy: computers, data banks, and dossiers. University of Michigan Press

18. Sweeney L (2000) Simple demographics often identify people uniquely. Health (San Francisco) 671:1–34

19. Dinur I, Nissim K (2003) Revealing information while preserving privacy. In Proceedings of the Twenty-second ACM SIGMOD-SIGACT-SIGART symposium on principles of database systems. New York, pp 202–210

20. Narayanan A, Shmatikov V (2008) Robust de-anonymization of large sparse datasets. In Presented at the security and privacy, 2008. SP 2008. IEEE Symposium on, 2008. pp 111–125

21. Ghinita G, Karras P, Kalnis P, Mamoulis N (2009) A framework for efficient data anonymization under privacy and accuracy constraints. ACM Trans Database Syst (TODS) 34(2):9

22. Ohm P (2009) Broken promises of privacy: responding to the surprising failure of anonymization. Social science research network, Rochester, NY, SSRN Scholarly Paper ID 1450006

23. Kryvinska N, Gregus M (2014) SOA and its business value in requirements, features, practices and methodologies. Comenius University in Bratislava, ISBN: 9788022337649

24. Kaczor S, Kryvinska N (2013) It is all about services—Fundamentals, drivers, and business models, The society of service science. J Serv Sci Res, Springer 5(2):125–154

25. Car2Go. Car2Go Wien Unternehmensprofil. Available: https://www.car2go.com/de/wien/unternehmen/. Accessed 07 Nov 2014

26. DriveNow Impressum. Available: https://at.drive-now.com/#!/impressum. Accessed 05 Jan 2016
27. Evfimievski A, Gehrke J, Srikant R (2003) Limiting privacy breaches in privacy preserving data mining. In Proceedings of the Twenty-second ACM SIGMOD-SIGACT-SIGART symposium on principles of database systems. New York. pp 211–222
28. Witten IH, Frank E (2005) Data mining: practical machine learning tools and techniques. Morgan Kaufmann
29. Kryvinska N (2012) Building consistent formal specification for the service enterprise agility foundation, The Society of Service Science. J Serv Sci Res, Springer 4(2):235–269
30. Dean J, Ghemawat S (2008) MapReduce: simplified data processing on large clusters. Commun ACM 51(1):107–113
31. Leavitt N (2010) Will NoSQL databases live up to their promise? Computer 43(2):12–14
32. Shvachko K, Kuang H, Radia S, Chansler R (2010) The hadoop distributed file system. In Presented at the mass storage systems and technologies (MSST), 2010 IEEE 26th Symposium on, pp 1–10
33. Aggarwal CC, Yu PS (eds) (2008) Privacy-preserving data mining, vol 34. Springer, Boston
34. Sweeney L (2002) k-anonymity: a model for protecting privacy. Int J Uncertainty, Fuzziness Knowl. Based Syst 10(05):557–570
35. Adam NR, Worthmann JC (1989) Security-control methods for statistical databases: a comparative study. ACM Comput Surv (CSUR) 21(4):515–556
36. Dalenius T (1974) The invasion of privacy problem and statistics production—an overview. Statistik Tidskrift 12:213–225
37. Agrawal V, Arjona LD, Lemmens R (2001) E-performance: the path to rational exuberance. The McKinsey Q, 31
38. Rotenberg M. Preserving privacy in the information society. Available: http://www.unesco.org/webworld/infoethics_2/eng/papers/paper_10.htm. Accessed 02 Jan 2016
39. EU Parlement. Data protection directive 95/46/EC of the European Parliament. Available: http://eur-lex.europa.eu/legal-content/en/ALL/?uri=CELEX:31995L0046. Accessed 02 Jan 2016
40. Chen FF, Adam EE (1991) The impact of flexible manufacturing systems on productivity and quality. IEEE Trans Eng Manage 38(1):33–45
41. Agrawal R, Srikant R (2000) Privacy-preserving data mining. ACM Sigmod Record 29: 439–450
42. Parameswaran R, Blough D (2005) A robust data obfuscation approach for privacy preservation of clustered data. Presented at the Workshop on privacy and security aspects of data mining, pp 18–25
43. Adam NR, Worthmann JC (1989) Security-control methods for statistical databases: a comparative study. ACM Comput Surv 21(4):515–556
44. Bohdalová M, Greguš M (2016) Value at risk with filtered historical simulation. Time series analysis and forecasting. Springer, Cham, pp 123–133
45. Charfaoui E (2008) Die Sprachfertigkeiten der Manager - der unabkömmliche Bestandteil ihrer Fachvorbereitung in der Informationsgesellschaft. Inovácie - podnikanie - spoločnosť. - Prešov : Vysoká škola medzinárodného podnikania, - S. 439-446. -ISBN 978-80-89372-03-4

Challenges in Management of Software Testing

Martina Halás Vančová and Zuzana Kovačičová

Abstract The chapter is focused on test management within projects of software development. The high quality of test management is a part of the success of IT projects. There exists a high quantity of recommended practices and techniques for test management, however, it is necessary to complete these recommendations by actual trends and practical experience, since every IT project is different in its nature and it is managed in a specific environment. The first part of the chapter deals with the theoretical background of IT testing. The second part specifies the research objective, methodology and research procedures. The last part is empirical, and it describes activities of test management carried out at the observed IT project.

1 Introduction

On a daily basis, we are surrounded by information technology. Either we talk about hardware or software, it is currently omnipresent. Naturally, there are people who every day create new types of software, by which they are trying to make our world more sophisticated and our daily activities simpler and more effective. The process of software development consists of many activities a`nd inseparable parts. Except for managing of the whole software development and programming, it is necessary to include quality assurance. Quality assurance in software development is carried out via testing processes, which can be further divided into different types or techniques. Software testing and its management is a crucial part of every software development mainly because it assures that software will fit the purpose for which it is created. Additionally, software testing enables a developing company

M. H. Vančová (✉) · Z. Kovačičová
Department of Information Systems, Faculty of Management,
Comenius University in Bratislava, Bratislava, Slovakia
e-mail: martina.halas@fm.uniba.sk

Z. Kovačičová
e-mail: zuzana.kovacicova@fm.uniba.sk

© Springer International Publishing AG, part of Springer Nature 2019
N. Kryvinska and M. Greguš (eds.), *Data-Centric Business and Applications*,
Lecture Notes on Data Engineering and Communications Technologies 20,
https://doi.org/10.1007/978-3-319-94117-2_7

to make sure that a product they are selling to a customer is of a high quality, without defects and but also it assures management that developers are doing their work correctly.

2 Current State of Software Testing in Slovakia and Abroad

Software testing is a crucial part of every software development. It is undeniable that this process cannot be excluded because it assures the overall quality of a developed software based on requirements.

Nowadays, we can say that software is omnipresent. It is an integral part of our business, systems that we use on daily basis, it is integrated into many mechanisms and electronic devices. Some software products can be bought directly in shops and installed into our personal computers, others are already embedded into products we buy [1].

Currently, we can claim that software development projects are facing a serious challenge. Very often costs for software development exceed the planned budget and, in the end, final software is not adding such a value for which it was created [2]. Therefore it is even more necessary to highlight need for high-quality IT management, which is also including IT test management.

2.1 Software Testing Defined

There have been several definitions of software testing created, while the most standard definitions are:

> The process of a system or component under specified conditions, observing or recording the results and making an evaluation of some aspects of the system or component [3].

> The process of analyzing a software item to detect the difference between existing and required conditions (that is, bugs) and to evaluate the features of the software items. [4]

> The process consisting of all lifecycle activities, both static and dynamic, concerned with planning, preparation and evaluation of software products and related work products to determine that they satisfy specified requirements, to demonstrate that they are fit for purpose and to detect defects. [5]

Definitely, we can sum it up into a simple form where we define testing as: *"comparing what is to what it should be and share the information obtained"* [6].

We all make mistakes and often it is good to be controlled in order to avoid or fix mistakes. It is similar also in case of software testing. Even if there is a developer with the best intentions to create software without any error, it is simply not possible because software is very complex, it processes lots of information and last but not least it is developed according to information provided by another person and mistakes can appear in analysis even before a developer starts to code.

Sometimes we have assumptions, sometimes we perceive things differently than other people meant and thus it is necessary to involve testing in software development to have an overview [6].

However, it is natural that a cycle of test scripts cannot guarantee that a tester will find all possible errors or mistakes within software, but at least he/she can find some of them and thus make software better and better. But even if we have the most motivated tester who wants to find all possible errors, it might not be possible (and also it would be probably too costly) [7].

2.2 Purpose of Software Testing

When we discuss development of software, we can come to conclusion that it is not necessary to involve testing and rely purely on check of developers and their unit or alternatively module tests. However, if there are more people involved in development of software, it means that there are several perspectives, thoughts, improvement ideas but also auditors who go through software through and through. In addition to this, testing adds value to software development despite it might seem to be an additional cost in the overall budget for software development but on it can naturally bring potential savings if errors are discovered as soon as possible, basically we never know how expensive errors can be if they happen to occur in production. Testing definitely improves reliability of a software product, and thus in the end it also helps to increase loyalty of customer towards the company and company´s goodwill itself. The main improvements, however, can be seen in the following main areas as they were defined [6]:

- The sooner defects or errors are discovered the less costly it is to fix it since future development will already be following the right path
- Crucial decisions about further development of software can be influenced by testing, a company can decide for a completely different solution, which would not be highlighted without interference of testing.

Additionally, software testing is not only about functionality. It includes many non-functional perspectives, for instance penetration testing, which is focused on verification of security of developed software and it includes various techniques aimed at decreasing of the probability that software can be attacked [8].

2.3 Black Box and White Box Testing

When talking about software testing we often encounter two types of testing: black box testing and white box testing, which are also often called as behavioral and structural testing respectively [9].

Black box testing is carried out according to given requirements or specification about how software should look like and how it should behave. It means that a tester is given a set of requirements for software and he/she verify if software is able to fulfil these requirements, but a tester does not have to know anything about internal structure of the application or how it works from the technical perspective.

On the other hand, there is white box testing, which is based only on internal structure of software. It means that it would be very difficult for a person who does not this software to test it because it often requires programming skills but also this person would not know what to expect from software [10].

2.4   Positive and Negative Testing

These two types of testing are opposite to each other, but both of them are necessary to be included in general testing of software. Positive test cases are designed to verify how software should behave, what it should do according to functional or technical specification and a tester is simply reviewing if software does what it is supposed to do. However, a certain part of test cases should be also covered by negative tests, which are supposed to verify is software does not do what it should not do [9].

Lyndsay [11] also highlights critics of negative tests. He claims, that negative testing is often focused on areas, which have never been mentioned in a design or functional analysis and thus a developer does not know that a system should not work like that and negative test cases are added on top of a defined project scope. Such cases can be understood as "*acceptable failure*". Naturally, it is also unnecessary to include negative tests if positive tests are already covering the same topic. Sometimes negative tests uncover defects from the business perspective while IT perspective is correct. It means that there is not a problem in coding but rather in a design [11].

2.5   Levelling of Testing

There are various approaches how to divide or level testing. According to Copeland [10] testing is divided into unit testing, integration testing, system testing and finally acceptance testing. However, Watkins [9] adds also system integration testing. In addition to this, Ammann and Offutt [12] add module testing.

There might be various definitions of testing levels and the most common are described in the following sub-sections.

2.6   Unit Testing

Unit testing is the first testing carried out directly by a developer, usually on a piece of code that he/she created. This testing is done at the lowest possible level of the whole application. As it was explained by Loveland, it is so-called "*man versus code*" testing [13].

This type of testing is very important because it enables to test individual parts of code separately without interactions with other modules, applications or systems and thus defects can be uncovered much easier and resolved faster. Unit testing often includes usage of additional programs or mocks for unit testing or various debugging systems. After a tester (in this case most often a developer himself/ herself) completes testing of one unit, he/she should follow up with testing of interrelated parts of code, which has already been unit tested. Unit testing usually involves review of basic operations done by a certain unit, verification of performance as well as reliability of this unit, checks of GUI, but also creating, updating or deleting of fields [9].

2.7 Integration Testing

The main aim of integration testing is to make sure that individual parts of software can be integrated without any defects and their interfaces can cooperate. Integration testing should be done directly after unit testing, however, in many cases it is included within unit testing without specific separation. Integration testing therefore should be done at the level of developers, while it is managed by a development lead (or sometimes a test manager). At the end of integration testing it is necessary to report that individual modules of developed software are integrated without errors as it is documented within a design [9].

Naturally, sometimes integration tests may be excluded especially when developed software is very small and it is not possible to carry out unit tests without having components already integrated, so thus in such a case unit and integration tests are the same [14].

2.8 System Testing

System testing is done by the testing team according to a certain functional or technical design explained in detail in a specification document, which should be approved by a customer (or a future business user of developed software). At this level a whole system is already tested, not only its individual components. During software testing, a test manager must be assured that all requirements of a customer will be covered in testing (usually there is specific mapping of requirements and test cases covered in a so-called requirement traceability matrix). However, in order to test software in the most realistic setting and environment, it is also necessary to use data, which are the closest to real business data (i.e. production data). For this purposes data anonymization is often necessary, which is nowadays carried out by anonymization systems and it can be understood as a relatively simple process [9].

As it was mentioned by Copeland [10], system testing already involves the final product or output of development, which means not only software but its correspondence with hardware, user guides, training information and so on. Additionally, the main aim of system testing is not only to verify overall functionality of software but also its usability, user friendliness, availability, capacity and also design [10].

In addition to this, there is often mentioned a specific type of system testing called *system and integration testing*. We already mentioned integration testing, but in this case, there is a slight difference. System and integration testing usually includes integration not only within the system but also outside the system—its inbound and outbound communication with the environment, i.e. other systems and applications. Cooperation with such system must be defined in specification and its developed based on that. It is very often challenging to test interfaces because systems and applications of third parties are usually outside of competences of a test manager and thus it is often replaced by mock applications [9].

System testing as well as acceptance testing are carried out according to specific test scenarios and test scripts. Firstly, a test scenario is high level overview of what should be done within a system, including business conditions, environment requirements and so on. Then, a test script is a more detailed process. It very precisely and specifically describes what exactly should be done within a system— every step that should be done by tester as well as what reaction should be expected from a system [15].

2.9 Acceptance Testing

The final level within the whole software testing is acceptance testing. This can be done only when unit testing, integration and system testing are accomplished. When this type of testing is successful it can be simply followed by customer's acceptance of a delivered system [10].

Successful testing is usually a direct pre-approval for Go Live. However, a certain pass rate is required. A pass rate is a ratio that shows how many of completed tests from the overall quantity were successful. This ratio is usually stated in a contract or in testing documentation [15].

2.10 Further Testing Techniques

Above we mentioned the most common types of testing, however, there exist a wide range of additional testing techniques such as regression testing, reliability testing, stress testing, performance testing, penetration testing, security testing, usability testing, volume testing and additionally also automated testing. It always depends on software and decision of test managers, which types of testing or testing techniques will be included. The more types of techniques, the higher the chance that software will be without any defects and it will serve the purpose in the best possible way. However, often it is useless to include many of mentioned tests because they are not required in case of smaller applications [9].

Types of testing and testing techniques also depends on how large a development project is. For very small applications it is often enough to carry our unit testing and simple integration testing. In case of medium applications, it is good to incorporate also system as well as acceptance testing. But in case of large applications it is mostly necessary to involve also regression testing, usability testing, performance testing and many others, usually requested by a customer in order to assure the highest possible quality.

2.11   Test Execution

Generally, typical manual test execution is based on requirements. Requirements are defined by a customer and then test cases are developed based on them by a test designer. Naturally, a test designer is also coordinated by a test plan, which is a high-level document capturing the whole process of testing. After test cases are prepared and a system is ready for testing, test execution commences. Every test execution is naturally connected with test results. This process is shown in Fig. 1 [16].

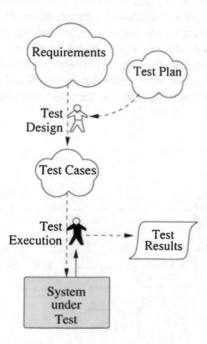

Fig. 1 Manual test execution [16]

In addition to this, waterfall process of test execution if often illustrated as V-model, which shows individual phases of testing and their relationship to other activities typical for software engineering. V-model shows that unit testing is on the same level as detailed design, then integration testing is based on preliminary design and finally system testing is based on requirements defined by a customer. However, there are many authors who modify V-model and naturally, V-model can be also specified on individual software development projects. Figure 2 shows V-model designed by Jorgensen [17].

There are various theoretical methodologies how to approach to test execution and its planning. In order to execute tests, it is necessary to allocate them to testers and plan test cycles or test attempts. Many IT projects apply the 30–50–20 ratio for test execution. According to this ratio, in the first testing cycle there are 30% of all tests pass and 70% are failed. Then those tests that were failed are taken into the second testing cycle, and it is expected that 50% of all tests (not only of those 70%) will pass. Then failed tests are taken into the third cycle where it is expected that 20% of all tests (or the remaining quantity) will pass. Thus, in the end, after three test cycles or testing attempts all 100% of planned tests are passed. In the end, it means that some tests are carried out only once and they immediately pass, some tests are carried out two times and they pass and some tests are carried out 3 times and only then they pass. According to this ratio, it is then easier for a test manager to plan time necessary for test execution as well as the quantity of resources. The approach is better explained in Table 1, where we can see that if we plan to test 100 test cases in three test cycles it is necessary to expect minimally 190 man-hours if we assume that completion of one test case takes one hours [15].

Completion of tests is affected by many factors and many of them are out of control of a test manager. Usually, progress in testing is achieved very slowly at the beginning, then after certain time a quantity of completed tests (either failed or passed) increases rapidly and then towards the end there is a small quantity of the

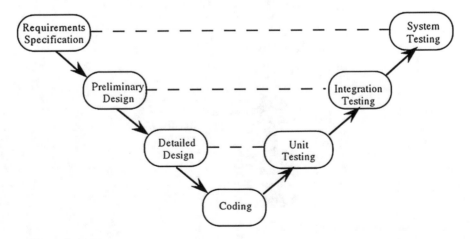

Fig. 2 V-model [17]

Table 1 Example of resource planning in testing according to the 30–50–20 ratio

Test cycle	Quantity of completed test cases	Passed test cases	Failed test cases	% of passed tests from overall quantity (%)	Man-hours expected (if 1 test case = 1 h)
1st test cycle	100	30	70	30	100 man-hours
2nd test cycle	70	50	20	50	70 man-hours
3rd test cycle	20	20	0	20	20 man-hours
Total	190	100	90	100	190 man-hours

Source Goodpasture [15]

most difficult test left and it is again very time consuming to complete them due to more serious defects or other facts which might be either in responsibility of a development team or outside the whole project. Below graph represents typical progress in test execution in the form of a so-called "S curve" [13] (Fig. 3).

When a test manager plans test execution, it is necessary to involve the whole management and also to gain commitment of a customer (whether internal or external) to be included in testing, at least in acceptance testing. A strategy for test execution should also include an approved quantity and content of test cases that will be completed, monitoring and reporting of progress, training and coordination of testing team according to a design testing process, etc. [9].

2.12   Management of Software Testing

Software development projects are often run based on well-known methodologies such as Project Management Body of Knowledge (PMBOK) or PRINCE2.

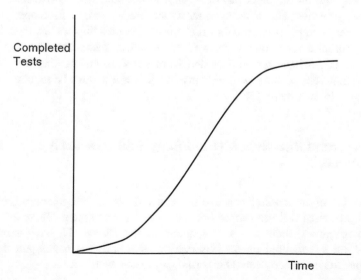

Fig. 3 Progress in test execution [13]

According to PMBOK software testing could be assigned under the topic Project Quality Management, which is focused on identification of quality requirements and standards as well as on their auditing and monitoring. These activities are done also during software testing, firstly requirements for software are analyzed and transferred into test scenarios, cases or scripts and then software is tested in order to find out if requirements are fulfilled. The main aim of software testing thus is to ensure ability to achieve customer satisfaction and to find out if developed software satisfies real needs [18].

Similarly, according to PRINCE2 methodology we can also assign software testing under the topic Quality, based on which project products should fit for purpose, which means to meet business expectations [19].

It is a best practice currently that software managers and project managers in software development should be certified by a certain project methodology, especially by one of the two mentioned. The methodology, however, is also very useful for test managers, because they are also part of a project team and best practices explained within these methodologies are very beneficial for successful management of testing processes during software development [20].

High quality of test management is necessary, because as it was highlighted by Gao et al. [21], testing can make from 40 to 60% of total costs for software development. According to this fact, it is very useful to include also automated testing and thus decrease requirements for human resources. Automated tests can replace human testers from repetitive manual testing operations, decrease duration of testing, increase quality of realized testing. To sum it up, in many cases it is less costly to develop a script, which will test a certain operation that needs to be tested several times than to hire a tester who will realize this test repetitively [21].

Naturally, the main aim of test managers as well as project managers it to keep testing as cost-effective as possible. There are various testing methods captured theoretically but on the other hand there are also challenges how to make testing even more effective. However, systems which are nowadays developed are very large and very complex and thus they make testing even a bigger challenge, because then they require more human as well as technical resources. The fact is that software testing will never be excluded from software engineering even if there would be hundreds of automated tests involved. It will always be required to test a system by a human being [22].

3   Research Objective, Methodology and Research Methods

The objective of our research was to observe a software development project and to describe processes that are carried out within test management. There are limited research resources about test management in the Slovak IT environment and therefore we are focused on the observation of a software development project based specifically in Slovakia. Our partial objectives were to:

- Observe and analyze creation of a high-level test plan
- Observe and analyze communication, coordination and management of human resources within a test team
- Observe realization of testing itself, analyze its parts, processes and activities
- Analyze defect management and its features
- Analyze opinion of a testing team about test management.

The research object was a software development project, which was aimed at development of web-based application consisting of several modules. The project was realized in the year 2015 and it consisted of 6 main phases, while software testing was one of them. Software testing was scheduled for 3 months, specifically from July to September 2015, however, the official phase of testing was postponed due to several reasons mentioned in the sections below. In addition to this, test management was present in the project since March 2015 until the end of the project, and thus our observation took 9 months, from March to November 2015 [29].

There were two test managers involved in the project, however, they were not working at the same time but they exchanged each other in August 2015 with 2 weeks hand-over period. The testing team contained also 8 testers with different background, age, skills and experience.

In addition to this we carried out two interviews with both test managers after completion of the project in January 2016. The interviews were focused on the observed project but also on general aspects of test management. Both interviews were structured, and they contained the same questions.

In order to find out opinions of members of the test team we realized internal online questionnaire survey. The questionnaire was filled in February 2016 by 8 testers who were member of the testing team during the observed project. The questionnaire survey consisted of 8 statements that were followed by five possible options: strongly agree, agree, neither agree nor disagree, disagree and strongly disagree. We asked respondents to express their opinion about the statements by choosing one of the possible options. In order to achieve maximal anonymity of the survey we did not ask statistical questions regarding age, gender or years of experience, because the testing team was so small and diverse that these simple questions would completely damage anonymity of the survey.

Observed processes were analyzed and described in the empirical part of this chapter and they were further supported by two interviews. Additionally, we analyzed results of the realized survey and it enabled us to provide a full picture of observed test management. Additionally, we used the method of comparison since we compared opinions of two test managers.

The interviews were realized during personal one-to-one meetings with test managers and outputs of the interviews were noted down and further used during the process of description of test management.

The questionnaire survey was distributed via emails to 8 former members of the testing team using the tool Google Forms. Results were analyzed in the tool and outputs of the survey are presented in the form of graphs in the empirical part of the chapter.

3.1   Description of the Observed Project

The project that we observed was managed according to principles of the project methodology PRINCE2. The methodology PRINCE2 provides best practices for managing of projects and it enables a project manager to coordinate a project in a structured manner. The main deliverable of the observed project was a web application consisting of several modules integrated together.

3.2   Project Organizational Structure

The project organizational structure was designed as follows (Fig. 4).

The organizational structure included the project board that consisted of an executive, senior supplier and senior users. The project executive was a representative of the customer who was responsible for overall success of the project and he was also responsible for allocation of financial as well as human resources to the project from the side of customer. On the other side, there was a senior supplier who was responsible for allocation of human resources on the project from the supplier side. There were three senior users included in the project board. During observation of test management, we came into contact with senior users for several times, because they were responsible for stating requirements for the developed system and they were also responsible for acceptance testing, which occurred in the final stage of the project. Every day execution of project activities was coordinated by the project manager and there were three team managers responsible for specific areas within the project—a test manager, managing business consultant and IT architect who was also a software development lead. The team managers were

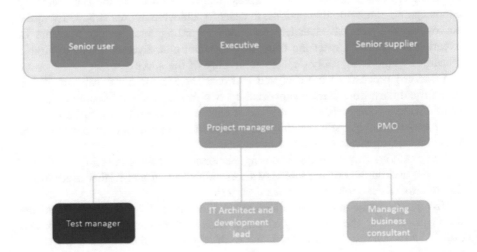

Fig. 4 Organizational structure of the project. *Source* Internal project documentation

responsible for work packages assigned to them by the project manager and they coordinated their teams—testing team, team of analysts and IT consultants and team of software developers.

3.3 Project Schedule

The project was realized from January until November 2015. The project consisted of the following stages.

As we can see in Fig. 5, testing processes were planned from July to September, but the test manager was included in the project since March because before testing processes itself it was necessary to plan the whole testing, including quality criteria, acceptance testing, contribution of customer on testing, etc. In addition to this, there were two test managers working at the project—not simultaneously but they exchanged each other during the project with two weeks transition period.

The first phase after the initiation of project was focused on analysis of customer's requirements. Requirements were documented and based on them acceptance criteria were developed, documented and approved. During this phase, it was not necessary to include the testing team because they will only further use outputs of the requirements analysis as introduction to their work on the project. Analysis of requirements was conducted by IT consultants, solution designers and business analysts in order to capture all needs of the customer and understand how the information system should look like and what functions it should offer. In addition to this, it was necessary to understand what should have be the added value of the solution. The analysis took 3 months and it required several workshops with senior users and future business users of the solution. Requirements were captured and toward the end of the analysis period they were reviewed by chosen representatives of customer (mostly by senior users and subject matter experts) and they were officially approved. It was very necessary to formally approve the requirements because they were inputs into subsequent project phases and processes—especially to testing processes. The phase was followed by detailed design of solution.

During the phase of detailed solution design it was necessary to prepare description of to-be state, thus to describe future function of the application. Primarily, this was the task of the analytical team. Discussion about the application was realized with customer, the phase required several workshops and brainstorming

Phase	JAN	FEB	MAR	APR	MAY	JUN	JUL	AUG	SEP	OCT	NOV
Requirements analysis											
Detailed solution design											
Solution implementation											
Testing											
Training and documentation											
Go Live and 30 day support											

Fig. 5 Project schedule. *Source* Internal project documentation

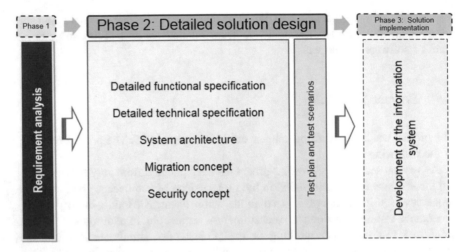

Fig. 6 Components of detailed solution design. *Source* Internal project documentation

sessions in order to make sure that all future function of the application will be understood, captured and prepared for the subsequent phase of development. The output of the phase was the document including detailed solution design of the information system. Within the design there were several areas included: detailed functional specification, detailed technical specification, system architecture, concept of migration and security concept. However, the phase included also planning of test management. It was parallel activity carried out by the test manager in order to capture the whole testing process and to assure its acceptance by the customer. Figure 6 shows all parts of the solution design and it highlights the fact, that planning of testing processes was carried out as a parallel activity, which should take into account all outputs of design completed in the phase 2.

Subsequently, there was the phase of solution implementation which was closely connected to the further phase of testing. As we mentioned before, testing was not present in the project only as it is shown in the schedule, but it was included since March, and also the phase of solution implementation already included testing processes (for example unit testing). After the phase of testing was successfully accomplished according to acceptance criteria, there was training of new users of the application and Go Live in November.

3.4 Test Plan

The test plan can be perceived as a so-called umbrella document. It is a high-level overview of all testing processes and procedures that were carried out during the project. There were several further test documents written during the project, but they were much more detailed and specific than the test plan. However, they all had to be in correspondence with the test plan. In addition to this, the test plan was further updated during the course of the project because there were several changes during the project and all these changes had to be incorporated into the test plan. The test plan was formally accepted by the customer, specifically by a chosen person responsible for testing processes from the side of customer. Naturally, this person was a counterpart to the test manager and it was necessary that she understands the overall concept of testing and moreover that she knows what kind of participation of customer's employees will be required.

Preparation of the test plan was an iterative process. It is necessary to highlight that it was not written from scratch, but it there were test plans from previous projects taken as a basis (or inspiration). In our interviews with the test managers we found out that they usually apply their practice from previous projects and they tailor their knowledge on a particular project environment. However, if there is a test manager hired on a project and he/she with insufficient experience it is necessary to provide him/her with testing documents from previous projects. In our case there was taken a test plan from a similar already completed project realized in the Czech Republic and on its basis the test plan was created, however, many parts of the inspiration document were changed or not used at all. It is a part of best practices in the observed company to use documentation from previous projects and to amend it. In addition to this, the company very often applies lessons learned from its previous projects. The test plan included areas explained in the following sub-sections.

3.5   General Approach to Testing

The test plan included the overall approach to testing as well as several specific areas. The introductory part of the test plan contained the V-model of testing. Based on our interview with the test manager, the V-model is often used in high-level testing documentations because it is very useful graphical expression of test processes and especially of links between individual processes carried out on a project.

The V-model that was created for the observed project is shown in Fig. 7. On the left side, the V-model shows how customer's requirements are decomposed into system functional requirements, further into system requirements and architecture of the system and then into very detailed design of individual components of the developed information system. Then there is development itself and it is directly connected with unit tests. On the right side, module tests are carried out after the unit tests, then system and integration tests and the last step are the user acceptance test. In addition to this, the V-model shows which processes of requirements analysis and decomposition are linked to particular testing processes and thus it was easier for the customer to understand the sequence of processes and also it was highlighted (in pink colour) when cooperation of the customer will be required. Additionally, the management summary included short introduction into topics that were further developed within the document.

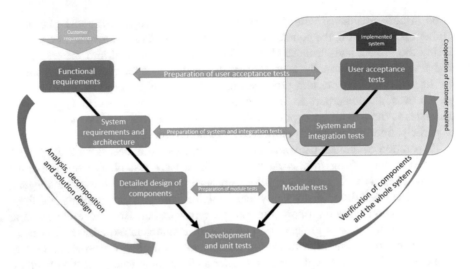

Fig. 7 V-model of testing processes using in the observed project. *Source* Internal project documentation

3.6 Objectives and Scope of Testing

It was necessary to highlight main objectives of the testing as well as its scope within the test plan. The test manager must have been assured that the customer understands what was going to be reached by testing and also what is included within the scope of testing and what is already out of the scope. To be specific, the scope of testing was only to test the information system itself and also its integration to another system on the customer's side but only to an extent, which is possible to be tested by the supplier. It means that if our application should interact with a certain customer's system, we were responsible only for testing of the interface but the whole end-to-end process must have been assured by the customer as well. The plan highlighted the fact that we, as the supplier, were responsible for outbound and inbound communication of our application but processing of information, receive and send processes of integrated systems were in hands of the customer. Further, it was also described which activities require cooperation of the customer. Specifically, the objectives were following:

- To assure that the implemented application corresponds with functional requirements of the customer
- To assure that the implemented application corresponds with non-functional requirements (security, performance, etc.)
- The scope of testing activities included:
- Preparation of test scenarios (required approval of the customer)
- Preparation of test cases (required approval of the customer)
- Preparation of test data (required inputs and cooperation of the customer)

- Testing of test environments
- Realization of test cases
- Creation of test protocols that will be accepted by the customer
- Test training for customer's tester who will be responsible for user acceptance testing
- Coordination of user acceptance testing (cooperation of the customer required).

Within the scope it was explained that testing will consist of two main areas: functional and non-function testing. **Functional testing** was divided into four parts:

- Unit testing—was in responsibility of the development team and it was carried out by automated tests and scripts that controlled quality and correctness of codes. It was responsibility of the software development lead to assure that unit test reports will be provided to the test manager.
- Module testing—to assure that individual functions of the application modules are working as it was required in the requirement analysis
- System and integration testing—realization of end-to-end test cases and review of the system as a whole. Interfaces with the developed application were tested as well.
- User acceptance testing—carried out by chosen customer's employees who were responsible to assure that the system fulfilled requirements and that it could be formally accepted.

Non-functional testing included the following parts:

- Performance testing—to assure that the system will be responsive and reliable in case of increased workload and stress
- Penetration testing—to assure that the system is without security flaws and that it is protected against potential attacks.
- Verification of readiness of test environments—to make sure that test environments are ready for testing and they meet requirements for realization of testing processes.

3.7 Description of Basic Testing Activities

The test plan included detailed information about processes that are typically carried out during software testing. It was necessary to describe them to make the customer familiar with methodology and specific terms that are related to testing. The chapter explains the common words such as end-to-end testing, test cases vs. test scenarios, smoke testing, user acceptance testing but also why it is necessary to use automated scripts during testing and why it is necessary to create a requirement traceability matrix. Individual activities of testing were also described in detail in order to prepare customer for them and to introduce common understanding of what is necessary to carry out if we want to assure the highest possible quality of the developed system. To be specific, the chapter included detailed description of how test scenarios and test cases will be developed, how and when testing will be carried out and how testing activities will be managed.

3.8   Communication During Testing

The chapter briefly explained communication within the project as well as within the testing team and towards the customer. It was necessary to establish formal communication rules because they enable the project team to understand hierarchy and also ways of communication. In addition to this it was explained how often formal testing meetings with will occur and when presence of the customer is required. Explanation of communication must have been in correspondence with the overall project communication and thus this chapter was reviewed and discussed with the project manager as well as with the customer.

3.9   Management of Testing

During our observation we were enabled to see overall management of testing processes during the project, which included various aspects. It is very complex coordination, because it requires managerial as well as technical skills and experience. Test management activities and areas should be explained in the test plan, however, there are many activities, which are not expected before and must be managed "on the go". During the project we were able to observe selection of a test management tool, procedure of creation of test scenarios and test cases, we were explained why it is necessary to have at least three environments within software development, we observed unit testing, module testing, system and integration testing as well as user acceptance testing. In addition to this we could have insight into penetration and performance testing. We also observed creation of defect management methodology and we could also see how to handle resistance of a customer to the newly developed application.

3.10   Selection of a Test Management Tool

During the observation of the project we were able to take part in the meeting, during which a test management tool was selected. It is a system that includes test cases, test cycles, recording of defects, their tracking but also overall statistics of testing processes. Currently there are many test management tools offered at the market. In our case, there occurred brainstorming during which the test management tool was selected. It was particularly necessary to carry out the brainstorming because the test management tool must have been agreed by the customer because the tool was also planned to be used during user acceptance testing. During the selection there were several criteria considered: a price, experience of the test manager with certain tools, experience of other team member with certain tools, services that a tool can offer and also a user-friendly interface. In the end there were two systems in the final round of decision-making—Redmine and JIRA. Redmine is a free tool for issue tracking and test management and therefore it was considered in order to save costs. On the other hand, JIRA was considered to have a friendlier user interface and also the customer as well as the project team had already had experience with this tool. Therefore, JIRA was chosen as a test management tool [25, 26, 32].

In smaller projects there is a tendency to avoid using of a test management tool and very often only simple tools are used—for example MS Excel. This is acceptable, if a project is small and if there are only few people included in testing. However, if there is a larger project it is necessary to have a specific test management tool where test cases as well as defects will be tracked. There are several reasons why it is was useful to choose the test management tool JIRA during the observed project:

- It is one-time cost. The company paid for licenses only once and but the tool can be used in several further projects.
- It offers user friendly environment and additional services such as time tracking, assigning of tasks to various users, creation of test cycles, controlling of test metrics and test statistics.
- It is a single source of truth. Every user has own access and if there are any changes it is traceable who and when made changes.
- Test cases can be assigned to specific people. Every user has a dashboard where he/she can see what tasks or test cases are assigned to him/her.
- Tracking how busy testers are. Statistics show how much time it takes to carry out a specific test and then quantity of tests per one user can be adjusted.
- Notifications. Users are notified if there are any changes, updates or questions added to their test cases or created defects and thus prompt reactions and communication are assured.

3.11 Creation of Test Scenarios and Test Cases

During the project the test team was responsible for preparation of test scenarios and test cases. Usually, these are written based on functional and technical analysis (often called as design) which is developed by solution designers, consultants and business analysts in cooperation with a customer. However, in case of our project, it was not so simple because due to several project changes there was not enough time for preparation of test scenarios and test cases by testers themselves, and therefore it was necessary to ask for cooperation of business analysts since they already knew the functional analysis and also how the final solution should look like.

However, our observation as well as subsequent interviews ensured us that it is always necessary to incorporate designers, consultants or business analysts into creation of test scenarios and test cases because at the moment of their creation they are the ones who have the best knowledge about the future system. Thus, it is at least good to let them review created test scenarios and cases before they are officially approved by a customer.

There are various approaches to definition of test scenarios and cases but in case of the observed project, test scenarios were a high-level view of particular function of the application and test cases were more detailed descriptions of individual steps that must have been carried out in order to accomplish correct function of the application. Many projects tend to write these scenarios and cases in very simple tools—such as MS Excel, but then it might be a bit clumsy during testing. In our case we used two types of information systems for writing and editing test scenarios and cases

- Test scenarios—they were derived directly from use cases of the system—only their denoting changed. This is a very good practice, because in the end, test scenarios must map real functionality of the system and nothing can be omitted. In our case the whole system design was developed in the program Enterprise Architect and also derivation of test scenarios was prepared here. Necessary feature of creation of test scenarios was their denotation. Test scenarios were denoted in the following way:

  - FT.01.001—Login
  - FT.03.001—Simple search in content
  - FT.03.002—Faceted search in content
  - NT.01.010—Language versions
  - IT.02.008—Download a document from SharePoint
  - E2E.01.003—Submission of a request to Central Evidence, etc.

  where FT stands for "functional tests", NT for "non-functional tests", IT for "integration tests" and E2E for "end-to-end tests". First number states a high-level group of functions (such as login/logout, creation of record, searching, change of data, validation…) and the last number is order number within the group. The test managers explained us during the interviews that in every project it is necessary to define specific denotation rules already at the beginning because they have to be followed in all analytical as well as testing processes and they enable team members to structure information but also their work in specific manner. In addition to this, names of test scenarios are often very long and very difficult to remember for people who are not included in their creation or in the project and thus it is better to have specific numbering which makes easier communication. Test scenarios should be always approved by a customer and only then we can approach to creation of test cases.

- Test cases—test cases were written in JIRA, which offers a perfect environment for test management. Test scenarios can be developed into several test cases but the most important part is to keep their denotation, as it was already mentioned above. Since test cases were deduced from test scenarios their denotation was following:

  - FT.01.001.001—Login—positive test
  - FT.01.001.002—Login—negative test
  - FT.03.001.001—Simple search in procurement database
  - FT.03.002.001—Faceted search in procurement database
  - NT.01.010.003—Review of English translation in headings
  - IT.02.008.001—Download a document from SharePoint
  - E2E.01.003.001—Submission of a request to Central Evidence, etc.

- Thus, the denotations of previously mentioned test scenarios are extended by the last number for every individual test case. In addition to this, in case of test cases it was necessary to state which information are necessary to be described and explained in every test case and subsequently prepare these fields in JIRA where test cases were recorded. Obligatory fields in test cases vary from project to project and in our case the following fields were obligatory:

  - name
  - an objective of a test case
  - difficulty/time estimate—expressed in minutes; necessary for calculation of time demand per one test case and to calculate time of a tester spent on certain test cases
  - prerequisites—what must be completed before a test case is carried out
  - test data—what data are necessary to carry out a test case—e.g. a filled database, pdf documents to be downloaded, etc.
  - test user—which user should carry out a test case—e.g. a regular future user, a system administrator, etc.
  - test steps—detailed description what should be done—divided into steps, steps should not be very detailed and if they are too detailed they should be further divided into another steps
  - expected results—what should the system do after a certain test step was carried out.

Before development of test cases in the observed project these obligatory parameters were not specified and thus they had to be added afterwards, which was rather complicated because there were more than 500 test cases created and all of them had to be manually adjusted. Thus, it is necessary for a test manager to have a list of typical properties of test cases, which may be required in software development projects, brainstorm them with a current project team and after information that should be obligatory for every test case are selected it is necessary to instruct a testing team to define them. But as we later found out during the interviews, IT projects are these days operating in very a dynamic environment and therefore

Fig. 8 Example of a created test case. *Source* Testing tool of the observed project

sometimes it might not even be possible to predict what information will be necessary in subsequent project phases and thus very often test cases have to be redesigned, updated or changed. When test cases are created they should be approved by a customer and only then testing should start. This is necessary because project human resources can spend their time on testing activity only when it is clear that created test cases are those that will be in the end accepted by a customer. Otherwise there would be waste of time spent on testing of incorrect test cases and this might strongly influence a project budget. Figure 8 shows an example of created test case in the system JIRA.

3.12 Test Environments

During the observation we were informed that according to best practices, environments at software development projects should be separated at least into three independent parts: development environment, testing environment and production environment. During the project this best practice was taken into consideration and there were two environments created at the beginning and before release into the production, there was also the production environment created. However, the testing and production environments were prepared with delay and thus it had impact on schedule of testing processes. In addition to this, the customer requested specific naming of the environments and it brought lots of confusion among project members and it required longer time to get used to the naming, especially when the naming was changed after several weeks of usage of the environments. The final naming was TEST_DEV, TEST_TEST and PROD. The environment TEST_DEV was used by developers and architects to create the system, they carried out unit testing in this environment. After software was ready for testing it was released into the TEST_TEST environment, where it was further tested by the testing team. It means that these environments were separated and while testers tested e.g. 1.08 version of the software, developers could continue with their work on 1.09 version where they, among the others, implemented also fixed features that were reported during testing. After the software was tested in the TEST_TEST environment (with a specific pass rate) it was further tested by the customer during the acceptance testing. During the project there were several discussions about how many environments are needed because in the final phases of the project the customer needed also pre-production environment for training of employees and some additional testing and final bug fixing. However, too many environments would be too expensive and time consuming for maintenance and therefore it was decided that the customer can use the TEST_TEST environment for trainings and their testing. This decision was reasonable because testing by the supplier at the TEST_TEST environment was almost finished and thus there were not so many complications to share one environment. It is also necessary to make sure that the correct version (latest) of the software is released at the environment.

3.13   Realization of Testing

Testing was realized with delay according to initial schedule of July to September. The delay was caused by two main factors:

- Delay of the whole development phase
- Delay of preparation of testing environments.

The delay was almost 3 weeks from initial schedule and it also meant that testing processes will be completed later than until the end of September. In fact, testing processes were carried out until the end of the project because there were several change requests approved within the project and thus it was necessary to realize further testing cycles.

3.14   Unit Testing

As we mentioned above, there were different types of tests carried out. Firstly, there were unit tests, which were carried out directly by developers and they were mostly coordinated by the development lead. The test manager was interested especially in outputs of this testing—in the form of unit test reports. After successful unit tests (in this case 100% pass rate was required) the test manager proceeded to realization of module testing. Developers were using a specific program for unit testing (which was previously approved by the development lead and the project manager). After processing the developed code in the program, they provided results of such unit testing to the test manager. Afterwards module tests could be started.

3.15 Module Testing

Since the developed application contained several separated modules, there were firstly test cases prepared for module testing were assigned to test cycles and then to individual testers. These were later in the testing phase followed by System and integration testing, which was focused on testing of the application as a whole. However, before start of realization of test cases assigned to a test cycle it was needed to carry out so-called smoke testing, which means to go through functions of application without any specific steps, click its buttons, fill fields, see if the testing environment is working properly, if the application is behaving properly and after smoke testing are perceived as correct a tester can approach to module testing itself. This process of smoke testing was carried out by testers themselves, it was not required to provide any documentation or confirmation that smoke testing was realized, but every tester was briefed to include smoke testing as a part of good practices within testing.

There were four releases and testing was planned after every release and after initial testing the phase of re-testing was scheduled. Individual testing phases were strongly dependent on timing of release and if there was delay in a particular release there was also delay in testing.

In every test cycle there were four possible states for every test case:

- Pass—when all steps of a test case were successful
- Fail—when at least one step of a test case was unsuccessful
- Work in progress (WIP)—when test steps were not completely finished and it was necessary to return to this test case
- Blocked—when a particular test step could not be realized due to various factors (e.g. the system not completely developed, waiting for external dependencies, etc.) and it was further necessary to return back to the test after blocker was removed/resolved.

This categorization implies that test cases marked as failed were necessarily connected with a certain defect, which had to be registered in JIRA. It means that there could not be any failed test cases without registered defects. At the beginning of testing it occurred many times that testers did not register any defect with the aim to wait few days and to see if this defect will be fixed. Therefore, there were lots of failed tests without any defect assigned. This situation was unacceptable, and testers were instructed to register any defect, even if it is minor. This is the point when it was necessary to remind testers that they should consider themselves as "auditors" of developers. Every tester had to be very careful about everything in the tested application, not only steps he/she was going through but also additional features of the application, layout, graphics—everything what was perceived as not typical, not according to specification, with not great design was necessary to be mentioned in defects.

Figure 9 shows a typical test cycle with several passed, failed test cases and one test case in progress. In the test cycle we can see that a tester was assigned 6 test

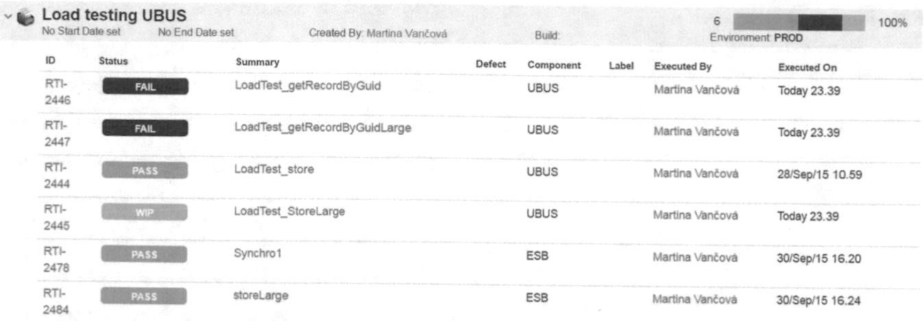

ID	Status	Summary	Defect	Component	Label	Executed By	Executed On
RTI-2446	FAIL	LoadTest_getRecordByGuid		UBUS		Martina Vančová	Today 23.39
RTI-2447	FAIL	LoadTest_getRecordByGuidLarge		UBUS		Martina Vančová	Today 23.39
RTI-2444	PASS	LoadTest_store		UBUS		Martina Vančová	28/Sep/15 10.59
RTI-2445	WIP	LoadTest_StoreLarge		UBUS		Martina Vančová	Today 23.39
RTI-2478	PASS	Synchro1		ESB		Martina Vančová	30/Sep/15 16.20
RTI-2484	PASS	storeLarge		ESB		Martina Vančová	30/Sep/15 16.24

Load testing UBUS — No Start Date set — No End Date set — Created By: Martina Vančová — Build: — Environment PROD — 6 — 100%

Fig. 9 Test cycle. *Source* Testing tool of the observed project

cases to be responsible for and they had to be completed until specific time set by the test manager. After the tests were completed the cycle was reviewed by the test manager and potential problems were discussed. In addition to this, every failed test was connected with at least one defect that was further handled by a developer. Developers were given maximally 5 working days to resolve defects, and then fixes of defects were deployed in the following sub-release (e.g. FR 1.2.4 where 4 denotes the number of sub-releases within the main release 1.2). If developers were not able to fix defects within 5 days, the status was escalated to the level of project management [23, 24, 31]. However, the methodology was given and developers were responsible for fixing defects according to their severity—i.e. critical and major defects must have been resolved primarily and only then medium and minor defects. In spite of planned 5 days turnaround, many defects were in the solving process for many days as it is shown in Fig. 10.

In the process of defect management, it was clearly shown how beneficial it is to use JIRA as a defect management tool. In many cases there were defects, which were repeated in several test cases and it was not necessary to write this defect again but rather use only its ID to assign it to another test case. It means that there was many-to-many relationship between test cases and defects—a defect could have been assigned to many test cases and vice versa—a test case could have many defects. If test cases were failed they were assigned to a subsequent test cycle. They could not be fixed within the same cycle. If test cases were blocked or marked as work in progress they could have been re-tested within the same cycle and their status could have been changed, however in reality those assigned as blocked were most often assigned to subsequent test cycles because their blockers were usually resolved together with a new release or sub-release.

3.16   System and Integration Testing

System and integration testing included such test cases that were testing integration individual modules within the application and also integration of the application to

Fig. 10 Defect resolution
time. *Source* Testing tool of
the observed project

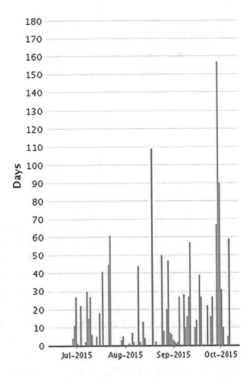

another systems (applications), which were outside the scope of the project. In many cases it was not possible to test this integration directly (even at the end of the project) and therefore these interfaces were replaced by so-called "mock applications". These applications were either simple applications developed by the supplier based on specifications of applications that should be integrated, but the most often they were simply substituted by SoapUI, which is used to test SOAP and REST services, on which the developed application was practically based. During testing through the program SoapUI it was necessary to have a certain technical knowledge and to closely cooperate with developers and therefore this part of testing was done only by tester of the supplier and it was omitted during testing with the customer (user acceptance testing). This decision was, naturally, discussed and initially agreed by the customer [27, 28, 30].

3.17   User Acceptance Testing

After completion of system and integration tests and final preparation of the application there was user acceptance testing (UAT) organized, which already included participation of representatives from the customer's side. After first session of UAT it was found out that it is necessary to assure participation of testers from the side of supplier because they helped to customer's testers how test cases should be done and therefore testing was faster. The first session of UAT testing basically showed that there should be one tester from the side of supplier assigned to one tester of the side of customer. The UAT did not include the whole scope of created test cases, it was stated in the initial documentation that it will include 40% of the whole scope of test cases. This specific percentage may vary from project to project and it always depends on the decision of a customer because a particular quantity of test cases may assure a customer about correctness of developed software. The scope of test cases for UAT was recommended by the supplier but finally approved and chosen by the customer.

Software development projects are operating in the very dynamic environment, which is naturally connected with changes and there were many changes present also at the observed project. For example, during the UAT it was decided that the scope of 40% test cases must be enlarged by E2E that assured the customer that the system is possible to be used through all its components up to integrated systems (that were outside the scope). UAT were a subject of official acceptance of the whole project, and therefore realization of every test case by a tester from the side of the customer had to be marked as passed and additionally signed by the customer. This step was very necessary to be done immediately after completion of testing in order to assure accuracy and current truth because, as it was mentioned earlier, software development projects often change, and change requests can have potential influence on UAT and then subsequently to formal acceptance.

3.18   Penetration and Performance Tests

Penetration and performance tests were outsourced to a sub-contractor previously approved by the customer. The decision for this step was based on wide know-how of the sub-contractor in this sphere and it was more beneficial (from financial as well as time perspective) to outsource these tests than to carry them out directly by the supplier. There were specific requirements for the sub-contractor about the tests that had to be carried out from the perspective of penetration and performance and their realization was explained in the extensive penetration and performance reports. The sub-contractor was given access to the developed application at the testing environment for a certain agreed period of time, during which penetration and performance testing had to be accomplished.

The methodology of penetration testing was based on the OWASP Risk Rating Methodology, according to which there was the risk matrix developed evaluating probability and impact of findings from penetration testing. Both factors, probability as well as impact were always evaluated by penetration testers within the scale of 0–9 and then risk was calculated by the formula probability x impact. Table 2 was defined within the project documentation and it shows overall severity of risk in penetration testing after impact and probability are evaluated. We can see that in case of low impact and low probability there is only note announced and given to

Table 2 Overall severity of risk in penetration testing

Overall severity of risk in penetration testing				
Impact	High	Medium	High	Critical
	Medium	Low	Medium	High
	Low	Note	Low	Medium
		Low	Medium	High
	Probability			

Source Internal project documentation

the test manager but in case of the high and critical findings it is necessary to proceed with corrective actions immediately because the application is in high risk to be easily attacked by hackers.

Penetration testing was focused on security of the developed application. In total there were 87 different automatized penetration tests realized (e.g. test account provisioning, testing for bypassing, password functionality, cookies attributes and many others) and there were 8 findings identified during the first round of penetration testing. These findings are shown in Table 3 and they were explained in detail and their risk was marked as low.

Afterwards, these findings were improved and the second round of penetration testing was carried out, during which only 8 mentioned findings were re-tested. It would not be meaningful to test all 87 penetration tests again since the majority of them was marked as passed already in the first round of testing and thus it would be too much time consuming. The second round of the penetration testing was without any finding. Penetration tests were also realized in the PROD environment after application was deployed in the final stages of the project. There were no findings identified.

Performance testing were focused on quality of the developed application in case there are too many users at once and too many operations and requests going on. There were specific requirements defined by the customer during the phase of

Table 3 Results of penetration testing

Section	Test ID	Penetration test name	Risk
2.1.8	OTG-INFO-008	Fingerprint Web Application Framework	Low
2.1.17	OTG-CONFIG-007	Test HTTP Strict Transport Security	Low
2.1.39	OTG-SESS-002	Testing for Cookies attributes	Low
2.1.40	OTG-SESS-003	Testing for Session Fixation	Low
2.1.46	OTG-INPVAL-001	Testing for Reflected Cross Site Scripting	Low
2.1.62	OTG-ERR-001	Analysis of Error Codes	Low
2.1.63	OTG-ERR-002	Analysis of Stack Traces	Low
2.1.64	OTG-CRYPST-001	Testing for Weak SSL/TLS Ciphers, Insufficient Transport Layer Protection	Low

Source Internal project documentation

analysis and the sub-contractor was testing these requirements by automatized tests as it was done also in case of penetration tests. There were several different aspects tested, which included for instance ability to process 2 millions of transactions per year, maximal size of a processed document cannot must be up to 20 MB, all 4000 expected future users is logged in at the same time, maximal response which does not repel users is maximally 1000 ms, etc. Performance tests did not identify any problems or defects, neither in the testing environment nor in the PROD environment.

Results of penetration and performance testing were summarized in forms of reports which were a key part of attachments during final acceptance of the project.

3.19 Defect Management

Defect management was an integral part of the whole management of testing. Defects that were created during the testing went through a certain life cycle and it was necessary for every interested party to have current information about actual status of created defects. Before the phase of testing started the test manager together with the development team designed a workflow for a life cycle of defects. This workflow was further implemented in JIRA and particular statuses of defects were created. This workflow is stated in Fig. 11.

In practice, when a tester found a defect he/she recorded it in JIRA and it was automatically set into the state "New". Then a tester, test manager or developer assigned this defect to a particular person who was responsible for its solution. Then the defect could go through the status "Need info" in case the developer

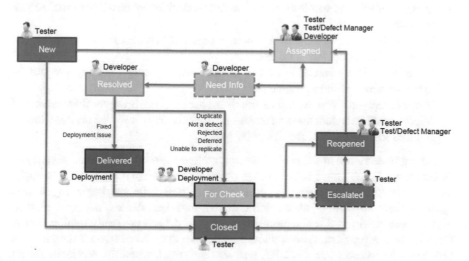

Fig. 11 Defect management workflow. *Source* Internal project documentation

needed additional information for solving the defect, then he/she put the defect into the status "Resolved". If the defect was resolved it could have several types of resolution:

- Fixed defect or Development issue—both solutions meant that a defect really occurred in the application and a developer fixed it
- Duplicate, Not a defect, Rejected, Deferred, Unable to replicate—a defect was not fixed; in case a defect was marked by a developer as Not a defect or Rejected a developer had to state one of the following reasons: Tester mistake, Wrong data, Wrong test script, Functionality not deployed, Deployment issue, Environment issue, Document issue, Different business expectation, other—thus everybody knew why a defect was not fixed and then subsequent actions could take place.

After a defect was resolved, it was put into status Delivered or For Check and afterwards closed by a tester if he agreed with a given resolution. However, if a tester did not agree he/she could either reopen the ticket (and then it would go through the whole life cycle once again) or escalate the ticket to the level of the test manager who further decided how to proceed with such a defect (usually discussion with the development lead followed or it was even escalated to the project manager).

After the interview with the test manager we discovered that it is very necessary to set specific rules at the beginning of testing because otherwise there can be chaos in processes. During the observed project it was strictly necessary to define all the rules immediately before testing processes started because individual team members were not based in the same building and thus communication was more difficult.

Within the project there were the following defect severity categories introduced and they were also enshrined in the project contract as well as the test plan:

- Critical defect—the whole system does not work, it is not possible to realize any function
- Major defect—the system as a whole work, but certain functions, which are necessary for its operation do not work
- Moderate defect—functions of the system are working but they are not in correspondence with requirements
- Minor defect—the system and its functions are working however there are small defects within its functions or cosmetic defects (e.g. how the system looks, grammatical errors, not user friendly, etc.).

During observation of the project we encountered several cases when severity of defects caused problems. First of all, we observed that internal testers tend to apply the severity "Moderate defect" almost in every defect (at the beginning it was approximately 80–90% of all created defects), and then the test manager tried to find the reason why is this situation happening—if defects are really moderate or if there is another problem. After a short analysis we discovered two reasons for this problem—first reason was that JIRA was set up to put the severity Moderate for all newly created defects by default and the second reason was that testers were not completely sure about differences between individual types of severity and it was necessary to repeat it. Thus, JIRA was adjusted to have no severity status when a defect was created and since it was required field testers had to choose among different types of severities.

Secondly, in spite of it was specifically explained in the test plan as well as in the contract, the customer did not always agree with severity of tests. This situation occurred also during the user acceptance testing and it was very difficult to explain why certain defects are not critical. It is natural that for a future business user it might seem that a defect is critical, however from the perspective of supplier it often cannot be understood as critical since the whole system was working correctly. There were several customer representatives present during the user acceptance testing and there were several discussion about severity of defects. For example, we come across the problem when testing data were not accurate and the information system did not provide correct data when a user needed to choose address of a legal entity—there were not all the cities in Slovakia imported correctly in the register and thus, the user could not record the legal entity with correct data. From the perspective of the user, this problem seemed to be critical because he could not execute his work task. However, from the perspective of the information system, it was only a moderate defect because the function of recording legal entities worked, but it did not meet customer's requirements. In addition to this, results from testing were always in the spotlight because statistics from testing gave the project manager overview about correctness of development and there were also specific performance indicators stated in the contract (as well as in the test plan) and they had to be met in order to achieve successful user acceptance testing. The indicators were as follows.

Table 4 shows that in case of Module, System and integration testing as well as User acceptance testing there cannot be any critical defects, otherwise testing could not be accepted, the whole session of testing must be repeated until we reach the point when there are no critical defects recorded. It implies that the test manager must be careful when the system is tested by customer's users and it is necessary to make sure that they are recorded with correct severity. Based on our observation we can conclude that it is good practice to enable a test manager to have overall power

Table 4 Formal acceptance of testing phases

Formal acceptance of testing phases	
Type of testing	Completion criterion
Module testing	90% of planned tests executed without any critical defect
System and integration testing	95% of planned tests executed without any critical defect and maximally 10 major defects
User acceptance testing	95% of planned tests executed without any critical defect and maximally 10 major defects

Source Internal project documentation

to change severity of defects in case they are not in accordance with the contract description. Naturally, it is afterwards connected with responsibility of a test manager for particular changes.

3.20 Communication Within the Testing Team

We observed and it was also confirmed during the interview with the test managers that when people work on projects they often think that all necessary information are already spread through teams and that everybody knows all news. However, it is very rare and cascading of information takes lots of time and therefore we found out that it is very important to repeat everything—rather repeat twice than be in "information darkness". There are decisions made on daily basis not only in the testing team but in the whole project team and teams should inform each other immediately. During our observing we noticed that crucial information was not flowing to the testing team especially from the development team. For example, very often the testing team was not informed when there were minor fixes of defects deployed to the testing environment and it made confusion among members of the testing team. Thus, there was a rule introduced that all deployments, even if it contains only very minor fixes has to be announced upfront to the whole team, not only the testing team.

In addition to this, during testing with the customer it was necessary to repeat testing methodology and certain processes more often, especially at the beginning of every testing session. It is natural, because customer's testers are not doing testing on daily basis, they were testing the application from the business perspective and they were not aware about testing methodology before it was explained in detail. For example, the most problematic aspect was describing details of found defects. There were several testing sessions with customer's testers and if they found defect they only briefly described it (mostly in one sentence) without specific details or screenshot. Then we came to the situation that developers were not able to simulate the situation or repeat the defect and they simply did not fix it, however, it was again reopened by customer's testers later and this whole process only lengthen fixing of it (and also increased time to resolution, which was one of the most important key performance indicators of the development team). This problem was resolved by the following steps:

- Testers were instructed how to create a detailed defect in JIRA at the beginning of every testing session
- Detailed instructions how to create and describe a defect were printed and put on a table of every tester
- The test manager was observing in JIRA newly created defects and immediately after a new defect was created he checked if it is created properly with all necessary information that are important for the development team.

After introduction of these measures we only seldom came to the situation that developers would not be able to understand or replicate a defect.

3.21 Resistance to the New Application

Observation of the testing processes enabled us to understand that the test manager (and often also testers who attend testing with a customer) are often put into the position where they need to face resistance of the customer's testers to use new application. This is one of the aspects of change management, and naturally, it should not be solved at the level of the test manager. The experience showed us that when people are used to working with a certain web application and then a new application is being introduced they are comparing it and it is natural feature of every change—users often do not like a new application and they are even rejecting to use it. In our case, this rejecting was also present during the testing processes because people who were rejecting the application were also rejecting to test it, they were complaining and they were also created many defects, which were not defects in comparison to the approved design of the application but rather to the application that they were previously using (i.e. they were comparing the newly developed application with the old application and did not take the design of the new application into consideration). Subsequently, this situation had to be escalated and discussed at the level of project management.

3.22 The Perspective of Testers

In our research we also focused on satisfaction of testers with management within the testing team, on weak areas within the project as well as on those that went well. We proposed 8 questions to 8 testers who were in the testing team, which led to the findings described below.

The first statement of the questionnaire survey was as follows: *Status meetings of the testing team were beneficial and their frequency was optimal.* Status meetings of the testing team were scheduled for Monday afternoon, they started at 1 pm and usually took from 1 to 3 h depending on the period in which the project was—i.e. at the beginning of the project and towards the end of the project they were shorter, but during the phase where most work was done they took up to 3 h because many things were discussed in detail and solutions for various issues were brainstormed. The question was included in the survey exactly because of the reason that some of the meetings took long time and thus we were interested in opinion of team members about them. Figure 12 shows that team members mostly agreed with the statement, which means that they were satisfied with benefits resulting from status meetings as well as with their frequency. Specifically, 37.5% (3 respondents) strongly agreed with the statement and the same quantity agreed with the statement. There was only one respondent (12.5%) who disagreed and no one who would strongly disagree. It means that set up of status meetings could be kept similarly also in further similar projects.

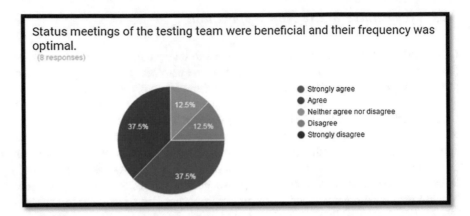

Fig. 12 Status meetings of the testing team. *Source* Own visualization of the survey

The second statement was explicitly focused on communication between the test manager and the testing team. We were interested in the fact, if the team considered quantity and timing of information coming from the test manager toward them as sufficient. The statement was proposed: *Communication of the test manager towards the testing team was timely, sufficient and accurate.* The results are very positive because 62.5% (5) respondents agreed with the statement and the rest 37.5% (3) strongly agreed with the statement (Fig. 13).

The objective of the third statement *Quantity of test cases assigned to me was optimal, I felt neither overloaded nor idle* was to find out if testers felt that they had enough work and at the same time not too much to be able to focus on its realization precisely. During test cycles there were test cases assigned to specific testers based on their estimated difficulty and time consumption. Test cases were evaluated

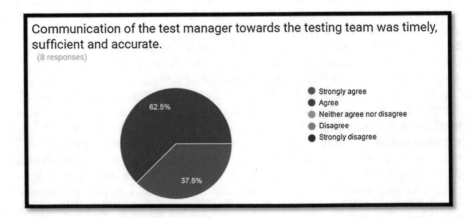

Fig. 13 Communication of the test manager towards the team. *Source* Own visualization of the survey

during their creation to low, medium and high, meaning that test cases marked low difficulty should be completed up to 15 min, those with medium difficulty up to 1 h and those with high difficulty up to 3 h. Based on this expected time consumption they were evenly assigned to testers. However, it was not always possible to precisely estimate real duration of realization of a test case and there were also another unexpected factors that influenced realization of testing (e.g. late deployment, blockers, integration with another systems, necessity to test with mocks, etc.) and therefore testing of one test case could take longer than it had been expected before. The survey shows that only two tester agreed with quantity of assigned test cases (12.5% (1) agrees, 12.5% (1) strongly agrees). 2 testers (25%) neither agree nor disagree and 4 testers were against the statement (25% (2) disagree, 25% (2) strongly disagree). The results imply that in future project there should be buffer included when test cases are assigned to testers. For example, a tester is able to complete 8 medium test cases within one working day, however, due to factors mentioned above there is a risk that he/she might not be able to complete all of them, therefore it is reasonable to include certain minutes as buffer for this unexpected factors, thus it might seem more realistic to assign 7 medium test cases and leave 1 h (12% of working time) for unexpected situations that may occur (Fig. 14).

In the fourth statement we focused on the fact, if the testing team considered tasks as explained sufficiently to be able to accomplish them. Findings for this statement are positive, and as it is shown in Fig. 15 testers either strongly agree (37.5%) or agree (62.5%) with the statement. This reflection shows that there was good communication about work that should be completed within the team.

The fifth statement was proposed as follows: *The detailed functional and technical analysis was sufficient for me during writing of test scenarios and test cases.* The functional and technical analysis was in the scope of the analytical team and it was used as input into processes of the testing team. However, it was considered as the most important input and therefore its quality was crucial.

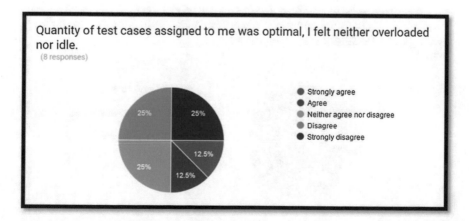

Fig. 14 Work assigned to testers. *Source* Own visualization of the survey

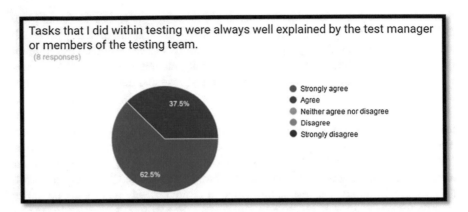

Fig. 15 Explanation of working tasks. *Source* Own visualization of the survey

We observed that during writing of test scenarios and cases there were required further discussions with designers and analysts of the solution and it was also necessary to include them into writing of test cases because it was very often unclear how the to-be state should look like.

Findings shown in Fig. 16 imply that members of the testing team were not satisfied with the analysis and it did not contain enough information for them to prepare test scenarios and cases easily. Only one respondent (12.5%) agreed with sufficiency of the analyses and the rest 25% (2) strongly disagreed and 62.5% (5) disagreed. Thus, the findings are showing that in future projects it is important to highlight necessity of more detailed functional and technical analysis and to be prepared for including solution designers and analysts into the process of writing test scenarios and test cases.

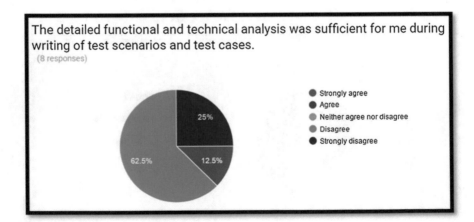

Fig. 16 Quality of analysis. *Source* Own visualization of the survey

The sixth statement *I liked using the tool JIRA for recording of defects and recording of realized test cases* was aimed at evaluating if the team members were satisfied with the tool chosen for test management and defect management. At the beginning of the project it was discussed what tool will be used, however, testers were not included in this decision making. 62.5% (5) respondents agreed with using JIRA as a testing tool, 25% (2) strongly agreed and one respondent (12.5) neither agreed nor disagreed. Findings are positive and we assume that many of them had not had any experience with different testing tools before and therefore they could not compare. Additionally, we discussed this tool during the interviews and also the test managers were satisfied with its functions as well as the friendly user environment. Thus we can recommend the tool for future usage in similar software development projects (Fig. 17).

Figure 18 shows positive opinions about the statement *Overall, I think that the testing processes were accomplished successfully.* The testers were generally satisfied with final accomplishment of testing processes, they were satisfied with the result and it is also positive reflection of management of the testing team. 62.5% (5) respondents strongly agree with the statement and 37.5% (3) respondents agree with the statement, which brings us to conclusion that activities of both test managers were set up correctly to lead the team to desired objectives.

As we mentioned earlier in outputs of our observation, there occurred several issues that were connected or influenced testing processes or the testing team itself. Therefore we proposed the last statement of the survey as follows: *Problems related to testing that occurred during the project were resolved successfully and timely.* Our aim was to find out if team members were generally satisfied with solutions of problems. Figure 19 shows that 37.5% (3) of respondents neither agreed nor disagreed with the statement, 12.5% (1) strongly agreed with the statement and 50% (4) of respondents agreed with the statement. The results imply that the team was in general satisfied with problem solving during testing.

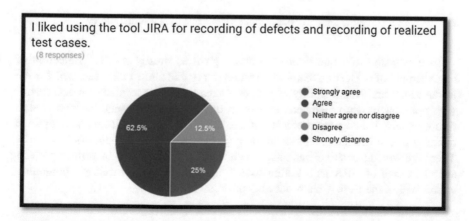

Fig. 17 Using JIRA as a testing tool. *Source* Own visualization of the survey

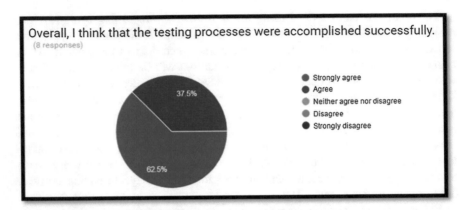

Fig. 18 Overall opinion of testers. *Source* Own visualization of the survey

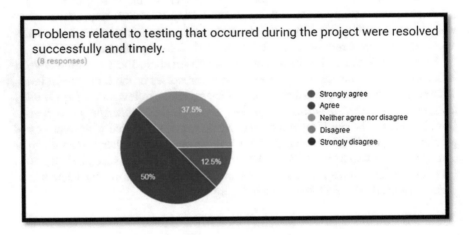

Fig. 19 Problem solving during testing. *Source* Own visualization of the survey

In conclusion, the questionnaire survey gave as mostly positive reflection of management of the testing team. The results imply that almost all areas, which were in the scope and responsibility of the test manager were coordinate in accordance with opinions of team members. However, one area which should be improved in future projects is assignment of test cases to testers in order to allocate them optimal quantity of test cases to enable them enough time to carry out their work properly. There are several positive findings, which may be applied also in further projects such as usage of JIRA as a testing tool, frequency of status meetings, communication within the testing team but also problem solving.

4 Conclusion

High quality test management is a necessary part of software development projects. It is undeniable that with effectively planned and managed testing it is possible to save costs and at the same time to assure required quality of a developed system. There are many theoretical approaches to test management and many sources of best practices. However, every project is different since it is surrounded by a different environment and consisting of different components. Thus, it is beneficial to enlarge sources of best practices by additional, which are based on practical experience on IT projects. Successful test management requires lots of experience and knowledge and its quality can be supported by relevant lessons learned from previously accomplished software development projects. The fact is that products of software engineering occur in everyday life of ordinary people and at the same time it cannot exist without testing. Therefore, it is beneficial for companies to apply strategies to improve its software testing processes in order to realize them more effectively and with added value.

In conclusion, test management is a set of very complex processes and it is a challenging area for proper coordination. Our research enabled us to identify several problematic areas that should be monitored in software development projects in order to achieve its success. Generally, there are many methodologies and best practices, which could be applied in test management, however, practice is always specific and there are always areas, which has never been covered by any authors. Thus, our outputs can be an addition to these best practices or lessons learned and they can enable test managers or project managers to be more successful during coordination of testing processes within software development projects.

References

1. Charatan Q, Kans A (2003) Formal software development. Palgrave MacMillan, New York. ISBN 0333992814
2. Majchrzak TA (2012) Improving software testing. Technical and organizational developments. Springer Briefs in Information Systems. Springer, Berlin Heidelberg. ISBN: 978-3-642-27463-3B
3. IEEE Computer Society: 610.12-1990 (2010) IEEE standard glossary of software engineering terminology. E-ISBN 978-0-7381-0391-49780471430209
4. IEEE Computer Society: 829-2008 (2008) IEEE standard for software and system test documentation. E-ISBN 978-0-7381-5746-7
5. ISTQB: Standard Glossary of Terms Used in Software Testing Version 3.01. [Online] 2015. [Cited 01.04.2016] Available on the Internet: http://www.astqb.org/documents/Glossary-of-Software-Testing-Terms-v3.pdf
6. Hass AMJ (2008) Guide to advanced software testing. Artech House Books, Norwood, US. ISBN 978-1-59693-285-2

7. Myers G, Sandler C, Badgett T (2011) The art of software testing. Wiley, Hoboken. ISBN 9781118031964
8. Allsopp W (2009) Unauthorized access: physical penetration testing for IT security teams (1). Wiley, Chichester. ISBN 9780470747612
9. Watkins J (2001) Testing IT. Cambridge University Press. ISBN 9780521795463
10. Copeland L (2003) Practitioner's guide to software test design. Artech House Books. ISBN: 9781580537919
11. Lyndsay J (2003) A positive view of negative testing. Workroom Productions Ltd. [Online] [Cited 26.04.2016]. Available on the Internet: http://www.workroom-productions.com/papers/PVoNT_paper.pdf
12. Ammann P, Offutt J (2008) Introduction to software testing. Cambridge University Press, Cambridge, UK. ISBN 978052188038
13. Loveland S, Shannon M, Miller G (2004) Software testing techniques: finding the defects that matter. Charles River Media, Boston. ISBN 9781584503460
14. Huang JC (2009) Software error detection through testing and analysis (1). Wiley, Hoboken. ISBN 9780470404447
15. Goodpasture JC (2012) Project management the agile way. Wiley, Hoboken. ISBN 9781604270273
16. Utting M, Legeard B (2010) Practical model-based testing. Morgan Kaufmann Publishers, San Francisco. ISBN 9780123725011
17. Jorgensen PC (2013) Software testing: a Craftsman's approach. CRC Press, Boca Raton, Florida. ISBN 9781466560697
18. Project Management Institute (2013) A guide to the project management body of knowledge (PMBOK® guide). Project Management Institute, Newtown Square. ISBN 978-1-935589-67-9
19. Axelos Limited (2014) Managing successful projects with PRINCE2. The Stationery Office, London. ISBN 9780113310593
20. Whitaker K (2010) Principles of software development leadership: applying project management principles to agile software development leadership. Course Technology Cengage Learning, Canada. ISBN 9781584505860
21. Gao J, Tsai HS, Wu Y (2003) Testing and quality assurance for component-based software. Artech House, Boston. ISBN 9781580534802
22. Bertolino A (2007) Software testing research: achievements, challenges, dreams. In: IEEE future of software engineering, pp 85–103
23. Charfaoui E (2007) Výuka odborného jazyka s využitím informačných systémov. Kompetence v cizích jazycích jako důležitá součást profilu absolventa vysoké školy. CJP UO, Brno S. 127–131. ISBN 978-807231-261-0
24. Franková P, Drahošová M, Balco P (2016) Agile project management approach and its use in big data management. Procedia Comput Sci 83:576–583. https://doi.org/10.1016/j.procs.2016.04.272
25. Kryvinska N (2012) Building consistent formal specification for the service enterprise agility foundation, the society of service science. J Serv Sci Res 4(2):235–269
26. Kaczor S, Kryvinska N (2013) It is all about services—fundamentals, drivers, and business models, the society of service science. J Serv Sci Res 5(2):125–154
27. Gregus M, Kryvinska N (2015) Service orientation of enterprises—aspects, dimensions, technologies. Comenius University in Bratislava. ISBN: 9788022339780
28. Kryvinska N, Gregus M (2014) SOA and its business value in requirements, features, practices and methodologies. Comenius University in Bratislava. ISBN: 9788022337649
29. Molnár E, Molnár R, Kryvinska N, Greguš M (2014) Web intelligence in practice. The society of service science. J Serv Sci Res 6(1):149–172

30. Kryvinska N, Auer L, Strauss C (2008) Managing an increased service heterogeneity in converged enterprise infrastructure with SOA. Int J Web Grid Serv (IJWGS) 4(4):440–460
31. Stoshikj M, Kryvinska N, Strauss C (2014) Efficient managing of complex programs with project management services. Springer, Global Journal of Flexible Systems Management, Special Issue on Flexible Complexity Management and Engineering by Innovative Services, pp 25–38. https://doi.org/10.1007/s40171-013-0051-8
32. Kryvinska N, Barokova A, Auer L, Ivanochko I, Strauss C (2013) Business value assessment of services re-use on SOA using appropriate methodologies, metrics and models. Int J Serv Econ Manag (IJSEM) 5(4):301–327. Special Issue on Service-centric Models, Platforms and Technologies, ISSN online: 1753-0830, ISSN Print: 1753-0822

The Role of Variety Engineering in the Co-creation of Value

Raoul Gorka, Christine Strauss and Claus Ebster

Abstract By integrating methods and models from systems science, the interdisciplinary field of service science has provided the basis for a possible radical shift in understanding the coordination mechanisms underlying global market dynamics. The *Variety Engineering Model* is found to support and extend the evolving framework of service-dominant logic by shedding light on the relational nature of interacting social agents, who co-create value by steering their behavior towards shared meanings through conversation. In our highly complex present-day world the main driver of balancing agents' complexity asymmetries is self-organization. Without adequate management, this self-regulation seldom produces socially desirable outcomes. We conceive the proposed systemic methodology as an effective guideline for supporting managers to coordinate this process towards common policies, which may foster sustainable structures.

1 Introduction

1.1 Relevance

The occurrence, speed and degree of penetration of technological achievements, especially in Information and Communications Technology (ICT), has led to a socio-economic shift towards a knowledge-oriented society and service-oriented economy. The interaction and interdependency of these forces fuel each other's growth, thereby further accelerating the pace of development [1–3]. The implications for organizations of already being "in the midst of a major evolutionary

R. Gorka · C. Strauss (✉) · C. Ebster
University of Vienna, Oskar Morgenstern Platz 1, 1090 Vienna, Austria
e-mail: christine.strauss@univie.ac.at

R. Gorka
e-mail: raoul.gorka@gmail.com

C. Ebster
e-mail: claus.ebster@univie.ac.at

© Springer International Publishing AG, part of Springer Nature 2019   179
N. Kryvinska and M. Greguš (eds.), *Data-Centric Business and Applications*,
Lecture Notes on Data Engineering and Communications Technologies 20,
https://doi.org/10.1007/978-3-319-94117-2_8

transition that merges technology, biology, and society" [4] are twofold: On the one hand, the changes concerning the continuing progress of ICT create new business opportunities by providing additional possibilities for service innovation and improvement (cf. e.g. [5]) on the other hand, managers of organizations are faced with increasingly complex situations they need to manage efficiently (cf. e.g. [6]). Both aspects require new models and a new understanding of social and economic dynamics [7, 8].

The forthcoming large-scale labor disruption is expected to radically change global workforce structures within the next years due to accelerated deployment of general-purpose robotics, machine learning, artificial intelligence, nanotechnology, biotechnology, genetics, and all their interrelations within the upcoming Internet of Things. In the context of social, political, and economic spheres, this technological revolution will lead to the redefinition of existing socioeconomic notions and organizing principles (e.g. labor, production, distribution, property, value) as well as change the human relation to socioeconomic coordination, which is currently operating "purely on profit-driven monetary logic without consideration for the complex and multi-dimensional spheres of human value" [9]. Current sociopolitical challenges (e.g. demographic change, acceleration of income and wealth inequality, financial crisis, political instability, etc.) require thorough consideration, comprehensive effort and resolute action. The study of service systems (i.e. service science) confirms this view by pointing out society's need for developing novel holistic models that would help define viable strategies for a sustainable future [1, 10, 11].

Recent developments in the fields of cloud computing and big data increasingly enable organizations to achieve higher performance and provide the basis for value creation, yet the potential for further improvement is still large. Enabled by digital technology new forms of collaboration have emerged, implemented as collaboration platforms and groupware. With this technology and its applications, it is possible to improve the participation of stakeholders by providing a platform for dialogue and mutual learning. This in turn enables organizations to improve anticipation and forecast of changes in demand and enhances their ability to adapt to uncertain shifts in competitive markets [12]. Consequently, the challenge for organizations in the next decades will be to improve their dynamic capabilities, i.e. the ability to develop and reconfigure key-resources and key-competencies in relation to organizational structure and processes [13, 14].

The emerging field of service science began to point out the central role general *systems theory* and the *viable system model* play in better understanding the processes by which global functioning of human society can be sustained [15]. These theories are found to offer a coherent methodology for interpreting the networks of service systems by proposing a new relational approach to manage interactions between these complex, adaptive, socio-technical systems [16, 17]. Aiming at better understanding the coordination mechanisms that enhance collaboration of interacting social agents, the Systems Approach, which is based on systems theory and cybernetics, is recognized by more and more disciplines as being a very fertile approach in helping managers cope with a growing number of organizational and ethical challenges [18].

Established by British cybernetician Stafford Beer during the second half of the twentieth century, the framework of *organizational cybernetics* provides an integrated approach called *variety engineering*, which is capable of tackling these challenges [19–27]. It aims at facilitating self-organization and self-regulation in organizations to improve their performance by providing managers with a methodology that helps to find better ways of co-creating value [28]. The Variety Engineering Model is the building block for the Viable System Model [24–26], which describes the necessary organizational structure for a service-system to survive in an unpredictably changing environment. This model is found to be at the core of the complex, urgent challenges society is facing [14, 29].

Based on this model, Espejo and Reyes [12] developed a template for complexity management strategies, which allows agents and actors (e.g. consumers and producers) to "learn from each other and co-create value through recurrent communications" [12, 14]. This observer-dependent methodology is able to improve the effectiveness of communications by providing a "meta-perspective", where actors become aware of their active participation in the larger ecosystem of mutual value creation.

1.2 Goals and Objectives

Despite the many contributions within the field of *service science* and *service-dominant logic (SDL)* to (co-)develop a framework that explains the processes by which producers and consumers interact to co-create value, only a small number of studies have developed a holistic account on the mechanisms underlying these interactions [30]. This work aims to contribute to a better understanding of the mechanisms that facilitate value co-creation in social systems by introducing the framework of *variety engineering*. The focus on systemic approaches for the management of complexity can help both researchers in the disciplines of service science as well as managers and other actors in organizations to reach a better understanding of the specific capabilities organizations need to achieve to enhance value co-creation.

This chapter is structured as follows. First, we deliver insight into the conceptual framework of *SDL* representing the foundation for service science. After introducing the science of *cybernetics*, the models and concepts provided by *organizational cybernetics,* which are considered relevant for improving value co-creation in social systems are explained and discussed. Then, the concept of *variety* as a measure of complexity is summarized and, on this basis, the model of *variety engineering* is introduced. As a further contribution, a second-order perspective on *Value Co-creation* is elaborated to support *SDL's* understanding of this inherently dynamic process.

Finally, we discuss how the model of *variety engineering* can be integrated into the concept of *value co-creation*; we thereby address the following two research questions: (i) How can *second-order cybernetics* contribute to a better understanding of the concept of *value co-creation* as put forward in SDL? (ii) What guidelines can the concept of *variety engineering* provide for effectively improving self-organization to improve value co-creation? The chapter closes with a summary and an outlook on further work.

2 Theoretical and Conceptual Background

2.1 Service-Dominant Logic: The Foundation for Service Science

The interdisciplinary field of service science is emerging within the larger context of current societal change. The accelerating pace of technological development has brought a fundamental shift in the context in which service is delivered and experienced. Advancements made in ICT have led to a proliferating variety of revolutionary services that change how consumers perceive themselves and their relationship to the world [11, 31, 32]. Parallel to the development and diffusion of digital technologies, which profoundly enhanced both producers' and consumers' communication capabilities and thereby their potential for mutual value creation, marketing science has seen a shift in perspective [14, 33, 34].

By integrating views from various disciplines (e.g. *consumer culture theory, agency theory, social network theory, viable systems approach, ecosystems theory,* etc.) the field of service science has been redefining the "traditional" concepts of marketing such as service, value creation, customer, market, and product [32, 35].

The models of marketing and management theory that prevailed in the last two centuries focused mainly on the efficient manufacturing of large quantities of homogenous outputs of tangible products (goods) in order to sell them to (passively receiving) customers at prices that allow short-term maximization of profits. The tangible good being the main unit of analysis in these models suited academics well in their pursuit of turning economics into a deterministic science that is based on linear cause-effect logic [36].

At the beginning of the 21st century, a paradigm shift occurred in marketing: the paradigm of goods-dominant logic, which inherently fragments the interrelatedness of consumers with producers, is more and more being replaced by a paradigm that focuses on the "togetherness between service provider and beneficiary" [34]. This shift towards a 'service-dominant logic'—where the focus moves from static transactions to dynamic relationships [36]—has serious implications on how exchange processes in the market arena are perceived and approached. The adoption of a service-oriented logic not only "embraces a focus on working together [...] to integrate resources [...] for mutual value creation" [34], but also "has the potential of shedding light on the role of exchange between and among service systems at different levels of analysis (e.g. individuals, organizations, social units, nations, etc.)" [37]. As a consequence, novel *SDL*-based models and concepts have emerged, such as servitization (cf. e.g. [38]) or service-oriented architectures (cf. e.g. [39]). For an overview of the development in the field of *SDL* cf [40].

Mainly influenced by systemic thinking, and thus being fundamentally inter-disciplinary, the framework of *SDL* has been advancing in reconceptualizing the basic notions of marketing. As distinctions in language (i.e. words) carry very specific connotations and an implied logic that are often incompatible with emerging conceptualizations, contributors of *SDL* have developed novel notions (see Table 1) enhancing understanding of this wider perspective [41].

Since the first introduction of *SDL* by [36], advancements have been made to further specify the foundational premises (FPs) and axioms on which the frame-work is being established (Table 2). Five of the currently eleven FPs have been given axiom status, as all other FPs can be derived from these [30, 37].

The foundational premises of *SDL* specify the core conceptual elements by which value co-creation is understood [37]. For instance, FP 6 and FP 9 as well as FP 11 imply that value co-creation always takes place in the larger context of

Table 1 Conceptual transitions from goods-dominant logic to service-dominant logic (based on [41])

Goods concepts	Transitional concepts	Service concepts
Goods	Services	Service
Products	Offerings	Experience
Value-added	Co-production	Value co-creation
Profit maximization	Financial engineering	Learning
Equilibrium systems	Dynamic systems	Complex adaptive systems
Supply chain	Value chain	Service ecosystem
Promotion	Integrated marketing communication	Dialog

Table 2 Axioms and foundational premises of service-dominant logic (based on [37])

Axiom 1	FP 1	**Service is the fundamental basis of exchange**
	FP 2	Indirect exchange masks the fundamental basis of exchange
	FP 3	Goods are a distribution mechanism for services
	FP 4	Operant resources are the fundamental source of strategic benefit
	FP 5	All economies are service economies
Axiom 2	FP 6	**Value is co-created by multiple actors, always including the beneficiary**
	FP 7	Actors cannot deliver value but can participate in the creation and offering of value propositions
	FP 8	A service-centered view is inherently beneficiary oriented and relational
Axiom 3	FP 9	**All social and economic actors are resource integrators**
Axiom 4	FP 10	**Value is uniquely and phenomenologically determined by the beneficiary**
Axiom 5	FP 11	**Value co-creation is coordinated through actor-generated institutions and institutional arrangements**

service networks and service ecosystems [16]. The collaborative nature of
service-for-service interactions between multiple stakeholders is reflected in FP 8.
The contextual nature of the inherently subjective construction of value is highlighted by FP 10. In Sect. 4 of this work, the concept of value co-creation is viewed
from a systems perspective, which further supports *SDL's* foundational premises.

2.2 Organizational Cybernetics

The term *cybernetics* comes from the ancient Greek word for steersman (kybernetes), a term first used by Plato to describe the art of effective government. Hence,
cybernetics is closely related to activities of steering and coordination and thereby
represents processes of regulation. In the middle of the 20th century, the mathe-
matician Norbert Wiener revived the term by using it to describe that all biological,
mechanical and social systems function and operate according to the same underlying laws and principles [42].

Initially, the field of cybernetics was concerned with the engineering aspect of
designing control structures in artificial machines. It had great influence on the birth
of many modern sciences, e.g. *control theory, information theory, computer science, artificial intelligence and artificial neural networks, system dynamics, cognitive science, computer modeling* and *simulation science* [43]. However, when
cybernetic thinking was applied in biology and the social sciences in the early
1970s, a new philosophy of science emerged. Expanding the traditional scientific
approach, in which "the properties of the observer shall not enter the description of
his observations" [44], the novel approach included the observer in the domain of
science.

The term second-order cybernetics as the "cybernetics of observing systems",
opposed to the (first-order) "cybernetics of observed systems" was coined by von
Foerster (1979, p. 7) [44]. Neurophysiological experiments provided the biological
foundation for this pioneering approach as it was shown that observations, independent of the observer's attributes, are not physically possible [45]. Influenced by
this meta-perspective on social systems, Stafford Beer applied the findings of
(second-order) cybernetics to the management of social organizations and institutions. Defining cybernetics as "the science of effective organization", he pointed to
the fact that there are fundamental organizational rules set out for all complex
systems, which, disobeyed, lead to instability or to a failure to learn, adapt and
evolve, and ultimately to a collapse and disintegration of the system as such [46].

In the late 1950s, Beer laid the foundations for a new second-order discipline in
the context of Operational Research, which he understood "as the use of trans-
disciplinary science in tackling ill-formed problems with no known solution" [25].
With this novel approach called Organizational Cybernetics, Stafford Beer tried to
model, make transparent and support the design of communication and control
mechanisms in social organizations. Beer created an extensive number of cybernetic models, with his Viable System Model (VSM) being the best known in the

management science community. This cybernetic model establishes the necessary and sufficient conditions for the viability of any human or social system. First defined in a set-theoretic model [20], the VSM was then revised using neuro-physiological terminology instead of mathematics [25] and in the final version was operationalized as a topological version of the original set-theoretic algebra [24, 26].

In Sect. 3 the model of *Variety Engineering* will be explained—to provide a basic view it is necessary to highlight selected main cybernetic terms and concepts that constitute the second-order approach to service systems. We have chosen seven items, i.e. System, Black Box, Feedback, Homeostasis, Ultrastability, Self-Reference, and Structural Recursion as ancillae.

- *System*: "a set of interrelated variables selected by an observer" [47]. In general, a system is understood as a mental construct providing observers with a device to order, communicate and explore shared perceptions; by naming them, observers bring systems into existence [12].
- *Black Box*: a cognitive device by which an 'outside' observer may describe an organization's transformation of inputs into outputs where the internal mechanisms performing this transformation are not accessible to observation. When a correlation between the input and the output can be established through time, the observer can model the (future) behavior of the black box and therefore "it is not necessary to enter the black box to understand the nature of the function it performs" [24].
- *Feedback* (negative, error-correcting): "the return of part of a system's output so as to modify its input", which is changed in a way that reduces variations in the output according to some specific purpose [26]. Feedback is one of the necessary mechanisms of self-regulation in all goal-seeking or purposive (teleological) systems; both in living organisms, artificial systems as well as in higher-level social systems (e.g. balance of supply/demand in economics).
- *Homeostasis*: the special case of a feedback system stabilizing the internal environment of a viable (living) system. The mechanism of a homeostat provides the system with the self-organizing and self-regulating characteristic to hold critical variables within physiological limits. (e.g. the control of blood temperature). This concept also denotes the tendency of a complex system to run towards a dynamically changing equilibrial state.
- *Ultrastability*: a new criterion emerging when describing complex systems that are self-organizing by homeostasis. This concept is relating to the capacity of a system to balance perturbations and adapt to changes in the environment that were unforeseen by the designer of the system.
- *Self-Reference*: a "property of a system whose logic closes in on itself: each part makes sense precisely in terms of the other parts: the whole defines itself" [26]; self-reference provides closure to an organizational system, thereby allowing it to conceive of itself as an integrated whole, a self having identity and purpose.
- *Structural Recursion*: a multidimensional, fractal organizational structure, where autonomous viable systems contain, and are contained in, autonomous viable

systems, each having self-regulating and self-organizing characteristics [24]. At lower levels of recursion, all viable systems are composed of autonomous sub-units of functionally differentiated viable systems (e.g. cells, organs, human beings, corporate divisions, branches of industry, states, etc.), each contributing to producing the larger viable system they are embedded in. One key feature of structural recursion for the management of complexity is that the structure is invariant at each level of recursion.

3 Complexity and the Concept of Variety

In the following, we discuss variety as a means to measure complexity, and the law of requisite variety, finally we elaborate on the *Variety Engineering Model*.

3.1 Variety as a Measure of Complexity

Organizational systems in the business or social domain are complex, the situations to be managed are mostly highly dynamic: the system is unfolding its richness over time, thereby constantly changing its output, which often turns out to be hard to predict. The system's internal mechanisms generate a proliferating variety that is to be handled by the manager. Likewise, in today's turbulent business arena, organizations are faced with the need to manage a large number of uncertain environmental changes. When decision-makers are unable to understand or explain all aspects impinging on the decision-process, they associate these "difficult-to-manage-situations" with complexity.

Complexity thereby is regarded to be an attribute ascribed by observers and thereby relates to the observer's ability to distinguish the parts and relations constituting the relevant system in the world. Following this line of thought, complexity does not reside in an outside world, but rather is determined by an observer using language to describe an outside world. In cybernetics, a system's complexity is defined as the number of distinctions an observer is able to make in his or her observational domain [12]. As a consequence, the question arises—as complexity is being shaped subjectively: how can this number of distinctions be determined?

A precise measure of systemic complexity was first proposed by the British cybernetician W. Ross Ashby in the 1950s. This simple measure of complexity is called *variety* denoting the number of different states or modes of behavior a certain system may adopt [48]. To measure the possible states of a system (i.e. its variety), the observer first has to draw a distinction (in language) to establish the system by differentiating it from its environment [49]. The system's border has to be set up according to the observer's purposes, which includes making boundary judgments—i.e. setting up the meaning of a proposition by establishing a specific reference system [50]. Measuring variety is thus a self-reflective process by which an observer constructs descriptions of possible states of a system.

To illustrate the process of measuring variety, consider the following: when thinking about the complexity of a standard lighting system—a bulb governed by a switch—we usually count two possible states: ON or OFF. However, as it sometimes happens, the bulb can burn out. This is then considered to be a different state from OFF, because flicking the switch doesn't light up the bulb; now three possible states are distinguished. If the bulb is replaced and there still is no light, we arrive at further differentiating additional states: the master switch may be turned off, the electricity bill hasn't been paid, the cable is broken, the bulb socket is defunct or the electricity generating power plant has blown up. In our small example two possible standard states have developed into eight possible states (cf. [24]).

This example shows the importance of including the observer when considering the variety of a system, as he is always a participant, an intrinsic part of the system and thereby sets the context from which it is observed. It also shows that the level of analysis must be accounted for when different purposes are to be considered. Therefore, by ascribing the system a specific purpose, the observer specifies its boundaries and thus determines its variety [24].

3.2 The Law of Requisite Variety

The essential theorem related to the concept of variety is Ross Ashby's Law of Requisite Variety [48], which states that only "variety can destroy variety". Ashby's law is considered to be as fundamental to management as the laws of thermodynamics are to engineering, or the law of gravity is to Newtonian physics. It states that, for a regulator to be effective, it needs to have at least as much variety at its disposal as the system to be regulated. In other words, "the larger the variety of actions available to a control system, the larger the variety of perturbations it is able to compensate" [43]. Indeed, this fundamental cybernetic law has important implications for every managerial situation: Management as the profession of control aims at a reduction of variety, but since the potential variety of disturbances impinging on the system is unlimited, managers have to enhance their internal variety to be flexible enough to react to unforeseen disturbances.

Perhaps the best known and most concise formulation of Ashby's law is the Conant-Ashby theorem, which states that "every good regulator of a system must contain a model of that system" [51]. This theorem makes clear that "the result of an organizational process cannot be better than the model on which the management of that process is based" [18]. In other words, the models that managers use in a decision process are fundamental to the quality of the attained results. This implies that the relentless rise of uncertainty and volatility in current business environment requires managers to use adequate models for the task, i.e. models that have requisite variety [52].

R. Gorka et al.

3.3 The Variety Engineering Model

One model, which offers effective strategies aiming at requisite variety is Variety Engineering as deployed in the *viable system model* [12, 24–26]. This approach aims at enhancing value co-creation by facilitating analysis and design of the communication structure between producers and consumers.

To understand the concept of variety engineering in the context of service systems—where consumers and producers interact to co-create value—the distinction of actors from agents is drawn. According to Espejo [28] "it is not until social agents coalesce around shared purposes (tacit or explicit) that they constitute themselves as actors of an emergent organizational system". The difference between actors and agents is thus defined by the contribution they make in producing the organizational system: Various social and economic actors interact in the creation and regulation of policies according to shared purposes and thereby produce, mainly through self-organization, the organization's structure. Agents in the environment (e.g. customers, suppliers, competitors) who interact with these actors become participants of the organizational system and thereby co-produce the organization's boundaries [14, 28]. This understanding of service systems is based on the notions of organizational closure and autopoiesis as developed by Maturana and Varela (1992) by which the boundaries of a system are being created by the system's own activities [53].

Figure 1 shows an organizational system with its management and embedded in an external environment. The variety of environmental agents is by nature much greater than the variety of the embedded actor's organizational system. Similarly, a management team has less variety than the organizational system, as it cannot have knowledge of the entire set of internal processes in detail; the variety of the challenges arising within the greater organizational context might by far exceed

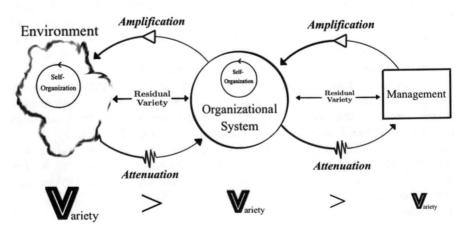

Fig. 1 Variety imbalance of three interacting systems Adapted from [12]

management's cognitive capabilities. Therefore, systems whose internal processes are not fully accessible by mere observing the system as an outside observer may be treated as a Black Box [47].

Ashby's law determines that, if actors (e.g. a multinational enterprise, a corporate division, or a project management team) wants to keep their organizational system under control (i.e. maintain dynamic equilibrium within its environment), they have to account for this variety mismatch by two strategies that have to be applied simultaneously: they have to either amplify their individual variety, i.e. develop more differentiated behavioral responses, or attenuate environmental variety, i.e. selecting an appropriate level of complexity they can deal with. According to Beer's *First Principle of Organization*,[1] these variety-balancing mechanisms emerge naturally as the variety asymmetries "are necessarily going to equate" [24].

However, in complex situations, these emergent variety amplifiers and attenuators show effectiveness of various degrees in leveling out variety mismatches. As can be illustrated by current developments in society and in the global ecosystem (e.g. climate-change, increasing scarcity of natural resources, demographic change, acceleration of income and wealth inequality, global oligarchy, financial crisis, political instability, migration, etc.), a misbalance potentially leads to social inequalities and high costs for the public. Consequently, in these situations, variety amplifiers and attenuators need proper management and regulation and should be designed to minimize damage to people and to cost [26]. In the *Variety Engineering Model* they are conceptualized by three types of variety operators:

- *Variety Amplification* is concerned with redesigning practices in a way that several relevant perturbations impinging on the management system can be taken care of simultaneously. With this strategy actors improve their behavioral repertoire and the resulting rise of individual complexity allows them to match a greater number of perturbations with the same number of responses. In other words, they can achieve "more with less" [28].
- *Variety Attenuation* deals with the categorization and classification of occurring perturbations in order to reduce situational complexity. Any mechanism or procedure that reduces the number of distinctions that management needs to pay attention to in a situation is an attenuator of variety. What aspects should be paid attention to depends on the purpose and performance criteria of the situation and on how much capacity the system has for self-organization. In this context it should be quoted that sheer ignorance is the "lethal variety attenuator" [26].
- *Variety Transducer* is another important aspect of the model. The mechanism of a transducer "leads across" communications between the environment, the organizational system and the management [26]. When information is passed between environmental agents and organizational actors the message transmitted has to cross a system's boundary. As each system has its own domain-specific

---

[1]"Managerial, operational and environmental varieties, diffusing through an institutional system, tend to equate; they should be designed to do so with minimal damage to people and to cost" [24].

language, the transmitted messages have to be "translated" (i.e. encoded or decoded) into the other system's particular mode of expression [24]. If the transduction process restrains the variety of the message, the information channel is not working properly, and effective communication is therefore inhibited [29].

One must note that both the agents in the environmental ecosystem as well as the actors that constitute the organizational system have themselves capacities to regulate and steer their behavior towards an agreed goal or purpose (see circles of self-organization in Fig. 1). This implies that the emerging processes of self-organization and self-regulation are capable of absorbing much of the variety that is produced by the interactions of agents and actors, thereby leaving management with only having to account for the unabsorbed residual variety [12]. The model of variety engineering aims precisely at the improvement of these self-organizing and self-regulating mechanisms, thereby reducing the relevant variety the organization and managers have to deal with. This is achieved through two possible modes of use [28]: (i) In a *diagnostic mode* the variety engineering model provides guidance to determine if the variety operators, that asserted themselves through self-organization, are effectively balancing the variety asymmetries between actors and agents. (ii) In a *design mode* variety engineering helps to formulate the necessary and sufficient variety operators that have to be implemented to achieve a desirable performance.

4 Systems Perspective on Value Co-creation

4.1 Paradigm Shift in the Concept of Market

As already anticipated by Toffler [54], the increasing trend towards a post-modern digital economy, where innovation and customization become the most thriving business drivers, led to a merging of producers' and consumers' boundaries. Facilitated by the wide range of advancements made in ICT, the traditional roles of consumers and producers are readjusted through various forms of new capabilities for collaboration, participation and customer engagement [55, 56]. Consumers today are more willing to provide information while at the same time being increasingly empowered to collaborate with producers in exchange for more customized products and services that better satisfy their needs, desires and preferences [35].

With the consumer's increased capability to participate in the value creation process through enhanced communication technologies, they can co-develop dynamically, together with producers, an active system of shared meanings [33]. These shared meanings are constituted by an ongoing process of reciprocal negotiation between producers and other stakeholders at a symbolic socio-cultural level [31]. As a result, value is constituted as the locus of these shared symbolic meanings [57]. This perspective resonates with *FP6* and *FP7* of *SDL* (see Table 2), which highlight the interactive nature of reciprocal value-propositions by producers and consumers.

4.2 Value Co-creation from a Second-Order Perspective

Considering these implications for market interactions, a deeper understanding of the mechanisms by which value is co-created is provided by *second-order cybernetics*. When viewing interactions between consumers and producers through the conceptual lens of the *variety engineering model*, the focus shifts to the conversational nature of service-systems constituting value co-creation through reciprocal value-propositions. In this context, Espejo and Dominici specified value co-creation as a dynamic process "through which the human constituents of the value system coordinate themselves in the creation of symbols and artefacts that become part of, and at the same time modify, the relational systems that create them" [33].

Consequently, both—consumers and producers—are conceived as participatory actors forming social networks within a global service ecosystem, which is treated as a black box. These actors are both involved in the dynamics that produce the black box's outcomes as well as being changed by these. Combined with *SDL's* conception of service ecosystems this implies that—as opposed to a first-order perspective, where social and economic actors stipulate each other's purposes—the second-order perspective provides a view where each participant is seen as an autonomous observer who enters the service ecosystem by stipulating his own purpose. By entering a dialogue, he changes his purpose and at the same time the purposes and orientation of all other actors in the entire global service ecosystem [44].

This circular relationship between market participants (producers and consumers) is what Maturana (2002) explains with the concept of structural coupling [58]. It states that, whenever there is a history of recurrent interactions between systems a structural congruence between these interacting systems emerges, by which they form each other's behavior by triggering internally generating structural changes that allow adaptation. In the case of a producer interacting with stakeholders in a market, this structural coupling is a learning process, where the result is the formation of shared meanings for products and/or services. The product is thereby "an embodied token underpinned by processes of value creation, which co-create, through recurrent producer–consumer relations, a stable form" [33].

Furthermore, SDL's conceptual transition from "products" to "experiences" (see Table 1) reflects it's close relation to the epistemology of *radical constructivism*, where knowledge is conceived as being actively constructed by an observer by his subjective interpretation of his experiences [59]. Heinz von Foerster's notion of an *eigenform* (1976) explains how structural coupling between producers and consumers leads to the emergence of stable points [60]. These *eigenvalues* and *eigenfunctions* represent the product as the recursive computation of shared meanings of congruent experiences. The product is thereby understood as a cognitive and symbolic entity, an emergent token of recursive interactions between producers and consumers. From this constructivist's perspective, a product (as a perceived experience) is necessarily different for every customer. Its value can even differ for one and the same customer over time, as the symbolic meaning of the subjective experience changes dynamically within a "co-ontogenic structural drift" [61].

The circular process of value co-creation resides in the relational domain of producer-consumer interactions and is the fundament for the creation of symbolic value. Hence, value co-creation occurs "when an object is seen as something perceived by an observer and socially valued, the perception being fluid and varying through time and space. [...] As such, the co-creation of value within a market is a continuous and dynamic process that needs to be fostered by continuous communication, negotiation, and action of all the agencies in the market arena" [62]. Designing an effective communications structure can be achieved by the adaptation and application of the *variety engineering model*.

5 Applying Variety Engineering to Value Co-creation

5.1 Effective Structures Through Variety Engineering

Applying a systems perspective to the study of value co-creation reveals that several theoretical and practical gaps are to be identified and need to be eliminated. Mainly because analytic-reductionist thinking still prevails in management practice, the *viable system approach*, which is based on the *variety engineering model*, is found to broaden the perspective on service systems, such as business or social organizations [16]. From the point of view of recent literature elaborating on the integration of second-order concepts into service research, this approach is found to be capable of interpreting the systemic dynamics of economic actors in the global business arena at a holistic level [15].

Figure 2 shows a homeostatic loop of structural coupling between a producer (low-variety side) and consumers (high-variety side). The producer is confronted with the high variety that is contained in consumers' value propositions (i.e. offerings and communications). Consumers themselves have to deal with the variety of product offerings and communications coming from the stream of the producer's value propositions. These propositions evolve with producer's dynamic capabilities and consumers' desires. The reflexive observing and communicating of both, producer's and consumers' relations, with each other constitutes the fundament for the emergence of value co-creation [33].

To achieve adequate performance according to a specified performance criterion (in our context the criterion is "effective value co-creation"), the scheme outlined in Fig. 2 is a design template for pairs of variety amplifiers and attenuators that have to

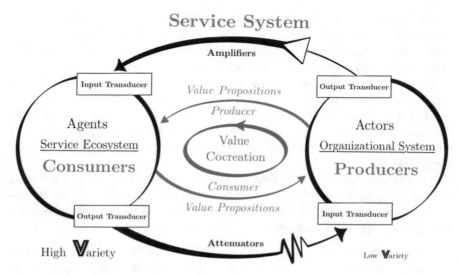

Fig. 2 Homeostatic loop in value co-creation [33]

be set up to deal with the residual variety that is left unattended by internal self-organizing processes. Both, the input and output transducers, need to be designed to match the variety that is being amplified or attenuated.

5.2 Variety Operators in the Global Service Ecosystem

In the context of global service ecosystem, these instances of variety operators provide guidelines to enhance value co-creation. In the following we discuss selected examples (c.f. [31, 33]).

Internet and Communication Technologies For the low-variety side, communications via mobile technologies increase producer's relationship capabilities to reach an increasing number of potential customers (amplification). The increasing variety of *platforms and applications* (transducers) that producers use to interact with customers require effective management for two reasons:

- The variety contained in consumers' value propositions may exceed producer's capabilities to implement these communicated customer demands.
- Too much variety in product offerings may overwhelm consumer's cognitive capabilities, which potentially leads to 'paradox of choice' and paralysis in consumers.

For the high-variety side, *online purchasing platforms* (transducer) with integrated search-engines facilitated by semantic web reduce the variety inherent in producer's value propositions by guiding self-organization. This helps customers choose from a proliferating variety of product offerings (attenuation).

Online collaboration platforms (transducer) allow customers to participate in the design process of products and services (amplification). These conversations constitute a balancing mechanism, by which producers learn to better meet the specific needs, desires and inner values of consumers with their value propositions.

Social media (e.g. blogs, social networks, virtual communities, apps for online-sharing, etc.) represent transducers for both producers and consumers.

Service Oriented Architectures and related concepts are to reduce or even eliminate boundaries among organizational entities (producers, supply chains, etc.) and thereby support for example the integration of producers in supply chains (amplification).

Artificial Intelligence tools and applications (attenuator) may help consumers as well as producers to interpret and translate online conversations. This would provide a conversations platform to negotiate meanings for services, thereby facilitating the emergence of 'consumer-tribes' and 'brands'.

Intermediaries As autonomous market agents form *self-organizing networks* in the service ecosystem, they absorb much of the relevant variety contained in consumers' value propositions.

Collaboration with environmental agents enhances the producers' response variety enabling them to reach a wider range of customers.

By instituting *distribution networks*, that connect e.g. retailers, distributors, and subcontractors, producers may increase (i) their flexibility to react on changes in the environment and (ii) their abilities to meet customers' demands.

Mass Customization If marketing research (attenuator) is able to identify specific demands and wishes of customers (e.g. via an online collaboration platform or by 'decoding' social media platforms) the producer can attenuate the variety necessary to meet the customer's demands by utilizing advanced manufacturing systems that use standardized components and procedures ('smart factory' in industry 4.0), and determine the latest possible point in the supply chain for customizing a product according the needs or wishes of a specific customer (postponement).

To effectively incorporate customer's requests for product design, the producer's organizational structure has to be flexible enough to manage the increasing volatility in trends (amplification).

Consumer Tribes When individual consumers have something in common, they interact to form culturally based tribal collectives; these consumer tribes assign products with specific symbolic meanings. If these idiosyncrasies are understood by the producer, they become susceptible to appropriate marketing practices, facilitating communication between the producer and the consumer tribal groups (attenuation).

By acknowledging the history, tradition, and culture of consumers the producer can better identify products that have the potential to create consumer tribes.

Brands Brands are creating awareness for the symbolic meaning of products and services (attenuation). Consumers see additional benefits from using the product (e.g. value-in-use, psychological-value, social-value).

Word of Mouth As actors and agents form social networks and relationships, they amplify producers' offerings and communications by word of mouth.

Institutions for Quality Assurance Organizations that manage the quality assurance of producer's value propositions by instituting subjective values and norms of customers are essential variety attenuators.

Universal Humanistic Value Sharing socially desirable meanings, values and norms (e.g. the Universal Declaration of Human Rights) may help producers and consumers to better reach consensus in their conversations. Redundancy in the service-system's structure provided by the agreement on basic humanistic and/or social values effectively increases the channel capacity, thereby making the communications of value propositions more predictable [63].

6 Conclusion

6.1 Synopsis

The proposed framework with its theoretical foundations in constructivist epistemology may be a starting point for developing enhanced models and improved approaches to overcome the limitations of traditional analytic-reductionist logic. With the integration of a second-order perspective into the understanding of the processes of value co-creation, the reference frame transitions towards a view of service systems as being constituted by the relational aspects of human observers, namely, their conversations. Consequently, service systems are understood as self-organizing, adaptive networks of communications among observers who participate in the service system by drawing distinctions and creating relations within it. This wider perspective shows the potential to transform our understanding of the actions needed to enhance value co-creation.

The perspective on self-organizing service systems as described by second-order cybernetics provides the framework of SDL assistance in establishing itself as a new paradigm. This chapter shows that organizational cybernetics could be used as a theory-driven underpinning for the emerging service science. The overview of the proposed concepts and models sheds light on some of the systemic interdependencies of actors and agents participating in service systems. Continuously co-developing in the evolutionary drift of structural coupling, variety operators assert themselves naturally via self-organization as long as lines of communication are maintained between service beneficiaries and providers in service systems. Improving the quality of these variety balances effectively requires holistic models that exhibit requisite variety according to Ashby's law. This implies, that a better understanding of these self-organizing patterns of interactions, in which value co-creation takes place, will lead to more differentiated models and—as a consequence—to better mechanisms and tools to guide and control the process of self-organization.

6.2   Further Research

The mere focus on the interactions taking place at the market level is found to be insufficient for better explaining and supporting the opportunities for value co-creation. To reach requisite variety of the model aiming at improving value co-creation, it is necessary to include the emotional, cultural, and cognitive-symbolic dimensions of human interaction. By enriching the design of decision-making models with the conceptual device of the Variety Engineering Model these dimensions could be taken into account. The resulting improvement of communications of value propositions enhances value co-creation by enabling shared meanings of all agents and actors to converge.

To enhance the possibilities for participation and collaboration arising with digital technology, variety operators have to be designed and installed to effectively improve value co-creation. It is likely that further development of the variety engineering model in the framework of service science may produce practical design tools and applications that help social actors to better guide the value co-creation process. One question that needs further elaboration in future research is: What measures are suitable for assessing the performance of the value co-creation process within a service system?

Systems thinking could be used to provide an open-minded, participative, pluralistic conversations-platform that creates an open dialogue between a wide diversity of actors in multiple human spheres. This could help redefine the purpose and orientation of the emerging human-technological society beyond current patriarchal institutions and neoliberal-capitalist forms. It provides a meta-perspective by which all social actors are perceived as closely intertwined and interdependent by their recursive communications, residing in the same coherent system.

With the proposed second-order view on service-systems the following question arises: Do we choose to think of ourselves as being *apart from* the world, deluding ourselves to think we can point to something independent from us 'out there', allowing us to be 'objective'; or do we want to use a model of us being a *part of* the world, realizing that our actions change us and everyone else, which would enable us to assume responsibility for our actions by consciously entering a co-evolutionary dialog with others?

References

1. World Economic Forum (WEF) (2016) The future of jobs: employment, skills and workforce strategy for the fourth industrial revolution. Retrieved from http://www3.weforum.org/docs/WEF_Future_of_Jobs.pdf. Accessed 16 June 2017
2. Kurzweil R (2001) The law of acceleration returns. Retrieved from http://www.kurzweilai.net/the-law-of-accelerating-returns. Accessed 16 Apr 2018

3. Kurzweil R (2005) The singularity is near: when humans transcend biology. Viking, New York
4. Gillings MR, Hilbert M, Kemp DJ (2016) Information in the biosphere: biological and digital worlds. Trends Ecol Evol 31(3):180–189
5. Stoshikj M, Kryvinska N, Strauss C (2016) Service systems and service innovation: two pillars of service science. Procedia Comput Sci 83(Ant):212–220
6. Kryvinska N, Strauss C, Zinterhof P (2011) Next generation service delivery network as enabler of applicable intelligence in decision and management support systems, Chapter 18. In: Besis N, Xhafa F (eds) Next generation data technologies for collective computational intelligence. Studies in computational intelligence. Springer, Berlin, Heidelberg, pp 473–502
7. Last C (2015) Information-energy metasystem model. Kybernetes 44(8/9):1298–1309
8. Heylighen F (2016) The offer network protocol: mathematical foundations and a roadmap for the development of a global brain. Eur Phys J Spec Top, 1–40
9. Last C (2017) Global commons in the global brain. Technol Forecast Soc Change 114:48–64
10. Maglio PP, Spohrer J (2008) Fundamentals of service science. J Acad Mark Sci 36(1):18–20
11. Stoshikj M, Kryvinska N, Strauss C, Greguš M (2015) Service science, service system and service innovation. In: Gummesson E, Mele C, Polese F (eds) The 2015 Naples forum on science: service dominant logic, network and systems theory and service science: integrating three perspectives for a new service agenda, SIMAS Lab di Salerno for Naples Forim on Service
12. Espejo R, Reyes A (2011) Organizational systems: managing complexity with the viable system model. Springer, Berlin
13. Teece DJ, Peteraf MA, Leih S (2016) Dynamic capabilities and organizational agility: risk, uncertainty and entrepreneurial management in the innovation economy. Calif Manag Rev 58 (4):13–35
14. Espejo R (2015) Good social cybernetics is a must in policy processes. Kybernetes 44 (6/7):874–890
15. Golinelli GM, Barile S, Spohrer J, Bassano C (2010) The evolving dynamics of service co-creation in a viable systems perspective, pp 813–825
16. Barile S, Lusch RF, Reynoso J, Saviano M, Spohrer J (2016) Systems, networks, and ecosystems in service research. J Serv Manag 27(4):317–360
17. Barile S, Polese F (2010) Linking the viable system and many-to-many network approaches to service-dominant logic and service science. Int J Qual Serv Sci 2(1):23–42
18. Schwaninger M (2000) Managing complexity: the path toward intelligent organizations. Syst Pract Action Res 13(2):207–241
19. Beer S (1959) Cybernetics and management. English University Press, London
20. Beer S (1962) Towards the cybernetic factory. In: Foerster H, von Zopf G (eds) Principles of self-organization. Oxford: Pergamon Press Limited. Reprinted in Harnden R, Leonnard A How many grapes went into the wine: Stattford Beer on the art and science of holistic management. Chichester: Wiley, pp 163–225
21. Beer S (1966) Decision and control: the meaning of operational research and management cybernetics. Wiley, London
22. Beer S (1968) Management science: the business use of operational research. Aldous Books, London
23. Beer S (1975) Platform for change: a message from Stafford Beer. Wiley, Chichester
24. Beer S (1979) The heart of enterprise. Wiley, London, New York
25. Beer S (1981) Brain of the firm. Wiley, Chichester
26. Beer S (1985) Diagnosing the system for organizations. Wiley, London, New York
27. Beer S (1994) Beyond dispute: the invention of team syntegrity. Wiley, London, New York
28. Espejo R (2015) Performance for viability: complexity and variety management. Kybernetes 44(6/7):1020–1029
29. Rosenkranz C, Holten R (2011) The variety engineering method: analyzing and designing information flows in organizations. IseB 9(1):11–49

30. Vargo SL, Lusch RF (2016) Institutions and axioms: an extension and update of service-dominant logic. J Acad Mark Sci 44(1):5–23
31. Dominici G, Yolles M, Caputo F (2017) Decoding the dynamics of value co-creation in consumer tribes: an agency theory approach. Cybern Syst 48(2):1–18
32. Ostrom AL, Parasuraman A, Bowen DE, Patricio L, Voss CA (2015) Service research priorities in a rapidly changing context. J Serv Res 18(2):127–159
33. Espejo R, Dominici G (2016) Cybernetics of value co-creation for product development. Syst Res Behav Sci 34(1):24–40
34. Greer CR, Lusch RF, Vargo SL (2016) A service perspective: key managerial insights from service-dominant (S-D) logic. Org Dyn 45(1):28–38
35. Dominici G, Basile G, Palumbo F (2013) Viable systems approach and consumer culture theory: a conceptual framework. J Organ Transform Soc Change 10(3):262–285
36. Vargo SL, Lusch RF (2004) Evolving to a new dominant logic. J Mark 68:1–17
37. Vargo SL, Lusch RF (2008) Service-dominant logic: continuing the evolution. J Acad Mark Sci 36(1):1–10
38. Kryvinska N, Kaczor S, Strauss C, Greguš M (2014) Servitization—its raise through information and communication technologies. In: Snene M, Leonard M (eds) Exploring service science, lecture notes in business information processing. Springer, Berlin, Heidelberg, pp 72–81
39. Kryvinska N, Auer L, Strauss C (2009) The place and value of SOA in building 2.0-generation enterprise unified vs. ubiquitous communication and collaboration platform. In: 2009 third international conference on mobile ubiquitous computing, systems, services, and technologies (UBICOMM), IEEE, pp 305–310
40. Kryvinska N, Olexova R, Dohmen P, Strauss C (2013) The S-D logic phenomenon— conceptualization and systematization by reviewing the literature of a decade (2004–2013). The society of service science. J Serv Sci Res, Springer, Col. 5(1):35–94
41. Lusch RF, Vargo SL (2006) Service-dominant logic: reactions, reflections and refinements. Mark Theor 6(3):281–288
42. Wiener N (1948) Cybernetics: or the control and communication in the animal and the machine. MIT Press, Cambridge MA
43. Heylighen F, Joslyn C (2001) The law of requisite variety. In: Heylighen F, Joslyn C, Turchin V (eds) Principia Cybernetica Web, Brussels
44. Foerster H (1979) Cybernetics of cybernetics. In: Communication and control, K. Krippendorff, Gordon and Breach, New York, pp 5–8
45. Umpleby SA (2016) Second-order cybernetics as a fundamental revolution in science. Constructivist Found 11(3):455–465
46. Beer S (1973) Fanfare for effective freedom: cybernetic Praxis in government. Retrieved from http://www.kybernetik.ch/dwn/Fanfare_for_Freedom.pdf. Accessed 16 June 2017
47. Ashby WR (1960) Design for a brain: the origin of adaptive behaviour, 2nd edn. Chapman and Hall, London
48. Ashby WR (1956) An introduction to cybernetics. Chapman and Hall, London
49. Spencer-Brown G (1969) Laws of form. Allen & Unwin, Australia
50. Ulrich W (2000) Reflective practice in the civil society: the contribution of critical systems thinking. Reflective Pract 1(2):247–268
51. Conant RC, Ashby RW (1970) Every good regulator of a system must be a model of that system. Int J Syst Sci 1(2):89–97
52. Pérez-Ríos JM, Velasco Jiménez I (2015) The application of organizational cybernetics and ICT to collective discussion of complex issues. Kybernetes 44(6/7):1146–1166
53. Maturana HR, Varela FJ (1992) The tree of knowledge: a new look at the biological roots of human understanding. Shambhala/New Science Library, Boston
54. Toffler A (1980) The third wave. Bantam, New York
55. Brodie RJ, Hollebeek LD, Jurić B, Ilić A (2011) Customer engagement: conceptual domain, fundamental propositions, and implications for research. J Serv Res 14(3)
56. Saarijärvi H (2012) The mechanisms of value co-creation. J Strateg Mark 20(5):381–391

57. Spohrer J, Maglio PP (2008) The emergence of service science: toward systematic service innovations to accelerate co-creation of value. Prod Oper Manag 17(3):238–246
58. Maturana HR (2002) Autopoiesis, structural coupling and cognition: a history of these and other notions in the biology of cognition. Cybern Hum Knowing 9(3–4):5–34
59. Glasersfeld E (2001) The radical constructivist view of science. In: Riegler A (ed) Special issue "The impact of radical constructivism on science", Foundations of science, vol 6(1–3), pp 31–43
60. Foerster H (1976) Objects: tokens for (Eigen-)behaviors. University of Geneva. Reprinted in Foerster H Understanding Understanding: essays on cybernetics and cognition. Springer, New York, pp 261–271
61. Maturana HR (1988) Reality: the search for objectivity or the quest for a compelling argument. Ir J Psychol 9(1):25–82
62. Dominici G, Yolles M (2016) Decoding the XXI century's marketing shift: an agency theory framework systems, vol 4(4)
63. Shannon C, Weaver W (1949) The mathematical theory of communication. The University of Illionois Press, Urbana IL

Enterprises Servitization for an Efficient Industrial Management a Survey-Based Analysis

S. Kaczor

Abstract The concept of servitization initially originates from the manufacturing industry which currently struggles with decreasing profit margins and therefore strives for a strategy to handle this. Consequently the enrichment of products by corresponding services appears to be a considerable approach, which in fact implicates a continuous movement towards a service economy. Accordingly, this paper strives to explain the concept of servitization in detail in order to declare its capabilities in being a powerful strategy to provide competitive advantage in manufacturing industries. This will be conducted by initially elaborating its notion and origins in literature as well as the characteristics and processes regarding value creation, which is consolidated as fundamentals of servitization. Subsequently, the driving characteristics separated by general environmental trends, financial-, strategic- and marketing-drivers are discussed in detail. The final section refers to the classifications of servitization and the actual options of implementation.

Keywords Service · Service science · Service-dominant logic
Servitization · Service economy · Value co-creation · Business models

1 Notion of Servitization

The concept of servitization is a term one can find in the manufacturing industry; it consequently deals with goods and products [1]. Though considering its denotation there has to be a relation to services as well. According to Vandermerwe and Rada [2], who described the concept as the process of creating additional value by adding services to products, finally were persuaded that:

S. Kaczor (✉)
University of Vienna, Vienna, Austria
e-mail: mis@fm.uniba.sk

© Springer International Publishing AG, part of Springer Nature 2019
N. Kryvinska and M. Greguš (eds.), *Data-Centric Business and Applications*,
Lecture Notes on Data Engineering and Communications Technologies 20,
https://doi.org/10.1007/978-3-319-94117-2_9

Modern corporations are increasingly offering fuller market packages or "bundles" of customer-focused combinations of goods, services, support, self-service and knowledge. But services are beginning to dominate. This movement is termed the servitization of business (see in [2]).

Recent literature adopted this concept to a competitive manufacturing strategy in order to establish entry barriers to lock-out competitors and to lock-in customers [3–6]. Besides the work of [2] also some other researcher attempted this topic deriving nominations like hybridisation, tertiarisation, service infusion or servicisation which left unaffected the dominance of servitization remaining as most cited and referred one in literature. Furthermore, the concept of a product-service-system (PSS) occurred which addresses exactly the same topic. In fact, both concepts strive for the same goals and are based on the same drivers and motivations as well [3]. Consequently the difference between both concepts lies in the perception of its ultimate result, which can be described by the following Fig. 1.

According to the definition of a PSS, the value for a common manufacturer is provided by the product itself, whereas every occurring service equates with additional costs. In order to achieve competitive advantage, to differentiate from and even to rise above competitors some product related services are offered to enrich the total offering. Finally the product and service merge to a product-service-system that provides its value solely in their combination [8, 9]. The transition within the concept of servitization goes a step further with the final transformation from a manufacturing company to a service provider [2], which corresponds to the concept of S-D logic [10]. Baines et al. [3] confirmed this conclusion by conducting an extensive research among 58 articles concerning PSS and servitization. Consequently they suggested refining the definition of servitization to encompass the PSS theme, by stating that: "*servitization is the innovation of an organization's capabilities and processes to better creates mutual value through a shift from selling products to selling PSS*" (see in [3]). Consequently the subsequent sections in this paper embrace the findings of literature concerning PSS as well as servitization in order to maximize credibility [1, 11].

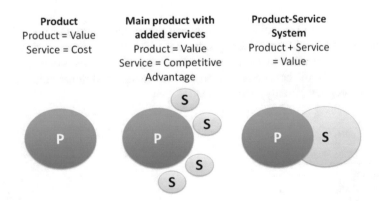

Fig. 1 Transition from a product to a product-service-system. *Source* own illustration according to [7]

2 Value Co-creation

Many industries identified the proposal of superior customer value as one of the most enduring and successful strategies to gain competitive advantage. Consequently, a lot of research papers are debating on the concept of value for a long time [12, 13]. The approach underlying this concept refers to the notion of value, which is conducted via the exchange of value among the involved parties. Accordingly, companies strive for establishing and maintaining a strategic buyer-seller relationship [7, 14, 15]. Concerning the topic of delivering and proposing services to add value, which are generally characterized by the fact that they have to be consumed in the same moment they are provided [16–18], the corresponding research [10, 19] referring to the concept of *value-in-use* has to be considered. Accordingly *value-in-use* is solely realized during the process of consumption [10]. Therefore, Woodruff [20] stated that:

> Customer value is a customer's perceived preference for and evaluation of those product attributes, attribute performances, and consequences arising from use that facilitate (or block) achieving the customer's goals and purposes in use situations [20].

Finally, recent academic research proposed that companies do not offer or provide value, but propose value. Thus, there is a need of contribution from the involved customer, who finally determines the value and consequently co-creates it [10]. Accordingly, companies store unrealized potential value in their products, which is finally executed through customers' co-creation in order to gain the benefit [13]. One has to be aware of the vital difference to the topic of co-production, which asks the customer to provide input to product design or self-service, and therefore appears in contrast to the notion of value co-creation [21–25]. In terms of servitization the desired value can vary across industries and the specific needs of customers, which may last from cost effectiveness, sustainable use of resources and operational improvement to less environmental burdens [26]. Consequently, the subsequent section deals with the drivers and motivations for the concept of servitization.

3 Drivers of Servitization

Every emerging approach or strategy evolves from several intentions and drivers. Consequently, the subsequent section discusses those driving servitization by splitting them into general environmental trends on the one hand and into internal drivers regarding finance, strategy and marketing on the other hand [1].

3.1 General Trends

Today's global manufacturing firms are facing highly competitive markets with, decreasing margins and revenues [3, 27]. According to the emergence of economies in the Middle and Far East which are able to provide low-paid and therefore cheap workforces, it is becoming harder if not impossible to compete solely on basis of costs. US manufacturers would have to cut their costs by 30% to meet the level of their Asian competitors, which is a threshold that seems not viable in developed economies [4, 5, 28]. Accordingly, several research suggest moving along the supply chain in order to innovate and create more sophisticated products and services to distinguish through additional value (cf. Porter and Ketels (2003) see in [4]). In fact, OECD identified a share of about 70% accounting for the total employments of the service sector [29], whilst approximately 30% of all manufacturing companies were conducting servitization [5]. Furthermore, [5] elaborated five underlying general trends from products to services to explain this development, which are shown in Fig. 2.

These five trends occur on different economic subdomains: (1) *products develop into solutions*, which means that solutions are rather supplementing products than replacing them [30]; (2) *outputs become outcomes* which especially refer to outcome-based contracts that guarantee for example the availability of a product [31]; (3) process oriented operational *transactions* develop into and are replaced by *relationships*; (4) *suppliers* develop into *network partners* and (5) *elements alter to eco-systems* [5].

3.2 Financial Drivers

The previous section briefly mentions financial benefits in context with a servitization strategy. It is a big issue for manufacturing companies to gain profitable

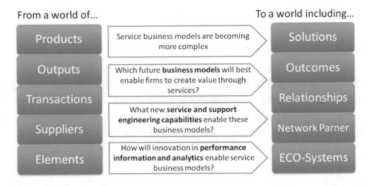

Fig. 2 General trends for the shift from products to services. *Source* own illustration according to [5]

revenues out of solely selling products, which relies to low and unstable profit margins [1]. Therefore, manufacturers are tempted adding services to their products in order to raise and stabilize margins [3, 32–34]. In fact, some companies performed well in applying this strategy and experience improved growth. As example for an early application General Electric Capital Services can be mentioned, they were exploiting higher margins and revenues out of services, which therefore accounted for 60% of their final profits [33]. The latterly prevalent case of complex engineered products accounts for another possible revenue stream in the manufacturing industry [35]. The resulting longer product life cycles raise the ratio of installed-base to new units, especially in the automotive, civil aircrafts and locomotive industry. Accordingly the ratio for automobiles is 13 to 1, those of civil aircrafts 15 to 1 and concerning locomotives even 22 to 1, which means that for every new build car already 13 operating ones already exist [36, 37]. In terms of financial benefits companies can establish stable revenue streams by charging for maintaining, repair and other product-related services [3, 33, 34, 38]. This push of considerable revenues downstream towards service support diminishes potential volatility originating of drops in sales and the inherent effects of mature markets and negative economic cycles [34]. In order to exploit these potential revenue streams one has to consider an appropriate pricing strategy and cost structure, concerning fixed/variable and direct/indirect costs as well as the choice between bundling or unbundling prices. Providing very specific and highly intense service offerings would equally raise the chance and possibility to gain higher financial benefits [33].

3.3 Strategic Drivers

Besides financial benefits as described above, the application of servitization in a manufacturing company may create strategic benefits. These are generally manifested by competitive advantage, which is carried out by the differentiation of manufacturing offerings [3, 34]. Consequently, services were identified as one of the most essential competitive factors in manufacturing, even support building of industry entry barriers [2, 33, 39]. In fact, it is difficult to imitate a well-intended, well-designed and tailored service, since it is less visible and mostly dependent on labor skills, which therefore creates a sustainable source of competitive advantage [3, 6, 34]. When establishing such competitive mechanism one has to consider that it is substantial to offer outstanding services to the customer to enable valuable co-creation. These services should directly enhance the value of the related product rather than being of general nature. This is due to the fact that, in a competitive market, consumers expect certain services as basics, which therefore would not provide any competitive advantage [33]. Higher strategic benefit could be gained by providing the more specific and intense service offering [3, 6, 34].

3.4 Marketing Drivers

Engaging in marketing opportunities literature shows that services likewise may account for several benefits. The services enrich pure goods offering, and consequently an appeal the customer [33], and therefore pushing the selling of products by influencing purchasing decisions [3, 34]. In a B2B context [33] derived three implications encouraging this pretention. Thus, a servitization strategy (1) *influences overall clients satisfaction*, whereas (2) *improving the adoption of the new product* and finally (3) *strengthening clients confidence* whilst (4) *improving supplier's credibility* [33].

The declaration for the increasing demand of services arises from the work of [2, 40, 41], who derived five main reasons for this evolution, consisting of the (1) connection of value to the use and performance of the system, (2) the desire for solutions instead of products or services, (3) gaining advantage of suppliers know-how, (4) the need for integrated and global offerings, and (5) the inherent customized relationships. These points consolidate the challenge in gathering the knowledge about the services customers require and desire, since some of them do not even know about their demand yet [33].

Considering the mentioned findings of [33], who elaborates on the degree of organizational intensity and service specification, he derived the evolution of servitization benefits which is shown in Fig. 3. Accordingly, the financial, strategic as well as the marketing benefit increases in relation to the extent of service provision growing even more whilst moving towards a solely service oriented company [33].

Fig. 3 Evolution of servitization benefits. *Source* [33]

Table 1 Typology of servitization benefits

	Offensive	Defensive	Neutral
Financial benefits	Increase total turnover	Stability of turnover	Extending life cycle of products
Strategic benefits	Differentiation of offering	Services are difficult to imitate	Sustainable environmental benefits
Marketing benefits	Aids understanding customer needs	Increase customer loyalty	

Source own illustration

The discussion above was structured according to financial, strategic, and marketing benefits, which were initially suggested by [3, 6, 33]. Another research conducted by [34] recommended the separation according to the intention of the competitive strategy, which tends to be offensive, defensive or neutral. In order to provide a valuable typology of servitization benefits two approaches were merged and illustrated in Table 1.

Furthermore, the emergence of powerful information and communication technologies (ICT) likewise enhanced the evolution and penetration of servitization strategies in manufacturing [42–44]. However, one has to be aware that these benefits exclusively depend on the mode of implementation of the servitization approach. Therefore, these technological advantages are incorporated in the subsequent section, which elaborates the options of servitization.

4 Conducting Servitization

Section 4 illustrates the approaches and possibilities concerning the implementation and execution of servitization. Accordingly the variety of services as well as their classification will be elaborated. Finally, the fundamental options of implementing servitization are depicted [1, 45].

4.1 Services and Value Offerings

According to the discussion in the Sect. 3, a company makes a value proposition to the customer who therefore might gain the benefits by co-creation, which finally is manifested in some kind of service [10]. Several studies dealing with servitization among manufacturing firms, showed that despite diverging manufacturing industries, a recurring pattern concerning the mode of services provided appears. Baines et al. [3] identified five fundamental services consisting of training, delivery, spare parts, repair, and customer helpdesks that occur most frequently. Furthermore, Tether and Bascavusoglu-Moreau [34] conducted a survey amongst UK manufacturers that elaborated 15 different services, which were ranked according to their extent. Comparing these two findings revealed that the top 5 services discovered by [34] perfectly matched the five ones discovered by [3]. The ranking of the 15 common services associated with servitization and introduced by [34] are illustrated in Fig. 4.

Concerning the choice of nature of the offered service provision as discussed above, one has finally to be aware of two fundamental considerations regarding the (1) requirements of infrastructure and skills and (2) the desired competitive advantage. Thus, the first service offerings illustrated in Fig. 4 generally implicate a huge and widespread infrastructure network to enable their execution among the market, which accompanies considerable initial investments. These preliminary fixed costs emerge regardless whether the service of delivery, customer helpline or maintenance is executed or not [46].

Additionally the service modes like delivery, repair, maintenance or warranty may appear in several industries as common ground among competition. Consequently these services still would still cause costs, however don't provide additional excitement for the customer and therefore equally don't provide any competitive advantage yet [3, 6, 33, 34].

Another research approach concerning service and value offerings is performed by [47] who focused on outcome based contract performance in B2B maintenance and repair services, examining value co-creation and its delivery. In order to provide the fundamental basis for the final act of value co-creation a certain

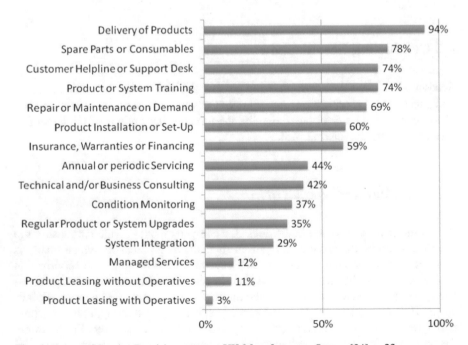

Fig. 4 Extent of Service Provision amongst UK Manufacturers. *Source* [34] p. 22

transformation of value has to take place. These three generic transformations are composed on the one hand by the (1) *transformation of materials and equipment* conducted by e.g. manufacturing, repair or installation throughout the supply chain or on the other hand by the (2) *transformation of information* like design or analysis through knowledge management. Finally also a (3) *transformation of people* may take place which therefore is characterized by training and the establishing of relationships [47]. Table 1 provides a synopsis on the extensive mapping regarding contextual offerings into value transformation. The study found that the majority of its processes are designed according to the transformation of materials and equipment. Nevertheless, the three transformations are interdependent, whilst executed jointly by the firm and the customer [47].

4.2 Classifications of Servitization

In this subsection an overview is provided on several classification schemes based on their implementation. One can distinguish three approaches concerning (1) the relation between *customization and integration* [48, 49] being in contrast with (2) the differentiation concerning *services supporting the product or the client* [50], and (3) one can differentiate concerning the *time frame* of the service execution [51].

One of the most critical impact factors refers to the degree of customization a company is capable to provide which has to be evaluated in relation to the extent of integration across its services applied [48]. Both variables consist of three characteristics, whereas no customization and no integration represent a possible degree of implementation. Accordingly, one level of customization is represented by *industry verticals* or *customer segments*, which come by with some differences in technical specifications, pricing and service levels. Furthermore, the degree of integration is crucial, since it reflects whether products and services within an offer create a value higher than the sum of their parts. In doing so there are two alternatives: (1) *commercial integration* standing for bundling of products and services into one transaction, and (2) *technical integration* based on physical interoperability [48]. The combination of these characteristics results in a 3 by 3 matrix [49]. Figure 5 shows the matrix together with exemplarily companies together their corresponding business models [11].

Another approach is pursued by [50] who draws a distinction between a service supporting the supplier's product or the client's action. Typical executions regarding the *support of supplier's product* are manifested for example by after-sales services like delivery and installation. In contrast, training services on the operation of the product are considered as *supporting the client's action*. Both approaches are heading for diverging goals, whereas the former one strives to ensure appropriate operation of the product and the latter one the extraction of knowledge about customer perceptions to work on subsequent product and service offerings [50].

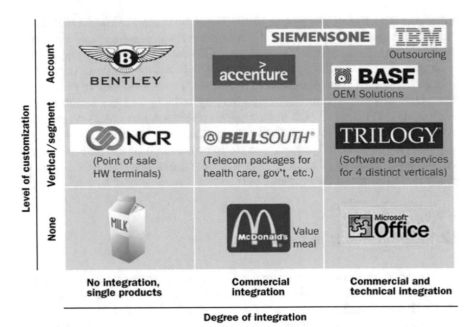

Fig. 5 Classification due to customization and integration. *Source* [49]

Literature likewise provides a classification regarding time frames of consolidating products with services. The pivotal point for this approach is manifested by the point-of-sale (POS) that yields the pre-purchase, at-purchase and after-sales time frame [51, 52]. Thus, during pre-sale within the manufacturing process it is classified as *product-service integration*, where services are linked to products while they are created. In fact, these inputs throughout the production process—concerning characteristics as well as physical composition of the product itself—are client-driven and therefore categorized as customization [52]. Considering the moment at-purchase or after-sales the appropriate terminology for this linking is *product-service packaging*. This packaging can be either conducted internally or by collaborating with external firms [51]. Referring to the automotive industry a typical paradigm occurs while purchasing a car and complementing it by financial and maintenance services [52]. The final process is characterized by a dominance of the service among the product and named as *product-service-bundling* [51]. The telecommunications industry provides therefore a well-established example by selling manufactured mobile telephones in conjunction with their mainly service offering concerning the access to telecommunication [52, 53].

The final perception concerning these classification schemes of product-services is the fact that providing excellence in services requires decisive knowledge on clients operations and activities. Therefore these classifications enhance the identification of potential benefits carried by appropriate value propositions which in turn provide competitive advantages [3, 6, 34].

4.3 Options of Servitization

Conducting a servitization strategy there occur several options for its implementation and the resulting business models. In fact, this classification differs considerably from the antecedent one (cf. Subsection 4.2), since it explains the structure and the nature of the business accounting for projected revenues [4]. Tukker [54] considered product-service systems (PSS); he finally defined three fundamental classifications manifested by *product oriented* PSS, *use oriented* PSS and *result oriented*. Since the topic of servitization is broader than PSS [36] enlarged this approach to finally five classes adding the *integration oriented* and *service oriented* approach. In the following these five options are described and discussed in detail (cf. Kryvinska et al. [11]).

The concept of providing services by reason of vertical integration is very common in the manufacturing industry and finally described as (1) *integration oriented PSS*, which was added by [36] to the initial approach of [54]. Typical applications of integration oriented PSS are retail and distribution services according to the case of the oil industry. These companies not only extract, refine and produce gasoline but also provide a extensive infrastructure for distribution and retail [28, 34]. This option is characterized by the transfer of ownership regarding the tangible product [36]. In contrast to the first option the (2) *product oriented PSS* offers a service directly related to the considered product, which finally also results in a transfer of ownership amongst the involved parties [34]. Thus, some basic services like disposal or warranty [55] as well as advanced ones like development and implementation services are conducted [36]. The corresponding evolution of the second option by shifting the focus even more towards the augmented service, finally results in a (3) *service oriented PSS*, which was added by [36] as well. The incorporation of the service into the product implicates that there is no choice for the customer whether he deploys the additional service or not. Accordingly the offer finally transforms into an inseparable bundle of product and service [56]. The (4) *use oriented PSS* constitutes a considerable change regarding the notion of the originally manufacturing business, since the ownership of the tangible product retains by the service provider [36]. For this reason a completely new business model has to be established to meet the need of solely charging for access instead of ownership among the product [7, 8]. This evolving relation and the requiring new business model scheme are illustrated in Fig. 6. The final characteristic concerning the options of servitization is manifested by the (5) *result oriented PSS*, which represents the most intense occurrence in terms of applying a servitization strategy for a manufacturing company. This option implies that the product is irrevocably replaced by a substituting service. The most popular example for this alternative is the introduction of voicemail services to replace individual answering machines [36].

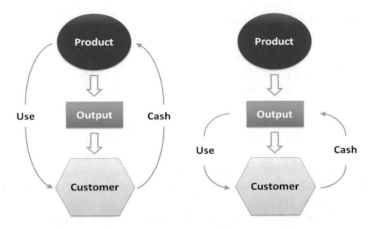

Fig. 6 Shift to access instead of ownership of products. *Source* own illustration according to [8]

Concluding it is important to highlight the differences between the initially introduced classification of ideal-typical servitization and the currently discussed options and implementation alternatives. Accordingly the classifications enhance the understanding about the nature of the service offered and the inherent requirements of the customer heading for potential benefits carried by appropriate value propositions to gain competitive advantage [3, 6, 34]. The options of servitization in contrast reflect on the fundamental idea about it referring to the extent of its application. Considering a manufacturing company applying an integration oriented servitization approach to gain competitive advantage, nevertheless remains mostly a manufacturing company. However, executing one of the other options of servitization pushes company's key business activities towards more service dominated ones. Whereas applying a result oriented servitization approach transforms the company ultimately to a service providing one. This fundamental idea and the corresponding transition of servitization are illustrated in the Fig. 7.

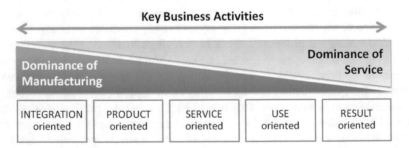

Fig. 7 Transition of servitization. *Source* own illustration

5 Summary

The concept of servitization is driven on the one hand by general environmental aspects like decreasing profit margins and revenues in manufacturing and equally internal drivers concerning financial, strategic and marketing aspects emerged on the other hand. Consequently higher margins on services tempt financial consideration whilst competitive advantage due to differentiation drives strategic aspects. Furthermore marketing issues are tackled since servitization pushes sales and tightens B2C-relationships resulting in a customer lock-in.

Although servitization exclusively deals with services, its origin and motivation is rooted in the manufacturing industry, since these services are augmented elements of tangible product offers. The aim is to co-create value with the customer in order to gain benefits by enriching its product offers with services. And, since the desired value co-creation is carried out by the underlying services, it is crucial for companies to know about theirs characteristics, modes and alternatives of implementation. Companies usually apply services like training, delivery, spare parts, repair and customer helpdesks, which in any case establish benefits for customers. Unfortunately these are considered in some industries as common ground, why they cannot provide any competitive advantage yet. In order to enhance the knowledge and understanding of customer's value perceptions one can classify services regarding the degree of customization in relation to its integration (cf. [1]). Servitization strategies can be consolidated to five options consisting of integration oriented, product oriented, service oriented, use oriented and finally result oriented product-service systems (PSS). Finally the options of servitization describe the transition from a pure manufacturing company to a service provider.

References

1. Kaczor S, Kryvinska N (2013) It is all about services-fundamentals, drivers, and business models. J Serv Sci Res 5:125–154. https://doi.org/10.1007/s12927-013-0004-y
2. Vandermerwe S, Rada J (1988) Servitization of business: adding value by adding services. Eur Manag J 6:314–324. https://doi.org/10.1016/0263-2373(88)90033-3
3. Baines TS, Lightfoot HW, Benedettini O, Kay JM (2009) The servitization of manufacturing: a review of literature and reflection on future challenges. J Manuf Technol Manag 20:547–567. https://doi.org/10.1108/17410380910960984
4. Neely A (2007) The servitization of manufacturing: an analysis of global trends. In: 14th European operations management association conference. Turkey Ankara, pp 1–10
5. Neely A, Benedetinni O, Visnjic I (2011) The servitization of manufacturing: further evidence
6. Oliva R, Kallenberg R (2003) Managing the transition from products to services. Int J Serv Ind Manag 14:160–172. https://doi.org/10.1108/09564230310474138
7. Pawar KS, Beltagui A, Riedel JCKH (2009) The PSO triangle: designing product, service and organisation to create value. Int J Oper Prod Manag 29:468–493. https://doi.org/10.1108/01443570910953595

8. Baines TS, Lightfoot HW, Evans S et al (2007) State-of-the-art in product-service systems. Proc Inst Mech Eng Part B J Eng Manuf 221:1543–1552. https://doi.org/10.1243/09544054JEM858
9. Goedkoop M (1999) Product service systems, ecological and economic basics
10. Vargo SL, Lusch RF (2004) Evolving to a new dominant logic for marketing. J Mark 68:1–17. https://doi.org/10.1509/jmkg.68.1.1.24036
11. Kryvinska N, Kaczor S, Strauss C, Gregus M (2014) Servitization strategies and product-service-systems. In: 2014 IEEE world congress on services, pp 254–260
12. Amit R, Zott C (2001) Value creation in E-business. Strateg Manag J 22:493–520. https://doi.org/10.1002/smj.187
13. Ng ICL, Nudurupati SS, Tasker P (2010) Value co-creation in the delivery of outcome-based contracts for business-to-business service. Advanced Institute of Management Research
14. Bastl M, Johnson M, Lightfoot H, Evans S (2012) Buyer-supplier relationships in a servitized environment: an examination with Cannon and Perreault's framework. Int J Oper Prod Manag 32:650–675. https://doi.org/10.1108/01443571211230916
15. Prahalad CK, Ramaswamy V (2004) Co-creating unique value with customers. Strategy Leadersh 32:4–9. https://doi.org/10.1108/10878570410699249
16. Parry G, Newnes L, Huang X (2011) Goods, products and services. In: Macintyre M, Parry G, Angelis J (eds) Service design and delivery. Springer, US, pp 19–29
17. Demirkan H, Spohrer JC, Krishna V (2011) The science of service systems. Springer, US
18. 27th ICAS Congress 2010: Paper ICAS 2010-7.7.1 Invited. http://www.icas.org/ICAS_ARCHIVE/ICAS2010/ABSTRACTS/823.HTM. Accessed 16 Nov 2017
19. Dohmen P, Kryvinska N, Strauss C (2012) "S-D Logic" business model—backward and contemporary perspectives. In: Snene M (ed) Exploring services science. Springer, Heidelberg, pp 140–154
20. Woodruff RB (1997) Customer value: the next source for competitive advantage. J Acad Mark Sci 25:139. https://doi.org/10.1007/BF02894350
21. Fang E (Er) (2008) Customer participation and the trade-off between new product innovativeness and speed to market. J Mark 72:90–104. https://doi.org/10.1509/jmkg.72.4.90
22. Ordanini A, Pasini P (2008) Service co-production and value co-creation: the case for a service-oriented architecture (SOA). Eur Manag J 26:289–297. https://doi.org/10.1016/j.emj.2008.04.005
23. Kryvinska N, Auer L, Strauss C (2009) The place and value of SOA in building 2.0-Generation Enterprise Unified vs. Ubiquitous communication and collaboration platform. In: IEEE, pp 305–310
24. Polaschek M, Zeppelzauer W, Kryvinska N, Strauss C (2012) Enterprise 2.0 integrated communication and collaboration platform: a conceptual viewpoint. In: IEEE, pp 1221–1226
25. Kryvinska N, Auer L, Strauss C (2011) An approach to extract the business value from SOA services. In: Snene M, Ralyté J, Morin J-H (eds) Exploring services science. Springer, Heidelberg, pp 42–52
26. Durugbo C, Bankole OO, Erkoyuncu JA et al (2010) Product-service systems across industry sectors: future research needs and challenges
27. Gebauer H, Fleisch E, Friedli T (2005) Overcoming the service paradox in manufacturing companies. Eur Manag J 23:14–26. https://doi.org/10.1016/j.emj.2004.12.006
28. Baines T, Lightfoot H, Peppard J et al (2009) Towards an operations strategy for product-centric servitization. Int J Oper Prod Manag 29:494–519. https://doi.org/10.1108/01443570910953603
29. Wölfl A (2005) The service economy in OECD Countries. https://doi.org/10.1787/212257000720
30. Sun HB, Mo R, Chang ZY (2009) Study on product service oriented enterprise servitization methods. Mater Sci Forum 626–627:747–752. https://doi.org/10.4028/www.scientific.net/MSF.626-627.747

31. Hypko P, Tilebein M, Gleich R (2010) Benefits and uncertainties of performance-based contracting in manufacturing industries: an agency theory perspective. J Serv Manag 21:460–489. https://doi.org/10.1108/09564231011066114
32. Falk M, Peng F (2013) The increasing service intensity of European manufacturing. Serv Ind J 33:1686–1706. https://doi.org/10.1080/02642069.2011.639872
33. Mathieu V (2001) Service strategies within the manufacturing sector: benefits, costs and partnership. Int J Serv Ind Manag 12:451–475. https://doi.org/10.1108/EUM0000000006093
34. Tether B, Bascavusoglu-Moreau E (2011) Servitization-the extent of and motivation for service provision amongst UK based manufacturers. AIM Res Work Pap Ser
35. Stoshikj M, Kryvinska N, Strauss C (2014) Efficient managing of complex programs with project management services. Glob J Flex Syst Manag 15:25–38. https://doi.org/10.1007/s40171-013-0051-8
36. Neely A (2008) Exploring the financial consequences of the servitization of manufacturing. Oper Manag Res 1:103–118. https://doi.org/10.1007/s12063-009-0015-5
37. Kumar R, Markeset T (2007) Development of performance-based service strategies for the oil and gas industry: a case study. J Bus Ind Mark 22:272–280. https://doi.org/10.1108/08858620710754531
38. Velamuri VK, Neyer A-K, Möslein KM (2011) Hybrid value creation: a systematic review of an evolving research area. J Für Betriebswirtschaft 61:3–35. https://doi.org/10.1007/s11301-011-0070-5
39. Schmenner RW (2009) Manufacturing, service, and their integration: some history and theory. Int J Oper Prod Manag 29:431–443. https://doi.org/10.1108/01443570910953577
40. Kryvinska N, Kaczor S, Strauss C, Greguš M (2014) Servitization—its raise through information and communication technologies. Exploring services science. Springer, Cham, pp 72–81
41. Dohmen P, Kryvinska N, Strauss C (2014) Viable service business models towards inter-cooperative strategies: conceptual evolutionary considerations. In: Xhafa F, Bessis N (eds) Inter-cooperative collective intelligence: techniques and applications. Springer, Heidelberg, pp 273–290
42. Belvedere V, Grando A, Bielli P (2011) The impact of information and communication technologies on operating processes—evidence from a survey. Research Paper | Bocconi University and SDA Bocconi School of Management. Google Search. https://www.google.at/search?q=Belvedere%2C+V.%2C+Grando%2C+A.%2C+%26+Bielli%2C+P.+%282011%29.+The+Impact+of+Information+and+Communication+Technologies+on+operating+Processes+-+Evidence+from+a+Survey.+Research+Paper+%7C+Bocconi+University+and+SDA+Bocconi+School+of+Management+.&ie=utf-8&oe=utf-8&client=firefox-b&gfe_rd=cr&dcr=0&ei=f28MWrOPHO3JXvvKmrAM. Accessed 15 Nov 2017
43. Belvedere V, Grando A, Bielli P (2013) A quantitative investigation of the role of information and communication technologies in the implementation of a product-service system. Int J Prod Res 51:410–426. https://doi.org/10.1080/00207543.2011.648278
44. Godlevskaja O, van Iwaarden J, van der Wiele T (2011) Moving from product-based to service-based business strategies: services categorisation schemes for the automotive industry. Int J Qual Reliab Manag 28:62–94. https://doi.org/10.1108/02656711111097553
45. Kryvinska N (2012) Building consistent formal specification for the service enterprise agility foundation. J Serv Sci Res 4:235–269. https://doi.org/10.1007/s12927-012-0010-5
46. Benedettini O, Clegg B, Kafouros M, Neely A (2010) The ten myths of manufacturing: what does the future hold for UK manufacturing? https://research.aston.ac.uk/portal/en/researchoutput/the-ten-myths-of-manufacturing(3eeca435-0aff-49b4-8e03-9d24b2359780).html. Accessed 15 Nov 2017
47. Ng ICL, Ding X (2010) Outcome-based contract performance and value co-production in B2B maintenance and repair service
48. Visintin F (2012) Providing integrated solutions in the professional printing industry: the case of Océ. Comput Ind 63:379–388. https://doi.org/10.1016/j.compind.2012.02.010

49. (2011) McKinsey study: solution selling is the pain worth the gain? In: Insight Demand Ltd. http://insightdemand.com/mckinsey-study/. Accessed 15 Nov 2017
50. Mathieu V (2001) Product services: from a service supporting the product to a service supporting the client. J Bus Ind Mark 16:39–61. https://doi.org/10.1108/08858620110364873
51. Almeida LF, Miguel PAC Product related services and the product development process—a preliminary analysis and research project outline, 15
52. Marceau J, Marceau PJ (2002) Australian Expert Group in Industry Studies Corresponding author
53. Becker A, Mladenowa A, Kryvinska N, Strauss C (2012) Evolving taxonomy of business models for mobile service delivery platform. Procedia Comput Sci 10:650–657. https://doi.org/10.1016/j.procs.2012.06.083
54. Tukker A (2004) Eight types of product–service system: eight ways to sustainability? Experiences from SusProNet. Bus Strategy Environ 13:246–260. https://doi.org/10.1002/bse.414
55. Cohen MA, Agrawal N, Agrawal V (2006) Winning in the aftermarket. In: Harvard Business Review https://hbr.org/2006/05/winning-in-the-aftermarket. Accessed 15 Nov 2017
56. Brax SA, Jonsson K (2009) Developing integrated solution offerings for remote diagnostics: a comparative case study of two manufacturers. Int J Oper Prod Manag 29:539–560. https://doi.org/10.1108/01443570910953621

The Future of Industrial Supply Chains: A Heterarchic System Architecture for Digital Manufacturing?

Corinna Engelhardt-Nowitzki and Erich Markl

Abstract Economic value networks are occasionally described as heterarchic systems. Hence, the application of supply chain management (SCM) practices means to be subjected to the fundamental principles of such systems—whether knowing and managing them consciously or not. However, this kind of knowledge has predominantly remained abstract and far from practical application. Besides, successful SCM relies on a fast and flexible information flow between the involved parties. Modern IT extends previous, at most hierarchical data structures and software systems by means of ubiquitous random access facilities. Comparable to the heterarchically negotiated structure of supply chain processes, these information systems are increasingly able to process data in reticulate structures, eventually even based on software agents with negotiation skills. This chapter characterizes basic principles of heterarchic systems. Accordingly, heterarchic, network-like IT-approaches are shortly discussed regarding the application in an SCM-context. Subsequently business implications for practical application are deduced regarding two questions: (1) 'What is heterarchy and why does it matter in SCM?' and (2) 'Could hierarchical layer models, as frequently used in SCM, informatics and automation, still serve for the purpose of achieving a holistic model of heterarchic systems in the context of digital manufacturing?'.

1 Introduction

Economic value networks are reported to be *complex, dynamic* and *adaptive heterarchic systems*, insofar as multiple heterogeneous and mostly legally independent companies (the actors or participants of a network) are tied together in highly interconnected and geographically wide-spread value creation processes

C. Engelhardt-Nowitzki (✉) · E. Markl
University of Applied Sciences Technikum Wien, Vienna, Austria
e-mail: corinna.engelhardt@technikum-wien.at

E. Markl
e-mail: erich.markl@technikum-wien.at

© Springer International Publishing AG, part of Springer Nature 2019
N. Kryvinska and M. Greguš (eds.), *Data-Centric Business and Applications*,
Lecture Notes on Data Engineering and Communications Technologies 20,
https://doi.org/10.1007/978-3-319-94117-2_10

[1–3]. However, the term 'heterarchic' is used diffusely and heterogeneously since years [4]. The concept of heterarchy was first introduced by [5], who disagreed to the previous concept that human brains are structured in a strictly hierarchical order, only being able to process in a sequential logic. Instead, he introduced the concept of a network-like nervous system, that enables parallel task processing. This concept was applied analogously in the field of economics and management, e.g., by Probst [6], who proposed the idea of fluctuating hierarchical relationships, that could adopt their structure according to situational requirements. In this context, the discussion of hierarchic versus heterarchic managerial structures needs to differentiate between a cross-company value network perspective (no central source of influence, thus only few hierarchical characteristics, unless the respective supply chain incorporates, e.g., a monopolistic supplier or customer) and a company-internal perspective (here, the extent of heterarchic attributes differs, depending on the implemented organizational structure).

Purely hierarchical systems have a unique root point that is the sole source of decision power, e.g., the owner or CEO of a company. All system elements, e.g., all company employees, are linked to this superior element, e.g. the respective CEO, as a direct subordinate or indirectly via several hierarchical levels—in this example, the organizational structure of the company. In a hierarchic organisation leadership is based on the managerial power and activeness of this superior role. In contrast to this, heterarchic structures rely on several stakeholders with distributed power who co-operate with each other in manifold ways. In this case, leadership also includes negotiation and decentral autonomy. Instead of a single source of power, decisions could be taken by either one or another organisational unit, or even within a joint team—depending on the problem to be solved in each case and the expertise required therefore. This is practically implemented by means of predefined rules and responsibilities. For example, the sales force of a company might decide on the question whether a certain customer is considered to be strategically important or not. However, the production units might take utilization decisions. Still a central management board or CEO could be the single point of power, having delegated this power to subordinated units. The resulting organisational constellation could be a hybrid between heterarchy and hierarchy, or could even be close to a hierarchical system. This depends on the question to what extent the principal would still be the central source of power behind operationally decentralized (i.e., delegated) decision responsibilities. The less influence the central principal has got over the involved decentral units, the more the organisational structure could be regarded to be a heterarchy. In the absence of a central and singular source of power, centralized (more or less authoritarian) decisions would be substituted by other decision mechanisms, e.g., negotiation, voting or team consensus. According to situational requirements, varying organisational units would co-operate with each other. Respectively, some authors assume the necessity of an overall hierarchical structure that could be supplemented with heterarchic structures on demand, in case a problem would require a high degree of flexibility and local autonomy [7]. This hybrid organizational constellation is called a '*heterarchic hierarchy*' [8].

Summarizing, the balance between both poles—pure hierarchy or pure heterarchy will depend on the actual distribution of property rights [9]. Practically spoken, an organizational structure with a high degree of decentralization, self-monitoring and autonomy can be assumed to be a predominantly heterarchic system [10]. Accordingly, the acting agents will be responsible for all activities and decisions, and will co-ordinate the major share of their information management and decision taking on their own account. Comparable to the heterarchically negotiated structure of supply chain processes, also the underlying information systems are increasingly able to process data in reticulate structures. This can, for example, be based on software agents with negotiation skills. Agent-based models could, for example, be used for the purpose of selecting supply chain partners, in particular enabling the analysis of agent behaviours throughout the bidding process [11].

From a *value network perspective*, in particular the availability of capable cross-company influence mechanisms has gained importance, as company success increasingly depends on suppliers, service providers and customers, as well as on other company-external parties, e.g. public authorities. This value network consists of several legally independent companies and can be considered to be a heterarchic structure by definition, as there is no central "supply chain authority". In this regard, SCM is not fully comparable to the management board or CEO of a company. Instead, companies are negotiating service and goods deliveries among each other, being determined by fair-scattered legal, political, cultural and social, technical, environmental or market-related influence factors.

Many companies have concentrated on their specific core capabilities, having externalized tasks with lower relevance or far distant from their core competences by means of *outsourcing* practices since many years [12]. As a consequence, the size and the boundary of singular companies have changed [13] towards more and smaller companies that are linked to each other within a specific supply network section by means of delivery contracts. Also, the overall network-wide proportion of tasks being coordinated hierarchically (inside a company) and tasks being coordinated through negotiation and contracting (company internally, or across company borders, at most within dyadic seller-buyer relationships) has shifted towards fewer make-decisions (i.e., hierarchical coordination) to the favour of more buy-decisions (i.e., market-based coordination; cp. in a supply chain context [14]. Furthermore, also the adequate amount of hierarchical coordination is a matter of discussion in company-internal organizational concepts, as more autonomous concepts like e.g. self-directed work teams, seem to handle a complex business better [15]. Altogether, it can be assumed that the *extent of a heterarchic distribution of power has increased* in current business environments.

In line with this assumption, the present chapter intends to contribute answers to the questions of (1) 'What is heterarchy and why does it matter in SCM?' and (2) 'Could hierarchical layer models, as frequently used in SCM, informatics and in automation, still serve for the purpose of achieving a holistic model of heterarchic systems in the context of digital manufacturing?'. The remainder of the chapter is as follows: Sect. 2 will briefly define SCM from a value network perspective with regard to the characteristics and principles of heterarchic systems. Subsequently, Sect. 3 shortly explains how hierarchical layer models are applied on the SCM and business process level as well as on the underlying information management perspective. Interestingly, the SCM- and the business process layers are undergoing similar shifts towards higher decentral autonomy and increasingly heterarchic structures, as the IT- and automation layers. Accordingly, Sect. 4 discusses the role of service-oriented architectures (SOA) and modern IT techniques, with a specific focus on current smart manufacturing and digital production concepts in industrial environments. We conclude in Sect. 5 with possible managerial implications of these converging trends for supply chain management that is based on a digitized manufacturing environment.

2 Heterarchic Supply Chain Structures for Dynamic Market Conditions

In the course of designing, planning, controlling and optimizing material-, information and monetary flows through a value network, companies have to perform several tasks, in particular supply chain (SC)-design, SC-configuration, SC-planning, SC-optimization and SC-execution (operation) in a collaborative way [16, 17]. In the course of fulfilling these tasks, each involved company takes more or less autonomous managerial decisions. The degree of autonomy depends on e.g., the mutual market power and reciprocal dependency. Thus, a value network can be assumed to be a heterarchic system [18]. Depending on factual market power, these companies have a different degree of influence on the network. This relative strength is fluctuating over time (see Fig. 1: all value network companies inside the dotted, grey shaded area are regarded to have a relevant influence on a company; however, this relevance might change over time: as well as new companies might arise and existing companies might disappear from this area of relevant supply chain parties).

This network structure represents a heterarchic system in the following sense: every company arranges a comparable picture with regards to its business, positioning its own operations (or eventually even more than one business field) in the central position of the network topology in Fig. 1. This might as well happen unconsciously, in case a company doesn't explicitly draw this structure, but only implicitly builds respective contracts with customers, suppliers and service providers. The resulting constellation is can't be as easy described as the frequently drawn assumptions that "supply chains compete with supply chains" [20], as this would assume a supply chain to be an acting entity. This is typically not the case, as it would require a central authority or extensive negotiation processes with the joint target of 'optimizing the supply chain as a whole entity'. Contrariwise, companies would rather enforce their individual objectives. Still, companies with mutual alliances could factually establish 'supply chain sections' (e.g., consisting of two,

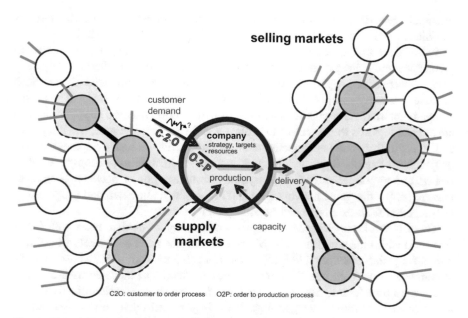

Fig. 1 Heterarchic supply chain structure [19]

three or sometimes more supply chain partners) via closely linked supplier-, customer-relationships or within collective companionship contracts. The closer the interdependent integration and managerial cohesion between them, the more the system will show a heterarchic behavior with syndetic supply chain sections (simplified supply chains).

Efficient and sustainable management requires the resolution of occurring conflicts in the face of limited resources and differing targets. However, the control and solution mechanisms that would support a manager to resolve these conflicts are rather different in hierarchical structures compared to heterarchic structures – no matter whether a certain conflict situation applies to a company-internal or to a cross-company setting or both. The more, a system can be considered to consist of comparably independent agents with strong self-interest, the more likely it is that decision rights and required information are distributed within the system. None of the acting agents can be sure to be completely aware of each other's intentions, objectives and operational action. As a consequence, authoritarian control ratio and related formal decision mechanisms (e.g., linear optimization algorithms within production planning) will fail to act. The same applies to situations characterized by information lacks—a rather usual state of things in highly dynamic and changing environments. As a consequence, managerial efficiency can be assumed to depend on mutual capabilities to exchange relevant information, to negotiate for working co-operation constellations and to resolve occurring conflicts. In doing so, one must be aware of the fact, that even the most efficient settings could be impaired through principal agent deficiencies (in particular hidden intentions, hidden characteristics and hidden action of agents, see for example [21] or more recent [22].

From the focus of each single company respective managerial methods, in particular supply chain management practices [23], have to be applied to adequately design, operate and optimize material-, information- and financial flows. Such practices usually imply integration and coordination issues. Thus, the company-internal value creation and the attached external material supply processes are aligned towards the objective to achieve a delivery performance, a selling price and a flexibility that is competitive in the respective industry segment and/or geographical region. Typical inhibitors are low predictability, incomplete information and opportunism [24, 25], asymmetric market power between companies that are part of the value network [26] and highly fluctuating demand progressions [17] together with extensive and volatile product variety [27].

There is evidence that the concept of heterarchy is neither been sufficiently discussed and understood in an SCM-context, nor has been adequately transferred into managerial practices [28, 29]. Unfortunately, concepts related to suchlike system theory oriented questions often are of abstract, sophisticated nature, which handicaps a transfer into practical usage [30]. In most cases no scientifically proven and practically feasible managerial guidance is provided regarding the question at how to achieve a well-working balance between opposing concepts, such as for instance central control versus decentralized autonomy [15, 31] or the exploration of innovations and new capabilities versus the efficient, reliable and standardized execution and improvement of existing processes [32] or reliable forecasting and planning versus a high instantaneous flexibility [33]. SCM typically describes such coherences as disparate dichotomies and—at best—provides segmentation criteria to differentiate adequate managerial policies.

Logistics and SCM theory have provided several conceptualizations regarding the nature of value networks and their attributes [3, 34] or [35], and have also investigated topological questions. For example, Lambert et al. [36] or Gosling [37] have provided generic value network models, emanating from a focal company and have subsequently modelled attached supplier and customer structures. However, practical circumstances often require a more elaborate consideration: simplified models for example often neglect the fact that each participating company establishes its own subjective value network conceptualization. This regards the valuation of the reciprocal importance within dyadic buyer-seller relationships: For example, a purchaser can rate a certain supplier as low priority 'C-type' supplier, whereas the supplier might perceive this customer as important 'A-type' customer (and vice versa). Therefore, a value network can be considered to be an aggregation of mutually differing value network notions [19] that are, besides, shifting over time. Further, the particular capability of each participating company to influence its environment according to its market power will be different among the network and may again vary over time according to developing technologies, logistics capabilities, selling and purchasing volumes or supported through further monopolistic (unique) or oligopolistic (rare) market constellations.

Typical *performance targets*, a company has to meet under such in-transparent and permanently changing conditions through the application of SCM concepts and practices are short lead and reaction times, a competitive economic position, a distinct operational flexibility [38] and a proficient capability to adapt internal processes and related supply flows to structural changes [39]. Relevant challenges for SCM arise from the fact that the coordination of economic value networks is reliant on accurate and exhaustive information [40, 41]. However, in practice the lack of available information and the mutual capability to exert influence upon each other are causing huge constraints, compared to the ideal state.

A further difficulty derives from the fact that value networks are regarded to be *complex adaptive systems* (CAS, see [31]). Whereas the original development of SCM concepts and practices took place under rather stabile economic circumstances, actual markets are showing turbulences and unpredictability [17]. Therefore, existing SCM methods and tools have to be developed further towards a point where they can be applied easily and provide for a fast and repeated execution within fluctuating conditions, but without losing explanatory power and validity due to oversimplification [42].

'Traditional' *reductionist approaches* to handle erratic changes are typically seeking to smooth turbulences and to best possibly predict residual unavoidable fluctuations. Respective control means require accurate and timely information to increase value network transparency [17]. Corresponding SCM practices are for instance order smoothing [43] or co-operative approaches like collaborative planning concepts or vendor managed inventory agreements. This is however stated to be inopportune, if applied in volatile environments—especially if there are notable impacts on the focal company that originate from tier-2 to tier-n network participants.

Thus, *holistic approaches* related to CAS and system theory have recently gained more attention also in a SCM context since long [44]. Respective approaches typically assume that a complex dynamic system is characterized through extensive relations and interdependencies, non-dynamic coherences, self-organizing behavior patterns and the occurrence of emergence phenomena. All these attributes also apply to economic value networks. Hence, a main paradigm shift in SCM is that the system behavior can't be necessarily explained from the analysis of its singular parts, but emerges either due to unexpected external impacts or as an inherent part of the internal system behavior that was neither observable nor predictable from the partial analysis of its single constituents [15]. In addition to simple linear coherences also circular causalities play a major role [45]. This is well-known in systems theory, time and again in principle discussed in SCM related theory, but has scarcely been operationalized for practical SCM transfer [30, 17]. One evident example for this is *Ashby's law of requisite variety* [46]: At first, universal sight, Ashby's proposition is obvious. At second, also generic principles can be deduced for SCM, such as 'Develop an adequate product variety towards your target markets: not too high, to avoid excessive efforts and complexity, but not too low, to avoid lost sales in favor of competitors.' However, the third and operationalizing step has remained unsolved: management approaches to quantitatively determine and achieve the 'adequate complexity or variety' have not yet reached a sufficient maturity [19].

These coherences have major consequences for SCM that have not yet found a broad propagation in corresponding methods and practices: reductionist partial analyses provide useful means, but may not always be expected to deliver exact and stabile evidence over a long time-span. They rather provide snapshots that are eventually clear and precise but possibly diffuse and faulty. Although the common assumption that logistics capabilities can be classified into demand-management and supply-management capabilities, together with an emphasis on interface and information management proficiency [47] is valid in principle, this approach doesn't go far enough in complex turbulent environments: An additional body of *dynamic capabilities*, such as for instance the transfer and practical application of decentralized theoretical approaches (e.g., the long-proposed theory of loosely coupled systems (for a reconceptualization of this concept, refer to [48]) into the context of value networks is strongly required. Further, it is important to develop practical means that support the evaluation and handling of dichotomies in the sense of 'dynamic, fragile balances'—e.g. between control and emergence, though a main difficulty lies in the fact that a suchlike emergent nature is not reducible to the characteristics of the network elements and therefore can't be predicted ex ante, but only becomes explainable ex post [15].

Altogether the theoretical base of SCM is contradictory and has not yet been developed to a point where it provides sufficient explanatory power and decision support regarding the design and operation of value creation flows in complex, volatile and unpredictable environments. In practice the proportion of hierarchically coordinated areas in value networks have decreased as a consequence of outsourcing. Further, also the adequate amount of hierarchical coordination is a matter of discussion in company-internal organizational concepts, as more autonomous concepts like e.g. self-directed work teams [15], virtuality concepts [41] agent-based modelling approaches [11], big data concepts [49] or concepts that involve elements of artificial intelligence techniques [42], seem to have the potential to handle a complex and unpredictable business better than conventional SCM- approaches that depend on planning reliability. Obviously managerial means beyond and supplementing traditional hierarchical and reductionist means are needed. In this context it is a noticeable observation, that the relevant body of literature has not only emerged recently: Most theoretical concepts have been introduced years ago (cp. the literature cited in this chapter as a characteristic—though on no account exhaustive—overview). Comparably, the related managerial challenges have been uncovered long ago. Still, operating answers aren't developed up to a sufficiently mature state. Obviously, *'hierarchy' is a useful structural principle under stable and predictable conditions*. However, recent dynamic and unforeseeable business conditions can't be handled well by means of a purely hierarchical perspective. Hence, several authors (for an overview see for instance [50] have proposed frameworks that are based on heterarchic strategies and principles. These contributions claim decision making to be far less centralized and hierarchical, but rather autonomous and linter-linked across independent companies compared to past business models and organizational structures. With regard to question (1) it can be stated that heterarchic concepts definitely matter within supply chain management, once only the issue of operationalizing the abstract concept with regard to concrete applications and competitive business advantages.

3 Hierachical Layer Models in SCM, Information Management and Factory Automation

An interesting learning opportunity originates from the evolution of *informatics*: With rising computational processing capabilities and hence, increasingly manageable application complexity, last recently modelling capabilities for data models have matured from first, dedicated relations between simple files to second, hierarchical databases (advantage: fewer file redundancy), later network models (advantage: more complex and flexible relationships, higher degree of semantics) and currently multiple types of specified complex concepts (advantages: e.g., adaptive data-structures, ability to process data from heterogeneous data sources). In order to analyze this coherence further, the next section of this chapters shortly discusses respective layer models in SCM, informatics and automation together with the evolution towards more heterarchic models. Figure 2 provides a short overview over the respective development.

A typical way of modelling in informatics is the use of *hierarchical layer models* (e.g., database architectures, network protocols, e.g., [51]). Elsewise, SCM is using

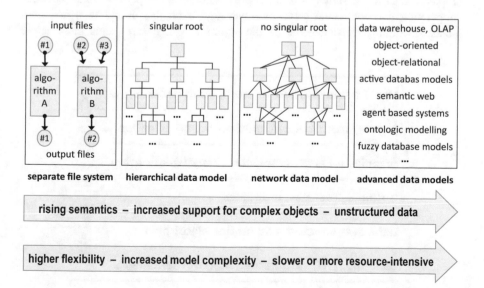

Fig. 2 Evolving data models within the field of informatics [51]

hierarchical modelling only occasionally. In previous contributions, we have proposed a generic hierarchical layer model with regard to enterprise and SCM (Fig. 3).

An important concept within suchlike layer models is the principle of *encapsulation*, which is however not fully applicable in a context that also claims to model human behaviours—not only deterministic computational coherences. Still, such models can serve well in order to reduce complexity, and can be applied as a descriptive architecture model for further formalization. There are obvious *similarities between data models and the organizational structures* with regard to the division of labour in companies and supply chains: a hierarchical topology allows for the rapid processing of well-standardized tasks. This is due to the singular centre of power and the respective uniform chain of commands in suchlike systems. Accordingly, also the communication means in use are unambiguous and well-defined. The disadvantage of hierarchical structures is the inherent difficulty to adapt, together with a low degree of flexibility. These characteristics can be identically observed on both layers—organizational structures and IT-structures. Contrarily, polycentric structures (such as heterarchic systems) are flexible and adaptive towards changes. They allow for fragmented requirement progression with higher performance. Besides, they show advanced capabilities to handle ill-structured problems. On the other hand, theses heterarchic systems are slower and less reliable when it comes to the execution of well-structured and properly defined standard tasks. For example, successful concepts on both layers, IT and SCM, seek to address opportunism problems through reciprocity and both have the inherent ability to solve "flash crowds" through distributed resource allocation [52]; In order to illustrate this assumption for an IT-example, one could refer to e.g., IT-based swarm algorithms. On a supply chain level, respective value network cooperation concepts could serve as illustrating examples: Here, local improvements are often disadvantageous at the superior system level, thus in principle

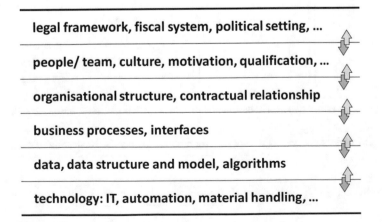

Fig. 3 Evolving data models within the field of informatics [51]

preventing their usage. However, in practice, the additional issues of human decision biases and distributed decision power have to be taken into account. This leads to the effect, that not always the most advisable decision is actually taken and consequently implemented.

Comparable to SCM, also within IT-based networks not only the network size, but primarily the proportion and density of relationships have risen. This goes hand in hand with an increased need for *dynamically configurable virtual networks* in the course of efficient resource sharing and a focus shift from resource sharing only to content transfer.

The same applies for *automation concepts*, that also have strictly applied hierarchical layer models in the past ([53]; for a more detailed overview see, for example, the contributions in [54]). Similar as SCM and informatics, also factory automation and industrial robotics have to handle increasing uncertainty and complexity. In this context, in particular industrial machinery and respective control devices (either industrial computers or programmable logic controllers—PLCs) have to be operated and optimized under these changed conditions.

The lowest layer includes equipment components (e.g., a pump, a valve, a drive, a measuring device, ...), sensors (either assembled to other components or independent) and actuators (e.g., claws, a drill spindle, a suction device, ...). On this field level (also device level), in particular the signal flow between these devices and the transformation of physical quantities into data signals and vice versa the triggering of (end-) effectors via respective data signals has to be managed.

The proximate control layer makes use of sensor information (having originated from the underlying device level) for the purpose of process control and generates control commands to be send to the particular devices (actuators) via the subordinate device level.

On plant level, comprehensive tasks like e.g., resource allocation, order scheduling or lot-size optimization are executed. Depending on respective optimization targets, the plant level tasks will influence process control tasks. For example, a robot welder might bond the work pieces at issue in a sequence that uses optimal setup times according to the instructions of a production planning system (plant level), in more detail instructed through a PLC (control level), and technically determined through signal flows from and to the affected sensors and effectors (device level) (Fig. 4).

Indeed, the hierarchical approach behind the automation pyramid enables the application of heterogeneous technologies within an automation system, e.g., a production plant. However, this approach increasingly fails to act in vague and ambiguous situations.

Currently, due to an increased decentralization of automation and robotics applications the decisional capabilities of automation systems are more and more transferred towards *distributed components*. Besides, the importance of vertical integration becomes a critical bottleneck, as risen business process and supply chain complexity are demanding a more efficient computational and automation support with less necessity for manual interventions. Especially, the amount and the heterogeneity of information, to be processed in this regard, have reached an

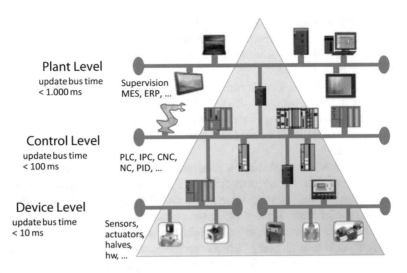

Fig. 4 Automation pyramid, modified from [53]

immense extent. Furthermore, in the absence of commonly agreed standards, heterogeneous devices (different control device manufactures, several technology generations and maturity levels, manifold operating systems and software applications, diverse country standards etc.) have to be integrated into a well-integrated system. As a consequence, different layer models from several disciplines—informatics, automation, SCM (and most probably further subject areas)—have to be integrated *interdisciplinary* beyond previous system borders in order to achieve an acceptable adaptability (or technically spoken re-configurability), flexibility and maintainability.

Moreover, current industrial systems need to be *resilient* on all aforementioned layers: As more and more unforeseen events might impair the system surprisingly, as well technical systems (informatics and automation) as organizational structures (company-internal and related to the cross-company supply chain context) have to be able to rapidly recover from internal failures and external disasters by means of being able to return to the previous, well-performing state or to achieve a new scenario, that is viable under mutated conditions [55]. This, however, requires network-like structures.

The *development from purely hierarchical architectures toward a network topology* has been repeated in several disciplines. The first to be mentioned in this context is the evolution of database models—at first hierarchical, later with (still hierarchically normalized) relational models, later evolving towards multi-dimensional architectures with enhanced distributed architectures like for instance online-analytical processing techniques (OLAP, e.g., [56]). However even in a distributed OLAP-datawarehouse with highly sophisticated locking and recovery procedures, the basic idea still relies on a hierarchical model. Second, the architectural design of IT-networks and respective communication protocols show a comparable progress that has started with the long-standing hierarchical OSI ref-erence model [57], and has developed towards more recent standards (in particular ISA88-95, NAMUR, TC-184 and ISO; see e.g., [58]). A first decentralisation step was reach in this field with the introduction of client-server architectures, and step by step advanced capabilities at the client-side. Also multiprocessor-systems have provided the capability of less hierarchical information processing in principle. However, these systems are even today widely operated by means of hierarchically control structures and software.

Only slowly, decentralized systems came into broader applications due to advancements in software engineering (e.g., neural network computing or agent-based systems, cp. e.g., [59]) and the increasing need to solve problems that couldn't be processed easily with common hierarchical systems. However, it must be stated, that these systems generally rely on a hierarchical core structure, when analysing it in the details: As well an x-based software relies on an operating system that is installed on the executing computational unit, and works with hierarchical principles. Moreover, recent simulation studies with regard to Industry 4.0 automation concepts have indicated that within the current state-of-the-art hybrid systems (i.e., a combination of heterarchy and hierarchy) still outperform fully decentralized systems with regard to production planning [60]. Obviously, today even a definitive heterarchy in the end still requires hierarchic roots—at least with regard to the current development stage of Industry 4.0 (also referred to as smart manufacturing, smart automation, digital factory and similar terms). Further research has to be done in order to better clarify this issue.

Last recently, concepts like the *Internet of things (IoT)* and in German-speaking countries industrial production paradigms like *Industry 4.0* show similar tendencies, although it is not yet fully conceivable, towards what technical standards, and to the favour of what novel architectures. The underlying assumption is, that a changed focus away from simple electric signal processing towards web-based semantic services is to be expected [61]. However, apart from the World Wide Web there is no such thing as one unique technology that drives the break-through in digitized automation systems, but rather a bundle of established and novel technologic options that have to be applied in a unique configuration according to situational conditions and requirements [62].

Without any doubt, *decentralization* is a major trend and is currently imple-mented in manifold ways, in particular through the use of *cyber-physical systems* (CPS, enabling real-time data exchange between relevant entities of the automation system [63]. Comparably, pervasive computing concepts are driving the number of (eventually autonomous) network nodes towards an immense amount of 'smart' devices. As these decentral entities are able to exchange data and to execute at least simple IT-processing tasks within a network architecture [64]. However, as long as there still remains a central source of authority—here a central server—the true paradigm change towards a heterarchy has not yet fully taken place. When reviewing decentralization concepts more closely, the concept is coupled with heterarchy only on a very loose base: decentralized real-time data processing also provides advantages for fully hierarchical and mixed ('hybrid') concepts, for example, when using conventional linear optimization techniques for the purpose of production planning [60].

At the same time many authors are assuming the resolution of purely hierarchical factory automation concepts, in particular at the middle layers of the automation pyramid: previously PLCs and SCADA systems have integrated superior and subordinated layers according to strictly hierarchical algorithms and access modes. Currently, newly emerging technical 'quasi-standards" like OPC/UA (unified architecture) allow for a more randomized access that may even omit singular automation layers and that at the same time enable higher semantic richness [65]. For example, an industrial robot could be directly controlled without using a PLC and respective middle-layer software.

This raises the question whether the observed trends are essentially contradictory. Revisiting the concept of layer models, this is not necessarily the case: referring to Fig. 1, for example the process level (e.g., looking at production planning and scheduling processes) could be controlled mostly hierarchically, as long as the planning premises are suitable. At the same time, a different layer could be organized more—or even fully—heterarchically, for instance, the communication between physical production devices (machinery, tools, sensors, transportation vehicles, etc.).

As a first tentative assumption that, however, would have to be investigated and validated within further theory-oriented and empiric research, we assume that hierarchical layer models might still have a profound eligibility for the following reasons: to begin with, they could support decision makers to conveniently segment between system partitions that are structured according to an in each case different extent of hierarchy (and respectively heterarchy). Moreover, the general advantage of hierarchical layer models, to partition a complex problem into smaller portions, thus reducing complexity and ensuring interoperability through consequent encapsulation, would remain a beneficial concept also in future systems that have an overall high degree of heterarchic coordination.

Furthermore, service-oriented architecture concepts (SOA) are becoming increasingly common in manifold environments, being applied as an integration approach that advances conventional industrial automation [66–68]. Among other fields of application, SOA-concepts are also applied in a supply chain context, intending to achieve agile reconfigurable process chains between companies [58, 68]. Hence, a company is enabled to react more rapidly and efficiently to occurring incidents, no matter whether they have a technical, ecological, political or economic root cause. Herewith, SOA and related services take advantages of an advanced capability to handle complex and heterogeneous systems. Basically, these systems have grown over years, and caused different technological and economic problems. The main idea is to connect the operative layer of real-time shop-floor tasks with high level services on company level in order to support process control and decision taking (Fig. 5).

Indeed, the hierarchical approach behind the automation pyramid enables the application of heterogeneous technologies within an automation system, e.g., a production plant. However, this approach increasingly fails to act in vague and

Fig. 5 A framework for the
service enterprise building
blocks interoperability [68]

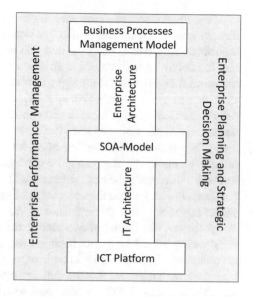

ambiguous situations. Frameworks like proposed in [68] allow for a consistent integration of at the one hand supply chain and business processes with at the other hand IT- and automation devices within the so-called 'service enterprise' in the 'service society' [69]. Moreover, also co-operation between value network partners can be facilitated by means of highly advanced services. This advanced services, in turn enable the evolution of heterarchic structures as well company-internal as across company borders.

The main idea behind SOA is the definition of all tasks, functions and business transactions as independent services. It is most important to optimize the condition-monitoring and the control of the underlying manufacturing devices, i.e., the physical manufacturing activities, in the course of for instance a production process, in order to support the required supply chain agility (and as well other supply chain objectives) best possibly. Here, heterarchic architectures offer promising improvement potentials by means of more autonomous and 'intelligent' devices with local decision capabilities [58]. Accordingly, critical success factors are not only the capabilities of the used physical devices (tools, sensors, actuators, measuring instruments, etc.), but also a proficient IT-integration and communication. In the current dynamic market, companies have to be able to collect and analyse unstructured information that is derived from widely distributed channels (i.e., multi-channel interaction with a high amount of customers who generate huge data volumes, see [70]). Subsequently, adequate conclusions, managerial decisions and activities have to be deviated across all layers—from the process level down to the field level. The underlying architecture either could be strictly hierarchical, as currently implemented in the majority of existing enterprise architectures on process-level, IT-level and automation level.

For example, an agile customer-to-order process could require the capability to directly impact a technical device that configures a 3D-printed part: The customer uploads a respective CAD- or printer-file or is allowed to choose from a database, that is offered by the company. This is a feasible business concept also within a hierarchical layer model, but might involve many abstraction layers, and hence be slow due to the high calculation efforts, when re-calculation through all layers strictly hierarchically. A more heterarchic architecture could skip layers that are not needed (because not influenced) for this operations. This however needs a well-substantiated understanding of respective technology impacts. For the (simplified) example of the customer-configured 3D-printed part production, for instance a different geometric structure will, e.g., influence cycle time and material consumption, and hence the order scheduling on super-ordinated layers; a different colour might not have this influence at all.

Altogether, developing theses architectures towards a more heterarchic structure is appealing on all layers as soon, as the existing, strictly hierarchical structure becomes too rigid to handle complexity and fuzziness. However, this requires a deep knowledge among the modelling experts as well regarding the users perspective (process level) as the concerned data structure(s), IT-infrastructure, automation and manufacturing technologies, including adjacent fields of knowledge like material sciences. Regarding question (1), we accordingly assume that heterarchic concepts have a high potential for the improvement of SCM concepts, but have to be developed with suitable cautiousness and diligence. Proceeding, it does not seem to be likely that heterarchy is going to completely substitute hierarchy according to the current state of the art knowledge. This means—concerning question (2)—that hierarchical layer models, as frequently used in SCM, informatics and automation, will still serve for the purpose of achieving a holistic model of heterarchic systems, at least in parts of the system or within highly standardized and predictable applications that don't require advanced capabilities.

4 The Role of Modern IT for the Management of Heterarchic Supply Chain Structures

Section 4 discusses the role of modern IT techniques, with a specific focus on current smart manufacturing and digital production concepts in industrial environments. As the previous sections have shown, interestingly, the SCM- and the business process layers are undergoing similar shifts towards higher decentral autonomy and increasingly heterarchic structures, as the IT- and automation layers.

Shroff [71] has proposed six elements that constitute the 'intelligent web': looking (looking 'at' information in terms of sensing and looking 'for' information in the sense of searching, especially within the web), listening and learning (i.e., retrieving the most relevant insights from this collective, digitized knowledge), connecting (these information fragments), predicting (what is likely to happen next in the relevant context) and correcting (in the sense of improving). Hence, we perceive major changes in the manner of data access—starting with humans who increasingly navigate through hyperlink-structures instead of sequentially and hierarchically organized information to the point of automatic data access on all levels and also between all levels of the automation pyramid. While stable situations (e.g. manufacturing objectives and conditions) can be handled satisfyingly with information systems that rely on a single entry point, dynamic situations with a huge amount of urgent decentral changes would overstrain conventional systems [60]. This leads to the following conclusion: theoretically spoken, the required capabilities (e.g., following the idea of looking, listening, learning, connecting, predicting and correcting, but also in the course of other capability taxonomies) can be designed independently from the underlying modelling principle—be it hierarchic or heterarchic, or be it a combination (a 'hybrid') of both. However, trying to take advantage from advanced technological possibilities, requires the in each use case applicable modelling approach.

Within digital manufacturing, respective considerations typically rest on the conventional automation layer concept [53]. At the device level, more and more perception points occur, due to the increasing types and amounts of sensors. Additionally, the growing extent of interconnectedness allows for more extensive access not even on machine level, but as well on the level of single components or tools, that are only a part of an industrial machine or a robot. Similar to the people layer, not only humans start to look for pieces of information in a hyper-link structure; the same applies to people and also to software systems that are looking for information about the actual condition inside a factory, or even inside a machine (technical or workload condition). At the same time the advancing development of standards (e.g., OPC/UA) allows for cross-level access, and hence more sophisticated logical interferences.

Shroff [70] mentions connectivity—as he calls it 'the ability to connect the dots'—as a next maturity step. In the context of human cognition this refers to logical reasoning; in the context of industrial manufacturing machinery, this relates to automated logic reasoning mechanisms. Again, such algorithms could be used on all aforementioned levels—be it the 'smart' combination of sensor data that has been retrieved from several sensor devices (i.e. intelligent sensor fusion within cyber-physical manufacturing, e.g. [72]) or be it the application of automated reasoning software applications for the purpose of appraising supplier performance [73]. Not only humans, but also connected devices on all levels—as well a singular sensor, a machine or a machine component, as a PLC-device or a MES-system (and so forth) are enabled to access an immense variety and volume of data via internet access. In a manufacturing context this can be limited to factory-internal devices only, but can also be extended to external internet resources.

As searching for relevant and meaningful data is a major challenge [70], digital manufacturing concepts like Industry 4.0 and similar seek to exploit the evolving technological options bet possibly. Accordingly, hierarchic architecture models come under criticism in some ways, despite the obvious advantage of unique control and accountability due to a single point of power and responsibility. On company- or plant-level, the alternative to hierarchic co-ordination has been discussed since long in the context of make-or-buy decisions, and so-called hybrid co-ordination forms, e.g., a joint venture or a franchising system [14, 21, 22, 25].

Nowadays, powerful IT-infrastructure and data processing algorithms have stimulated the same discussion analogously also at lower technology levels, down to the device-layer: comparable to human agents that execute managerial tasks on behalves of a superordinate principal [24], software agents could make use of the advanced data access capabilities and take decentral decision 'on behalves of' a smart production system. Comparably to a supply chain or enterprise scenario, where more than one participant owns a certain extent of influence which allows for at least some autonomous decisions based on respective organizational roles or contracts, also technical devices are enabled to de-centrally influence the manufacturing system to the favour of better decision quality, faster decision taking, higher flexibility and reduced transaction cost. De-central decision taking requires knowledge and the ability to judge. As soon, as technical systems are able to judge, the same structural principles can be used, as on the organisational layer. One doesn't have to wait for the IoT or the internet of services, in order to explore this assumption: as well a simple control loop (e.g., for temperature control) without any learning abilities, and with few sensors (often only one temperature sensing device) is a first example for de-central power that alleviates superordinate layers—here humans that do not need to control the thermal conditions of a e.g., a building manually.

A further factor of immense importance for the discussed progression is the availability and processability of huge amounts of SCM- and manufacturing-related data. This includes as well company-internal from all automation pyramid layers, as company-external data from customers, suppliers or public sources. Whereas well-structured data can be processed comparably easily with conventional algorithms (given that a sufficient processing capacity is available), huge challenges have to be seen in the processing of ill-structured data from heterogeneous sources, which is assumed to be the disproportionally higher share of relevant data in a SCM context [49]. Big data processing and data-driven strategies are a major success factor in digital SCM and manufacturing. According to the mentioned six elements of looking, listening, learning, connecting, predicting and correcting, a significant enabler is the achievement of richer semantics out of collected data sets. Again, hierarchic layer models play a major role in the course of systemizing this advanced data analysis and processing approaches.

Though being hierarchically structured at first sight, a more detailed look onto the framework in Fig. 6, shows the high variety of data sources, data entities as well as the highly diversified approaches for data acquisition, processing, visualization, transmission and finally application. All these steps across the lifecycle of manufacturing data implicate manifold biases and inconsistencies [74] and illustrate both —the promising opportunities and at the same time the urgent need for advanced data modelling capabilities. Typical application objectives are assigned to the plant- and process-level, e.g., production planning, quality tracing, manufacturing process optimization and material replenishment. Remarkably, the corresponding

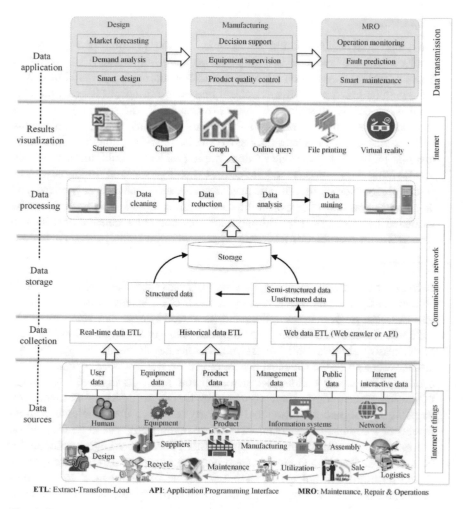

Fig. 6 Systematizing manufacturing data, cp. [74]

implementation procedures are linked to the subjacent technological layers. This requires a holistically integrated vertical communication between all layers. Purely hierarchic architectures can be assumed to be unable to accomplish this task. Altogether, these considerations lead to the conclusion that SCM challenges have to be analysed from a horizontal perspective on the process-level, whereas from a vertical perspective on subordinate control- and device-levels.

Wang [75] draws similar conclusions and claims the need to combine decentralized authority with a concentrated business focus as key driver for the evolution of hierarchy towards heterarchy. Based on a case-study, Wang accordingly recommends the following propositions to design an effective organisation in a complex and dynamic environment:

- *Proposition 1*

  First, decisions taking should be assigned to the position where they are resolved with the best available quality. The more 'intelligent (i.e., capable of decision taking) a system on a lower level of the automation pyramid becomes, the less important the strictly hierarchical line of commands, as the subordinated devices can decide autonomously instead of 'having to ask' superordinate layers for appraisal and permission. Does that at the same time mean, that hierarchy will get obsolete with increasingly smart devices? We assume the opposite, as this first proposition also suggests that simple problems would not unavoidably require this high amount of de-central intelligence and autonomy. The assumption rather is, that an adequate segmentation between centralized and de-central allocation of responsibility has to be achieved. Ideally, this can be modelled by means of a hybrid approach that combines conventional system elements with autonomous (software-)agents within a heterogeneous network.

- *Proposition 2*

  Second, it is advantageous to continuously advance the ability to decide autonomously also on lower levels. In the fields of SCM, enterprise organization and business process optimization, this is a well-known fact since long. On technical layers, the future development has only begun with current IT-processing abilities. Still, we assume, that the development will proceed in a similar way also on the technical layers. However, it is important to consider the economic benefits of building an increased amount of intelligence into a manufacturing system or factory: comparable to the necessity of consequently focusing human and organisational company-resources upon the core competencies of a company, also 'technical factory intelligence' has to be focused on the gaining of competitive advantages. This generates a huge challenge for current smart factory and digital manufacturing concepts, as the current experience regarding the linkage of technical 'low-level intelligence' to the comprehensive business strategy layer is limited to currently available use cases and business scenarios.

- *Proposition 3*

  Third, in an organizational context de-central intelligence seems to be applicable in particular in the case of complex, scattered and quickly changing environments [75]. Taking into account the aforementioned structural similarities and analogies between the different layers and the phenomena that have been discussed in the previous sections of this chapter, this is likely to be applicable also on technical levels

  Altogether, modern IT can be expected to be a strong enabler that allows to repeat previous developments from linear and purely hierarchical control structures and organisational principles towards more heterarchic characteristics within system layers that previously didn't allow for a distinct degree of de-central autonomy.

With regard to question (2), hierarchical layer models are not becoming obsolete in the course of this development. Quite the contrary, the explicit linkage of de-central manufacturing intelligence to a well-aligned and comprehensive business focus seems to be a critical success factor for the purpose of achieving a holistic model of heterarchic systems in the context of digital manufacturing.

5 Managerial Implications

In view of the aforementioned theory-based considerations regarding hierarchic and heterarchic organisation structures, SCM has to face a 'systemic paradigm' shift. Current geographically distributed, complex, dynamic and unforeseeable business conditions have generated the urgent need for managerial means that are capable of handling these intricate and fragmented conditions. At the same time, it is to an increasing degree possible to augment also low (i.e., technical) layers with 'automated intelligence' be it the generic ability to execute software procedures, the elementary capability to store data (e.g., regarding an object's own characteristics and actual condition) or the advanced ability for reasoning, (machine) learning or agent-based and negotiation-driven automated task execution.

For the field of SCM this leads to the conclusion, that from a current point of view heterarchic structures offer favourable improvements potentials (question (1)). Despite the advances to be expected from heterarchic structures, most likely a 'hierarchic backbone' will remain necessary or at least desirable in terms of efficiency in order to achieve a holistic model of heterarchic systems in the context of digital manufacturing (question (2)). According to the amount and variety of existing research from multiple disciplines, this chapter doesn't claim to present an exhaustive review or survey. Still, we have concretized several essential aspects regarding the more extensive application of heterarchic principles in order to advance SCM practices. Henceforth, we assume the necessity of future research with a respective focus.

References

1. Johannessen JA, Hauan A (1993) Linking network organization to holographic design: future industrial organization. Kybernetes 22(4):6–23
2. Hedlund G (2003) The hypermodern MNC—a heterarchy? In: International Business. Aldershot (u.a.): Ashgate Dartmouth, pp 379–399
3. Sachan A, Datta S (2005) Review of supply chain management and logistics research. Int J Phys Distrib Logistics Manage 35(9):664–705
4. Hedlund G (1993) Assumptions of hierarchy and heterarchy, with applications to the management of the multinational corporation. In: Ghoshal S, Westney DE (1993) Organization theory and the multinational

5. McCulloch Warren S (1945) A heterarchy of values determined by the topology of nervous nets. Bull Math Biophys 1945(7):89–93
6. Probst GJB, Mercier J-Y (1993) Organisation. Strukturen, Lenkungsinstrumente, Entwicklungsperspektiven. 1. Aufl. Landsberg/ Lech: Verlag Moderne Industrie
7. Scholz C (1997) Strategische organisation. Prinzipien zur Vitalisierung und Virtualisierung. Verlag Moderne Industrie, Landsberg/ Lech
8. Blecker T, Kaluza B (2004) Heterarchische Hierarchie. Ein Organisationsprinzip flexibler Produktionssysteme. Klagenfurt: Inst. für Wirtschaftswissenschaften (Diskussionsbeiträge / Universität Klagenfurt, Institut für Wirtschaftswissenschaften, 2004/01)
9. Andersen Birgitte (ed) (2006) Intellectual property rights. Innovation, governance and the institutional environment. ebrary, Inc, Edward Elgar, Cheltenham, U.K., Northampton, Mass
10. Windt K (2008) Ermittlung des angemessenen Selbststeuerungsgrades in der Logistik – Grenzen der Selbststeuerung. In: Peter Nyhuis: Beiträge zu einer Theorie der Logistik. Springer, Berlin, pp 349–372
11. Braggins Don, Ilie-Zudor Elisabeth, Monostori László (2009) Agent-based framework for pre-contractual evaluation of participants in project-delivery supply-chains. Assembly Autom 29(2):137–153
12. Quinn James B, Hilmer Frederik G (1994) Strategic outsourcing. Sloan Manag Rev 35 (4):43–55
13. Bustinza OF, Arias-Aranda D, Gutierrez-Gutierrez L (2010) Outsourcing, competitive capabilities and performance. An empirical study in service firms. Int J Prod Econ 126 (2):276–288
14. Williamson OE (2008) Outsourcing. Transaction cost economics and supply chain management. J Supply Chain Manag: A Global Rev Purchasing Supply 44(2), 5–16
15. Hodgson GM (2000) The concept of emergence in social sciences. Its history and importance. Emergence 2(4), 65–77
16. Christopher M (2004) Logistics and supply chain management. Strategies for reducing cost and improving service, 2nd edn, reprint. Financial Times/Prentice Hall, London
17. Crum M, Christopher M, Holweg M (2011) "Supply Chain 2.0". Managing supply chains in the era of turbulence. Int J Phys Distrib Logistics Manag 41(1), 63–82
18. Ahlert K-H, Corsten H, Gössinger R (2009) Capacity management in order-driven production networks—A flexibility-oriented approach to determine the size of a network capacity pool. Int J Prod Econ 118(2):430–441
19. Arthofer K, Engelhardt-Nowitzki C, Feichtenschlager H-P, Girardi D (2012) Servicing individual product variants within value chains with an ontology. In: Jodlbauer H, Olhager J, Schonberger RJ (eds), Modelling value. Selected Papers of the 1st International conference on value chain management. Physica Verlag, Heidelberg, pp 333–354
20. Antai I (2011) Supply chain vs supply chain competition. A niche-based approach. Manag Res Rev 34(10), 1107–1124
21. Williamson OE, Masten, Scott E (1999) The economics of transaction costs. E. Elgar Pub (An Elgar critical writings reader), Cheltenham, UK, Northampton, Mass. USA
22. Williamson OE, Masten Scott E. (2016): Transaction cost economics. An Edward Elgar research review. Edward Elgar Publishing Limited (International library of critical writings in economics), Cheltenham, UK
23. Li Suhong, Ragu-Nathan Bhanu, Ragu-Nathan TS, Subba Rao S (2006) The impact of supply chain management practices on competitive advantage and organizational performance. Omega 34(2):107–124
24. Jensen MC, Meckling WH (2009) Theory of the firm. Managerial behavior, agency costs and ownership structure. In: The economic nature of the firm: a reader. Cambridge Univ. Press, Cambridge (u.a.), pp 283–303
25. Coase RH (2012) The nature of the firm. In: The roots of logistics: a reader of classical contributions to the history and conceptual foundations of the science of logistics. Springer, Berlin (u.a.), pp 317–333

26. Liu X, Çetinkaya S (2009) Designing supply contracts in supplier vs buyer-driven channels. The impact of leadership, contract flexibility and information asymmetry. IIE Trans 41(8), 687–701
27. Wan X, Dresner ME, Evers PT (2014) Assessing the dimensions of product variety on performance. The value of product line and pack size. J Bus Logistics 35(3), 213–224
28. Schein E (2009) Concrete examples needed. People Strategy 32(1):8
29. Schein EH, Schein Peter (2018) Organisationskultur und Leadership, 5th edn. Verlag Franz Vahlen, München
30. Dekkers R, Sauer A, Schönung M, Schuh G (2005) Complexity based approach for collaborative enterprise networks. In: Michael Zäh und Gunther Reinhart: 1st International Conference on Changeable, Agile, Reconfigurable and Virtual Production (CARV 2005). München: Utz, pp 1–12
31. Surana A, Kumara S, Greaves M, Raghavan UN (2005) Supply-chain networks: a complex adaptive systems perspective. Int J Prod Res 43(20):4235–4265
32. Gibson CB, Birkinshaw J (2004) The antecedents, consequences, and mediating role of organizational ambidexterity. Acad Manag J 47:209–226
33. Boin A, Kelle P, Whybark DC (2010) Resilient supply chains for extreme situations: outlining a new field of study. Int J Prod Econ 126(1):1–6
34. Chen IJ, Paulraj A (2004) Towards a theory of supply chain management. The constructs and measurements. J Oper Manag 22(2), 119–150
35. Croom Simon, Watt Adrian (2010) Managing operations improvements through relational capabilities in the context of small-firm networks. Int J Logistics Res Appl 3(1):83–96
36. Lambert DM, Cooper MC, Pagh JD (1998) Supply chain management. Implementation issues and research opportunities. Int Jrnl Logistics Management 9(2), 1–20
37. Gosling T (2003) The simple supply chain model and evolutionary computation. In: CEC 2003. The 2003 Congress on Evolutionary Computation. The 2003 Congress on Evolutionary Computation, 2003. CEC '03. Canberra, Australia, Dec. 8–12, 2003. Congress on Evolutionary Computation. Piscataway, IEEE, N.J, pp. 2322–2329
38. Stevenson M, Spring M (2007) Flexibility from a supply chain perspective. Definition and review. Int Jrnl of Op Prod Manag 27(7), 685–713
39. van Hoek R (2004) Adapt or die: transforming your supply chain into an adaptive business network. Supply Chain Manag An Int J 9(1):118–119
40. Holweg M, Pil F (2008) Theoretical perspectives on the coordination of supply chains. J Oper Manag 26(3):389–406
41. Caridi Maria, Crippa Luca, Perego Alessandro, Sianesi Andrea, Tumino Angela (2010) Do virtuality and complexity affect supply chain visibility? Int J Prod Econ 127(2):372–383
42. Jain V, Benyoucef L (2008) Managing long supply chain networks. Some emerging issues and challenges. J Manufact Technol Manag 19(4), pp. 469–496
43. Balakrishnan A, Geunes J, Pangburn MS (2004) Coordinating supply chains by controlling upstream variability propagation. Manufact Serv Oper Manag 6(2):163–183
44. Tielebein M (2006) Decentralized supply chain management: a view from complexity theory. In: Thorsten Blecker, Nizar Abdelkafi: Complexity management in supply chains. Concepts, tools and methods. E. Schmidt (Operations and technology management, 2), Berlin, pp 21–35
45. Tarride Iván Mario (2013) The complexity of measuring complexity. Kybernetes 42 (2):174–184
46. Ashby WR (1968) Variety, constraint, and the law of requisite variety. Modern systems research for the behavioral scientist: a sourcebook. pp 129–136
47. Mentzer JT, Min S, Michelle Bobbitt L (2004) Toward a unified theory of logistics. Int J Phys Distrib Logistics Manag 34(8):606–627
48. Orton JD, Weick KE (1990) Loosely coupled systems. A reconceptualization. The Acad Manag Rev 15(2), 203
49. Papadopoulos T, Gunasekaran A, Dubey R, Altay N, Childe SJ, Fosso-Wamba S (2017) The role of big data in explaining disaster resilience in supply chains for sustainability. J Clean Prod 142:1108–1118

50. Chakravarthy B, Henderson J (2007) From a hierarchy to a heterarchy of strategies. Adapting to a changing context. Manage Decis 45(3), 642–652
51. Engelhardt-Nowitzki C, Kryvinska N, Strauss C (2011) Strategic demands on information services in uncertain businesses. A layer-based framework from a value network perspective. In: 2011 international conference on emerging intelligent data and web technologies. 2011 international conference on emerging intelligent data and web technologies. Tirana, 07.09.2011–09.09.2011. IEEE, pp 131–136
52. Paul S, Pan J, Jain R (2011) Architectures for the future networks and the next generation Internet. A survey. Comput Commun 34(1):2–42
53. Calvo I, Pérez F, Etxeberria-Agiriano I, de Albéniz OG (2013) Designing high performance factory automation applications on top of DDS. Int J Adv Rob Syst 10(4):205
54. Low, K-H (ed) (2007) Industrial robotics. Programming, simulation and applications. plV pro-Literatur-Verl. (ARS, Advanced robotic systems international), Mammendorf
55. Ehrenhuber I, Treiblmaier H, Engelhardt-Nowitzki C, Gerschberger M (2015) Toward a framework for supply chain resilience. Int J Supply Chain Oper Resilience 1(4):339–350
56. Mucksch H, Behme W (eds) (2000) Das data warehouse-Konzept. Architektur - Datenmodelle - Anwendungen. 4, vollständig überarbeitete und, erweiterte edn. Gabler Verlag, Wiesbaden, s.l
57. Zimmermann H (1980) OSI reference model—The ISO model of architecture for open systems interconnection. IEEE Trans Commun 28(4):425–432
58. Cândido G, Barata J, Colombo AW, Jammes F (2009) SOA in reconfigurable supply chains. A research roadmap. Engineering applications of artificial intelligence 22(6), 939–949
59. Mařík V, McFarlane D, Valckenaers P (eds) (2003) Holonic and multi-agent systems for manufacturing. In: First international conference on industrial applications of holonic and multi-agent systems, HoloMAS 2003, Prague, Czech Republic, September 1–3, 2003; proceedings. International Conference on Industrial Applications of Holonic and Multi-Agent Systems; HoloMAS. Berlin, Heidelberg: Springer (Lecture notes in computer science, 2744)
60. Meissner H, Ilsen R, Aurich JC (2017) Analysis of control architectures in the context of industry 4.0. Procedia CIRP 62:165–169
61. Zuehlke Detlef (2010) SmartFactory—Towards a factory-of-things. Ann Rev Control 34 (1):129–138
62. Bildstein A, Seidelmann J (2014) Industrie 4.0-Readiness: migration zur Industrie 4.0-Fertigung. In: Thomas Bauernhansl, Michael ten Hompel, Birgit Vogel-Heuser: Industrie 4.0 in Produktion, Automatisierung und Logistik. Anwendung, Technologien und Migration. Aufl. 2014. Wiesbaden: Springer Fachmedien Wiesbaden GmbH, pp 581–597
63. Broy M (2013) Engineering Cyber-physical systems. Challenges and foundations. In: Aiguier M, Caseau Y, Krob D, Rauzy A (eds) Complex systems design management. Proceedings of the third international conference on complex systems design management CSD&M 2012. Springer, Berlin, Heidelberg, pp 1–13
64. Kolomvatsos K (2018) An intelligent, uncertainty driven management scheme for software updates in pervasive IoT applications. In: Future generation computer systems 83, pp 116–131
65. Rohjans S, Uslar M, Appelrath HJ (2010) OPC UA and CIM. Semantics for the smart grid. In: IEEE PES transmission and distribution conference and exposition, 2010. T & D 2010; 19–22 April 2010, Ernest N. Morial Convention Center, New Orleans, Louisiana. IEEE PES T&D 2010. New Orleans, LA, USA. Power & Energy Society; IEEE PES Transmission and Distribution Conference and Exposition; T & D. IEEE, Piscataway, NJ, pp 1–8
66. Kryvinska Natalia, Gregus Michal (2014) SOA and its business value in requirements, features, practices and methodologies. (SOA reference models and schemes; criteria and features in defining of SOA business value; methodologies in defining and measuring of SOA business value), 1st edn. Univ. Komenského, Bratislava
67. Xhafa F, Kryvinska N, Gregus M (2017) Models and management of elasticity and openness. Towards flexible organizations. Glob J Flex Syst Manag 18(1), 1–2

68. Kryvinska Natalia (2012) Building consistent formal specification for the service enterprise agility foundation. J Serv Sci Res 4(2):235–269
69. Kaczor Sebastian, Kryvinska Natalia (2013) It is all about services-fundamentals, drivers, and business models. J Serv Sci Res 5(2):125–154
70. Molnár E, Molnár R, Kryvinska N, Greguš M (2014) Web intelligence in practice. J Serv Sci Res 6(1):149–172
71. Shroff G (2015) The intelligent web. Search, smart algorithms, and big data. Oxford University Press, Oxford
72. Cai Y, Starly B, Cohen P, Lee Y-S (2017) Sensor data and information fusion to construct digital-twins virtual machine tools for cyber-physical manufacturing. Procedia Manufact 10:1031–1042
73. Sepúlveda JM, Derpich IS (2014) Automated reasoning for supplier performance appraisal in supply chains. Procedia Comput Sci 31:966–975
74. Tao F, Qi Q, Liu A, Kusiak A (2018) Data-driven smart manufacturing. J Manufact Syst
75. Wang F (2010) The evolution of hierarchy toward heterarchy. A case study on Baosteel's managerial systems. Front Bus Res China 4(4), 515–540

Analysis of GDP and Inflation Drivers in the European Union

Kitty Klacsánová and Mária Bohdalová

Abstract This chapter describes different effects of inflation and GDP on the European Union countries pointing out the costs attendant on high and low rate of inflation. To improve the understanding of GDP and inflation differences there were applied statistical methods. They verified some of the big economic crises in the history. Nominal GDP developments yield similar results in both the shorter (1995–2015) and longer period (1967–2015), suggesting that only two common principal components are needed to explain a significant amount of variance in the data. Concerning the inflation, we have investigated the co-movements and the heterogeneity in inflation dynamics across the analyzed countries over two periods (1967–2015 and 1994–2015). The findings indicate that there are three substantial common principal components explaining 99.72% of the total variance in the consumer price indexes in the period from 1994–2015, that can be related to the common monetary policy in the euro area. Finally, the chapter employed multiple linear regression models.

1 Introduction

GDP as a comprehensive scorecard of a country's economic health is one of the most important and widely researched economic indicator. It represents the market value of all goods and services produced by an economy during a given period [14]. Economists are interested in measuring the overall amount of goods and services produced in a country over a given period, which is not affected by the changes in

K. Klacsánová · M. Bohdalová (✉)
Department of Information Systems, Faculty of Management,
Comenius University in Bratislava, Odbojárov 10, Bratislava, Slovakia
e-mail: maria.bohdalova@fm.uniba.sk

K. Klacsánová
e-mail: kitty.klacsanova@fm.uniba.sk

© Springer International Publishing AG, part of Springer Nature 2019
N. Kryvinska and M. Greguš (eds.), *Data-Centric Business and Applications*,
Lecture Notes on Data Engineering and Communications Technologies 20,
https://doi.org/10.1007/978-3-319-94117-2_11

prices. Real GDP considers the impact of inflation, allowing to compare the economic output of two countries or two different periods and to evaluate, whether a certain economy is expanding or contracting [15].

In literature, many authors are interested in exploring or forecasting GDP growth. For example [19] used panel data approach to explain the real GDP using the correlation with employment. Sorić [21] has discovered the time-varying impact of consumer confidence on GDP growth. His empirical analysis was based on a dataset of 11 new EU Member States. Taş et al. [24] analyzed the relationship between economic growth and macroeconomic indicators of the EU countries using panel data approach. They used 10 years data and they examined the effects of eleven macroeconomic indicators on GDP. The forecast of the German GDP is reported in [17, 18]. These authors propose alternative methods for forecasting quarterly GDP with monthly factors. Another approach to predict French GDP used Bec and Mogliani in their paper [1]. These authors investigate forecast combination and information pooling, in the context of France's GDP prediction in real time with monthly survey opinions. Rusnák [16] describes a Dynamic Factor Model to predict Czech GDP using multiple historical data over the period from 2005 to 2012. Bjørnland et al. [2] examined the existence of a systematic influence in the co-movement across countries to improve the forecast accuracy at the national level. They used dynamic factor model and their results show that exploiting the informational content in a common global business cycle factor improves the forecast accuracy across a large panel of countries. Soares et al. [20] used wavelet analysis to analyze inflation dynamics convergence in the Euro area. They discover that before Euro adoption the synchronization pattern was stronger in the 11 countries that first adopted the Euro as after Euro adoption. Development of GDP and inflation is also necessary for service-oriented enterprises as was described in [7, 9, 12, 13, 22, 23].

The main purpose of this chapter is to analyze the nominal GDP and inflation to point out the co-movements and the heterogeneity in its values. We researched the first member states as well as all the countries of the EU together by employing statistical methods. The paper contains descriptive statistics and principal component analyses for both the longer and shorter period. The results indicate how many principal components are needed to keep the original variability of the data. They also reveal the problematic countries and years. Additionally, in the longer period (1967–2015) we employed linear regressions. The established regression models for France and Luxembourg enable to explain the relationship between these countries nominal GDP and the principal components.

The main objective of this chapter is to analyze the changes in inflation and GDP, in order to identify the differences between the selected EU countries. To achieve our primary goal, we set the following partial goals:

• Describe the developments of inflation and GDP in the examined periods by descriptive statistics.
• Identify the main factors using the PCA analysis, which express the inflation and GDP in the analyzed EU countries.

- Compare the interdependence of the EU countries that were not affected by changes in their political system as well as the interdependence of the countries in Central and Eastern Europe, who became independent states after the end of the socialist regime.
- Identify problematic countries, based on the results from the PCA analysis.
- Reveal the problematic years based on the results from PCA analysis.
- Determine the linear regression models for France and Luxembourg to be able to assess the relationship between their inflation/GDP and the principal components in the period between 1967 and 2015.

2 Inflation

The growth of the price level has been one of the main subjects of economic researches during the development of the economic thinking. In the period of mercantilism, economists J. Locke and T. Mun found a positive correlation between the amount of money in the economy and the price level. The views of economists are broadly consistent with the fact that the rate of inflation in the long run depends on the growth of money supply, but we also should consider the growth rate of real output. The most frequent topics in the field of inflation are its causes, the impact on individual subjects and the ways to remove the negative trends and the associated costs.

During inflation the value of money declines and for the same amount of money it is possible to buy fewer goods and services. Increases in the price level also causes increased incomes. However, real income will not be affected by the changes in the price level in case that it changes proportionally in the same way with the prices.

Inflation represents an increase in the economy's overall price level, leading to a decline in the purchasing power of money. There are various reasons why faster growth of aggregate demand compared to the growth of aggregate supply causes inflation. The imbalance of the aggregate demand and supply can be linked to the state's budget deficit, the increase of the interest rates and the foreign demand.

Inflation can be associated with deflation, which is a continuous decline in the price level of goods and services. It is important to distinguish it from the phenomenon of falling prices in one economic sector, which is called disinflation. In addition, deflation can also have a positive impact on the economy, when lower prices increase the real income.

There are two reasons why a high level of inflation is not perceived as a desirable target for central banks. The first is that it is obviously difficult to reduce nominal wages. Deflation trap occurs when low levels of aggregate demand causes the price levels to fall, leading to a negative inflation. In this situation, the central bank's aim is to stimulate the economy by decreasing the nominal interest rates. The combination of low interest rates with a deflation results positive real interest rates, which is not the best solution to stimulate the demand in the private sector. The demand will continue to decline, so deflation does not require the central bank's active co-operation.

2.1 Measuring the Inflation

Inflation is an increase in the overall price level. It causes a decline in the real value and the purchasing power of money. Inflation is measured by the rate of inflation which is expressed as a percentage change in the price level. We distinguish various price indices that are used to monitor the price level in the economy according to Čaplánová and Martincová [5] as Consumer Price Index (CPI), GDP deflator, Producer Price Index (PPI) and Harmonized Index of Consumer Prices (HICP). Each of them reports the price level from a different perspective.

The most common method is based on the consumer price index. An analysis of the consumers' behavior identifies the goods and services that an average consumer buys in the certain economy. This includes information about the daily consumption, long-term consumption as well as the regular monthly payments. The consumer basket is put together by assigning weights to the goods and services according to their importance in the consumer's budget. The next step in calculating the CPI is to find out the prices of the selected items in different time points. Usually, the prices of these products are observed monthly in a wide range of stores. At first it is necessary to calculate the price of the consumer basket in each year. After determining the base year, the price index can be calculated by dividing the value of the consumer basket in a given year by its value in the base year and then multiplying it by 100. The calculated CPI serves to determine the inflation rate, which is defined as the percentage change in the price index compared to the previous period.

$$\textit{Inflation rate in the year } t = \frac{CPI(t) - CPI(t-1)}{CPI(t-1)} \times 100$$

As wrote Gerdesmeier in [6], inflation expressed by a certain price index is only an approximate measure of the economic situation. Over the time, consumers try to replace more expensive products with cheaper ones and the consumer basket becomes no longer representative. Another challenge is how to include the product quality developments in the price index. An increase in the price of a product due to an improvement in its quality is not considered to be the cause of inflation because it does not reduce the purchasing power of the money.

The CPI index is used to monitor the price changes in the short term. Any increase in the consumer prices is interpreted as an increase in inflation. This price index does not take into account the price increases due to improvements in the quality of the products. This means that the estimated rate of inflation can be higher than its actual value. It is important to mention that an increase in the nominal prices is similarly followed by an increase in the nominal wages. At the end, the real price of the products remains unchanged. The methodology of putting together and the constant updates of the consumer basket play an important role in revealing the inaccuracies.

The Harmonized Index of Consumer Prices, initiated by EUROSTAT, serves to point out the differences in national consumer habits. HICP is calculated differently from the other price indices. First, it is calculated from a smaller consumer basket. It considers only the current consumption expenses. For example, it excludes the expenses related to the cost of living in family houses. The main difference between the HICP and CPI is the calculation of weights. Additionally, the HICP includes the expenditures of the countries households. Expenditures of expats living in the country whose HICP is measured are also included in [11]. This index does not involve the spending of the households based outside their home country. In contrary the CPI considers only the expenditures of the national households excluding expat households.

3 GDP

In the past, to measure the overall economic activity economists needed to combine a wide range of information, for example data about iron shipments or sales in the department stores. As Blanchard and Johnson wrote in [3], the national income accounts were not reported before the Second World War. Similarly, the first release of the aggregate output data was in 1947 in the US. Of course, we can find earlier data but these had been compiled retrospectively.

GDP is one of the most frequently observed economic indicator that measures the economic well-being of a country. It represents the market value of final products and services produced in a country over a given period. GDP data are predominantly reported quarterly or annually. To avoid seasonality, quarterly data are usually cleared from seasonal fluctuations [14]. Intermediate goods that are subsequently used in the next stages of production are not considered when calculating the GDP. According Blanchard and Johnson [3], there exist the following three approaches to calculate the GDP:

- GDP is the value of all final goods and services produced in a country over a given period.
- GDP is the value added in the economy.
- GDP is the sum of each subject's income in the economy over a given reference period.

By Blanchard and Johnson [3], the main components of the GDP are consumption (C), investments (I), government spending (G) and finally the net export (NX), which is the difference between export and import.

$$HDP = C + I + G + NX$$

Consumption, the largest part of the GDP involves the goods and services purchased by households, for example food, clothing and medical services. Their decisions depend on various factors. One of them is the disposable income excluding the taxes. The following equation indicates that consumption is a function of disposable income

$$C = C(Yd)$$

If the disposable income increases, then a household will own more money that they can invest, resulting their consumption to increase. This and similar equations economists consider as a behavioral because it explains the consumers demeanor.

The private domestic investments also form a large proportion of the GDP, including asset purchases. As an example, they occur when firms expand their capacity by buying plant, property and equipment. These investments do not include purchases of different shares on the secondary market, since they are not considered as an investment [4].

Government spending consist of expenditures for numerous purposes, for example wages for employees in the public sector, investments in the infrastructure expected to generate benefits in the future. Transfer payments e.g. unemployment or child benefits are spending that are not intended to conduct transactions of goods and services, but instead they serve as a transfer of money. In a macroeconomic context they can be explained as negative taxes [14].

Export represents the purchase of goods and services produced in the home country by foreign subjects. Conversely, import is a purchase of goods and services originating from another country by domestic subjects. The difference between export and import is called net export or trade balance. There can occur two situations based on the trade balance. If export exceeds the country's import, it results a trade surplus. Otherwise, higher import than export is followed by a trade deficit as is wrote in [3].

Another term which we introduce is the GDP deflator, used to monitor the prices changes in the long term. The data are reported on a year-on-year basis. It is an aggregate price index, including all the goods and services produced in the economy during a given year. GDP deflator is appropriate when compiling time series from the price changes as well as comparing countries over a long period [5].

The GDP deflator is calculated using both the nominal and real GDP:

$$HDP\ deflator = \frac{Nominal\ HDP}{Real\ HDP} \times 100$$

In the basis year the value of the GDP deflator is equal to 100 as the values of the nominal and real GDP in that year are identical (see [4]).

4 Data and Results

We analyzed the annual CPI index and the nominal GDP denominated in USD. The data were obtained from the World Bank's (2016) [25], publicly available database. For the data analysis we employed the statistical software SAS Enterprise Guide.

The EU countries were divided based on the changes that took place in their political system. The developments of these macroeconomics indicators were examined during two periods. In the case of GDP, the first analyzed period was from 1967 to 2015. During this period, we analyzed the first member countries (except Germany), i.e. countries, that were not affected by socialism and the countries that joined the EU before 1995 inclusive. This analysis included the following 14 countries of the EU: Austria, Belgium, Denmark, Finland, France, Greece, Ireland, Italy, Luxembourg, the Netherlands, Portugal, Spain, Sweden and the United Kingdom. The second period spans from 1995 to 2015, where we analyzed the development of the nominal GDP in each country of the EU. Concerning the inflation, the analyzed period from 1967 to 2015 includes the original member states and the countries, which joined the EU before 1995 excluding Germany, the United Kingdom and the countries which were one of the succession states of the socialist regime. Additionally, this analysis contains Malta and Cypress. The period between 1994 and 2015 includes all the countries of the EU.

Principal Component Analysis (PCA) [8, 10] was used to find main driver in GDP and CPI of the EU members states in this chapter. We discover some principal components (PCs) for each sample data. Linear regression models are given for selected countries to predict their GDP and inflation based on PCA results.

4.1 Nominal GDP During the Period from 1967 to 2015

Table 1 displays the mean and standard deviations for each country. The findings indicate that over the analyzed period the highest average nominal GDP was recorded in France and the United Kingdom had the highest standard deviation. In addition, Luxembourg had the lowest average nominal GDP along with the lowest standard deviation. In this context we should consider the fact that the impact of price changes is not included in the nominal GDP. There are two reasons behind higher nominal GDP. First, it occurs because of increased production volume, and second, the inflation is followed by an increase of current market prices. We cannot fully presume that Luxembourg moves slowly, lagging behind the countries.

However, the mean rate of inflation in Luxembourg was among the lowest in the analyzed countries. Its value reached 3.6%. The country may have had lower nominal GDP due to slower growth of the price level compared to the other countries. We would be able to confirm this assumption, if we looked at the real GDP that is not affected by price growth.

Table 1 Descriptive Statistics of Nominal GDP for the First EU Member Countries (except Germany). Data period from 1967 to 2015

Country	Mean	Standard deviation
AUT	1.82E+11	1.39E+11
BEL	2.26E+11	1.65E+11
DNK	1.50E+11	1.10E+11
FIN	1.18E+11	8.70E+10
FRA	1.28E+12	9.00E+11
GRC	1.24E+11	9.93E+10
IRL	9.06E+10	9.39E+10
ITA	1.03E+12	7.36E+11
LUX	2.04E+10	1.99E+10
NLD	3.77E+11	2.88E+11
PRT	9.97E+10	8.37E+10
ESP	5.86E+11	5.01E+11
SWE	2.42E+11	1.67E+11
UK	1.25E+12	9.70E+11

Source Own processing base on OECD data

Now that we have studied the co-movements, as well as the heterogeneity in the nominal GDP across the EU we perform principal component analyses. To explain a significant amount of the total variability in the data we determined two principal components. Table 2 displays the eigenvalues of the covariance matrix established from the nominal GDP data in decreasing order, the proportion of explained variance and the cumulated explained variance by each principal component. The first principal component interprets 98.84% of the variance in the data, while 99.61% of the variance is interpreted by both components. Figure 1 shows that only 2 principal components are needed.

The covariance matrix's eigenvectors indicate that the first principal component represents the following countries: France, the United Kingdom, Italy and Spain. The second principal component has the highest positive loadings on France and Italy reflecting their variability of the nominal GDP. Concurrently, the second principal component acquires high negative loadings on United Kingdom (-0.8006), Ireland (-0.085) and Luxembourg (-0.001) (see Table 3).

The component pattern profiles are displayed on the Fig. 2, showing a strong positive correlation between all the countries and the first principal component. Its

Table 2 Eigenvalues of the Covariance Matrix of Nominal GDP for the First EU Member Countries (except Germany). Data period from 1967 to 2015

	Eigenvalue	Difference	Proportion	Cumulative
1	2.72E+24	2.69E+24	0.9884	0.9884
2	2.10E+22		0.0076	0.9961

Source Own processing base on OECD data

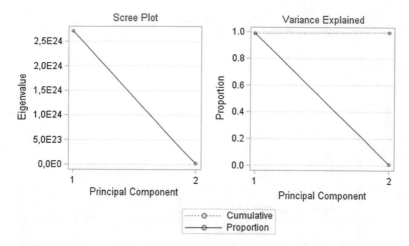

Fig. 1 Number of Principal Component versus Eigenvalue of Nominal GDP for the First EU Member Countries (except Germany). Data period from 1967 to 2015. *Source* Own processing base on OECD data

Table 3 First two principal components of Nominal GDP for the First EU Member Countries (except Germany). Data period from 1967 to 2015

Country	PC1	PC2
AUT	0.084	0.055
BEL	0.099	0.069
DNK	0.066	0.032
FIN	0.0524	0.0302
FRA	0.5447	0.4169
GRC	0.0588	0.0343
IRL	0.055	−0.085
ITA	0.4442	0.3649
LUX	0.0117	−0.001
NLD	0.174	0.0651
PRT	0.0505	0.0065
ESP	0.3014	0.1629
SWE	0.1	0.0554
UK	0.5845	−0.801

Source Own processing base on OECD data

values are close to one. Additionally, between the second principal component and the countries there is a very weak correlation, either negative or positive. Figure 3 shows a pairwise component pattern plot, which is also helpful for determining the correlations between the countries and the principal components.

This analysis highlighted an isolated year, 2009 (see Fig. 4). It can be associated with the global economic and financial crisis which has resulted in a gradual slowdown, and higher rate of inflation. In 2007, when the recession started, there

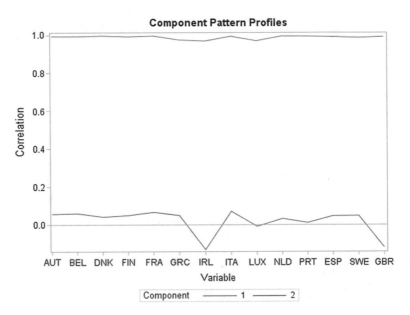

Fig. 2 Component Pattern Profiles of Nominal GDP for the First EU Member Countries (except Germany). Data period from 1967 to 2015. *Source* Own processing base on OECD data

had been already a crisis on the real estate market in the United States. The European Union started to experience its consequential impacts in 2008. As a gradual slowdown occurred, expectations of lower economic growth and higher inflation rate raised. The performed analysis confirmed this phenomenon.

In the next part of this analysis we will discuss the estimation of nominal GDP in France and Luxembourg. We use the two principal components as independent variables in the regression models. The results for France and Luxembourg are presented in the Tables 4 and 5. Considering the values of the Adjusted R-Square statistic, the model predicting the French GDP is better. Adj. R-Square is higher for France (0.9984) than in the case of Luxembourg (0.9404).

There is a positive linear relationship between the French GDP and the principal components as the estimated regression coefficients acquire positive values. If the value of the first principal component increases by one unit, the GDP in France will increase by 8.98E+11 USD on average.

Simultaneously the value of the second principal component remains unchanged. Based on the regression results, both regression coefficients are statistically significant, indicating a significant impact of the principal components on the GDP in France. Additionally, the model meets the iid. assumptions. It explains 99.85% of the French GDP. 0.15% of the country's variability in the GDP is due to other factors than the development of the GDP in the analyzed countries.

From the regression model we received an important information, which is presented in the Table 5. If the GDP of the countries represented by the second principal component increases by one unit, then the GDP of Luxembourg will

Component Pattern

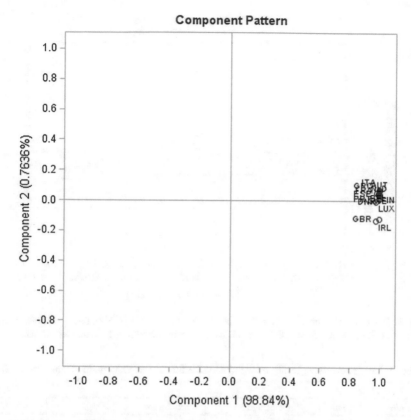

Fig. 3 Component Pattern of Nominal GDP for the First EU Member Countries (except Germany). Data period from 1967 to 2015. *Source* Own processing base on OECD data

decrease on average by −2E+08 USD. Additionally, only the first principal component has a statistically significant impact on Luxembourg's GDP. While the model as a whole is statistically significant, the residuals do not meet the assumptions of normal distribution. The residuals have a systematic tendency. The Durbin Watson's statistic equals to 0.079, showing a positive autocorrelation

4.2   Nominal GDP During the Period from 1995 to 2015

This section contains the analysis of the nominal GDP data of each EU member states over a period of 21 years, from 1995 to 2015. The lowest average GDP and the lowest standard deviation of the data was recorded in Malta. In contrary the highest average GDP and the highest standard deviation had Germany (see Table 6).

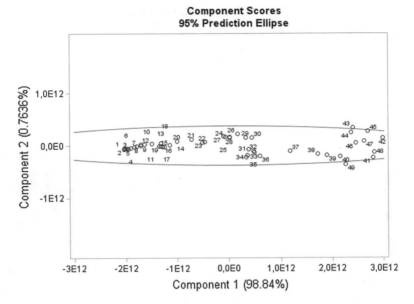

Fig. 4 Component Scores of Nominal GDP for the First EU Member Countries (except Germany). Data period from 1967 to 2015. *Source* Own processing base on OECD data

Table 4 Linear regression results of Nominal GDP for France. Data period from 1967 to 2015

Analysis of variance					
Source	DF	Sum of square	Mean square	F value	Pr > F
Model	2	3.89E+25	1.94E+25	15,124	<0.0001
Error	46	5.91E+22	1.28E+21		
Corrected	48				
Parameter estimates					
Variable	DF	Parameter estimate	Standard error	t value	Pr > |t|
Intercept	1	1.28E+12	5.12E+09	250.04	<0.0001
PC1	1	8.98E+11	5.17E+09	173.53	<0.0001
PC1	1	6.04E+10	5.17E+09	11.67	<0.0001

Source Own processing base on OECD data

The results from the PC analysis (Table 7, Fig. 5) showed, that two principal components explain 98.63% of the original data variability, while the first component contributes to the total variability by 95.41%. The second principal component reflects only 3.22% of the variability.

The first eigenvector (first principal component) of the covariance matrix acquires the highest loadings on the following variables: Germany, France, Italy, Spain and the United Kingdom (see Table 8). Given that the first and second eigenvector has high loadings on United Kingdom, both principal components well reflect the country's GDP over the analyzed period.

Table 5 Linear regression results of Nominal GDP for Luxembourg. Data period from 1967 to 2015

Analysis of variance

Source	DF	Sum of square	Mean square	F value	Pr > F
Model	2	1.80E+22	8.98E+21	379.43	<0.0001
Error	46	1.09E+21	2.37E+19		
Corrected	48	1.90E+22			

Parameter estimates

Variable	DF	Parameter estimate	Standard error	t value	Pr > |t|
Intercept	1	2.00E+10	6.9E+08	29.37	<0.0001
PC1	1	1.9E+10	7.00E+08	27.55	<0.0001
PC1	1	−2.00E+08	7.00E+08	−0.23	0.8193

Source Own processing base on OECD data

Table 6 Descriptive Statistics Nominal GDP of Nominal GDP for all EU Member Countries. Data period from 1995 to 2015

Country	Mean	Standard deviation	Country	Mean	Standard deviation
AUT	3.15E+11	9.14E+10	ITA	1.73E+12	4.44E+11
BEL	3.84E+11	1.13E+11	LVA	1.83E+10	1.04E+10
BGR	3.30E+10	1.87E+10	LTU	2.71E+10	1.51E+10
HRV	4.28E+10	1.70E+10	LUX	3.91E+10	1.67E+10
CYP	1.79E+10	6.89E+09	MLT	6.68E+09	2.53E+09
CZE	1.39E+11	6.72E+10	NLD	6.57E+11	1.97E+11
DNK	2.56E+11	7.21E+10	POL	3.31E+11	1.56E+11
EST	1.44E+10	8.18E+09	PRT	1.83E+11	5.28E+10
FIN	1.99E+11	6.04E+10	ROM	1.10E+11	6.82E+10
FRA	2.14E+12	5.83E+11	SVK	6.27E+10	3.01E+10
DEU	2.92E+12	6.67E+11	SVN	3.58E+10	1.28E+10
GRC	2.19E+11	7.30E+10	ESP	1.06E+12	3.81E+11
HUN	9.72E+10	4.03E+10	SWE	3.96E+11	1.23E+11
IRL	1.81E+11	7.53E+10	UK	2.22E+12	5.80E+11

Source Own processing base on OECD data

Table 7 Eigenvalues of the Covariance Matrix of Nominal GDP for all EU Member Countries. Data period from 1995 to 2015

	Eigenvalue	Difference	Proportion	Cumulative
1	1.52E+24	1.47E+24	0.9541	0.9541
2	5.15E+22		0.0322	0.9863

Source Own processing base on OECD data

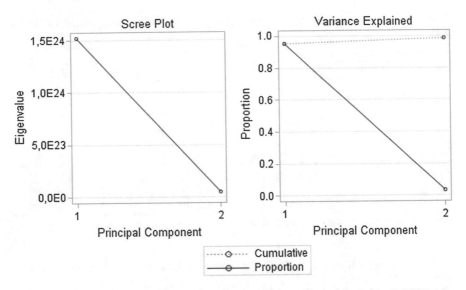

Fig. 5 Number of Principal Component versus Eigenvalue of Nominal GDP for all EU Member Countries. Data period from 1995 to 2015. *Source* Own processing base on OECD data

Table 8 First two principal components of Nominal GDP for all EU Member Countries. Data period from 1995 to 2015

Country	PC1	PC2
AUT	0.0734	−0.0438
BEL	0.0908	−0.0495
BGR	0.0148	−0.0055
HRV	0.0136	0.0000
CYP	0.0054	−0.001
CZE	0.0535	−0.0141
DNK	0.0583	−0.0157
EST	0.0065	0.0017
FIN	0.0488	−0.0046
FRA	0.4708	−0.1899
GER	0.5282	−0.5304
GRC	0.0527	0.0163
HUN	0.0322	0.0183
IRL	0.0581	0.0869
ITA	0.3544	0.0128
LVA	0.0083	0.0026
LTU	0.012	0.0023
LUX	0.0131	−0.0023

(continued)

Table 8 (continued)

Country	PC1	PC2
MLT	0.002	0.0001
NLD	0.1591	−0.0197
POL	0.122	−0.0305
PRT	0.042	0.0084
ROM	0.0541	−0.0096
SVK	0.0241	−0.0001
SVN	0.0103	−0.0011
ESP	0.303	0.0541
SWE	0.0967	−0.0264
UK	0.4435	0.815

Source Own processing base on OECD data

Figure 6 displays the interdependence between the analyzed countries and the principal components. It indicates that there is a strong interdependence between each country and the first principal component. Additionally, the correlation between the countries and the second principal component is weak, either positive or negative.

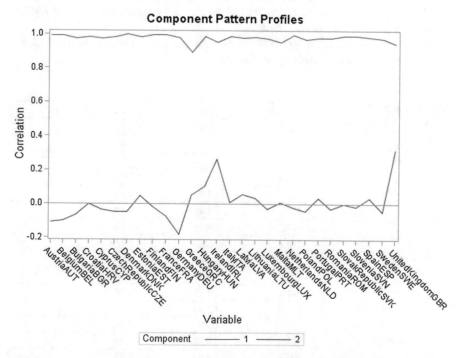

Fig. 6 Component Pattern Profiles of Nominal GDP for all EU Member Countries. Data period from 1995 to 2015. *Source* Own processing base on OECD data

4.3   Inflation During the Period from 1967 to 2015

In the Table 9 we can see the mean and the standard deviation for each country. The values of the average inflation in the countries were between 3 and 5%, except in Greece, Ireland, Italy, Portugal and Spain. In these countries the average inflation reached 6–9%. According to the input data Greece may have had higher average inflation due to the fact, that from 1973 to 1993 it reported its inflation on the level 12–20%. The highest variance of the data had also Greece and Portugal. In contrary the lowest variance of the data had Austria.

The overall variability of the input data is 355.56. Table 10 displays the values of variance for each component, which represent the eigenvalues of the covariance matrix, the proportion of variance explained by each component and the cumulated explained variance for up to four principal components. We determined four principal components, which explained a significant amount, more than 95% of the total variability (Fig. 7). The first principal component explains 80.36% of the variability in the original data. The other three components keep less variance than 10%.

Based on the values of the covariance matrix's eigenvectors we determined the following:

- The first principal component represents the most the southern countries of the EU (Portugal, Greece, Italy and Spain) as the first eigenvector has the highest positive loadings on these countries.
- The second eigenvector acquires the highest positive loadings on the variable Italy and the lowest loadings on Greece.

Table 9 Descriptive Statistics of CPI for the selected EU Member Countries. Data period from 1967 to 2015

Country	Mean	Standard deviation
AUT	3.37	2.11
BEL	3.73	2.92
CYP	4.16	3.32
DNK	4.76	3.69
FIN	4.96	4.52
FRA	4.48	3.97
GRC	9.61	8.14
IRL	6.01	5.91
ITA	6.43	5.84
LUX	3.6	2.68
MLT	3.27	3.18
NLD	3.47	2.61
PRT	9.4	8.56
ESP	6.94	5.79
SWE	4.62	3.95

Source Own processing base on OECD data

Table 10 Eigenvalues of the covariance matrix of CPI for the selected EU Member Countries. Data period from 1967 to 2015

	Eigenvalue	Difference	Proportion	Cumulative
1	285.7150	252.7744	0.8036	0.8036
2	32.9406	18.9576	0.0926	0.8962
3	13.9830	7.4931	0.0393	0.9355
4	6.4898		0.0183	0.9538

Source Own processing base on OECD data

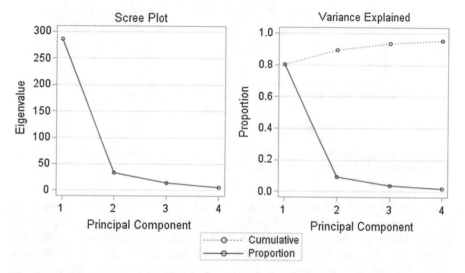

Fig. 7 Number of Principal Component versus Eigenvalue of CPI for the selected EU Member Countries. Data period from 1967 to 2015. *Source* Own processing base on OECD data

- The third principal component supremely reflects the inflation in Malta, Greece and Cyprus.
- The fourth principal component represents the following countries as the fourth eigenvector acquires the highest positive loadings on Belgium and Luxembourg as well as high negative loadings on Spain and Malta.

Figure 8 shows the component pattern profile, where we can see a strong positive correlation between the first principal component and all the countries. Compared to the other three components, not all the countries are positively correlated with them. Strong positive correlations exist between the Netherlands and the second component as well as between Malta and the third component. We can point out the relationship between Greece and the second principal component or between Malta and the fourth component as an example of a negative correlation. Figure 9 displays a pairwise component pattern profile plot. Additionally, this plot

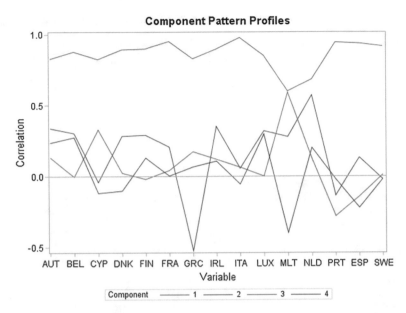

Fig. 8 Component Pattern Profiles of CPI for the selected EU Member Countries. Data period from 1967 to 2015. *Source* Own processing base on OECD data

illustrates the correlations between the countries and the principal components. We can see that year 1975 is an outlying value from the second principal component. Greek inflation has fallen by 13.5% this year compared to year 1974, while Cyprus and Portugal have also fallen by more than 7.5%. The year 1980 is an outlier for the third principal component due to the increase in inflation in all countries compared to the previous year except for Portugal and Spain, which recorded a fall in the price level. However, each country showed volatile inflation, with the exception of Austria and the Benelux countries (Belgium, Luxembourg and the Netherlands).

The second part of the analysis focuses on the estimation of inflation in France and Luxembourg. Each principal component is an independent variable in the regression models. The results are reported in the following Tables 11 and 12. The estimated regression coefficients are positive, therefore there exists a positive linear relationship between the specific countries' inflation and the principal components. If the value of the first component increases by one unit, then the inflation in France will also increase by 0.2236 on average. In the next step we analyzed whether the regression coefficients are statistically significant. At the significance level 0.05 only the first two regression coefficients have a significant impact on the inflation in France. The findings indicate that the inflation in this country had been mainly influenced by the inflation in the southern countries of the EU (Portugal, Greece, Italy and Spain) and Ireland. Additionally, the model meets the iid. assumptions and

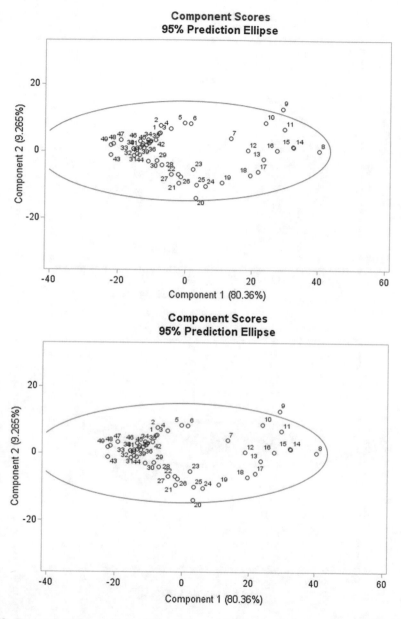

Fig. 9 Component Scores with Prediction Ellipse of CPI for the selected EU Member Countries. Data period from 1967 to 2015. *Source* Own processing base on OECD data

is also statistically significant. The estimated model explains 95.24% of the variability in the inflation of France. The remaining 4.76% variability is caused by factors other than the changes in the value of CPI in the analyzed countries that we did not include in the model.

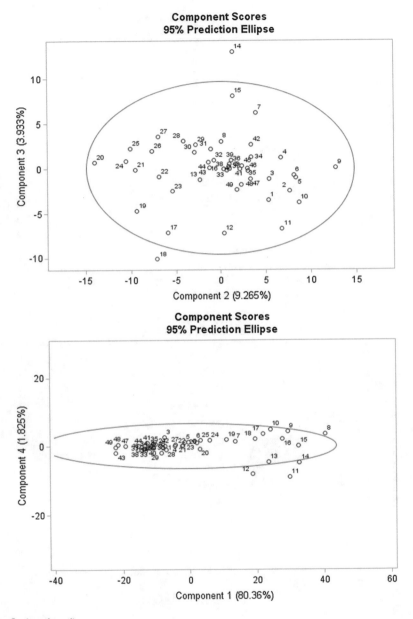

Fig. 9 (continued)

From the prediction of Luxembourg's inflation, it is obvious that the CPI index will increase predominantly, if the value of the fourth principal component increases by one unit. Furthermore, the inflation in Luxembourg will increase the least, if the value of the third principal component increases by one unit.

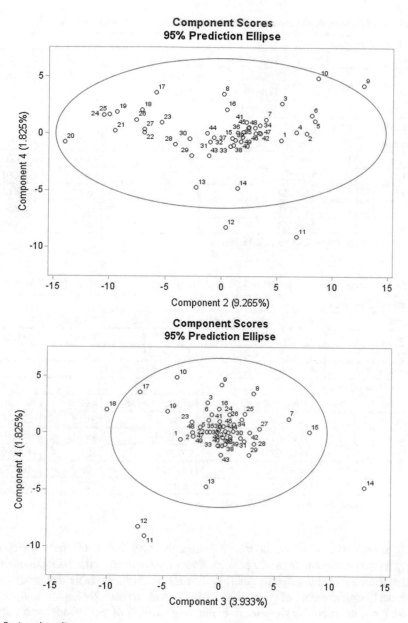

Fig. 9 (continued)

The previous facts indicate that an increase in all the principal components causes likewise an increase in the average rate of inflation in the country. Each country included in the analysis affected Luxembourg's inflation adversely. We can state that the changes in the inflation rates of the countries initiate an increase in the

Table 11 Linear regression results of CPI for France. Data period from 1967 to 2015

Analysis of variance

Source	DF	Sum of square	Mean square	F value	Pr > F
Model	4	718.8818	179.7204	220.31	<0.0001
Error	44	35.8939	0.8158		
Corrected Total	48	754.7756			

Parameter estimates

Variable	DF	Parameter estimate	Standard error	t value	Pr > |t|
Intercept	1	4.4789	0.1290	34.71	<0.0001
PC1	1	0.2236	0.0077	28.99	<0.0001
PC2	1	0.1424	0.0227	6.27	<0.0001
PC3	1	0.0451	0.0349	1.29	0.2028
PC4	1	0.0059	0.0512	0.12	0.9087

Source Own processing base on OECD data

Table 12 Linear regression results of CPI for Luxembourg. Data period from 1967 to 2015

Analysis of variance

Source	DF	Sum of square	Mean square	F value	Pr > F
Model	4	316.88551	79.22138	121.21	<0.0001
Error	44	28.75891	0.65361		
Corrected Total	48	345.64442			

Parameter estimates

Variable	DF	Parameter estimate	Standard error	t value	Pr > |t|
Intercept	1	3.59736	0.11549	31.15	<0.0001
PC1	1	0.13541	0.00690	19.61	<0.0001
PC2	1	0.14903	0.02033	7.33	<0.0001
PC3	1	0.00089646	0.03121	0.03	0.9772
PC4	1	0.31193	0.04581	6.81	<0.0001

Source Own processing base on OECD data

average inflation of Luxembourg. Between the CPI index of the country and the principal components exist a positive linear relationship as the estimations of the regression coefficients are also positive. The results from the table demonstrate the statistical significance of the first, second and fourth principal component. In addition, the model is significant, explaining 91.68% of the overall variability of the data and meets the iid. assumptions too.

According to the values of the Adj. R-Square the model for France is better, while it explains 94.81% of the variability in the input data compared to the model for Luxembourg, where 90.92% of the variability is interpreted.

Table 13 Descriptive Statistics of CPI for the all EU Member Countries. Data period from 1994 to 2015

Country	Mean	Standard deviation	Country	Mean	Standard deviation
AUT	1.9	0.78	ITA	2.29	1.22
BEL	1.88	1.04	LVA	7.22	9.02
BGR	65.28	224.16	LTU	8.64	17.01
HRV	7.73	22.27	LUX	1.97	0.9
CYP	2.26	1.77	MLT	2.39	1.04
CZE	3.91	3.4	NLD	2.02	0.81
DNK	1.96	0.74	POL	7.34	9.13
EST	8.13	11.31	PRT	2.45	1.49
FIN	1.53	1.13	ROM	29.58	41.42
FRA	1.46	0.73	SVK	5.31	3.87
DEU	1.5	0.67	SVN	5.61	4.98
GRC	3.55	3.05	ESP	2.59	1.46
HUN	8.65	7.53	SWE	1.17	1.14
IRL	2.15	2.28	UK	2.04	0.99

Source Own processing base on OECD data. Period from 1994 to 2015

4.4 Inflation During the Period from 1994 to 2015

In this section we analyzed the CPI index of the 28 EU member states over a period of 22 years from 1994 to 2015 (Table 13). During this period the highest average inflation had Bulgaria (65.28%) and Romania (29.58%). On the contrary the lowest average values of the CPI were recorded in Sweden, France, Germany and Finland.

The results of the PC analysis (see Table 14, Figs. 10, 11, 12, and 13) show, that three principal components are necessary to explain 99.72% of the variability in the original data. While the first component contributes to the total variability by 96.33%, the remaining two components by less than 5%. Based on the eigenvectors of the covariance matrix we determined the countries that are largely represented in

Table 14 Eigenvalues of the covariance matrix of CPI for the all EU Member Countries. Data period from 1994 to 2015

	Eigenvalue	Difference	Proportion	Cumulative
1	51230.7264	49612.4973	0.9633	0.9633
2	1618.2291	1433.8253	0.0304	0.9937
3	184.4037		0.0035	0.9972

Source Own processing base on OECD data

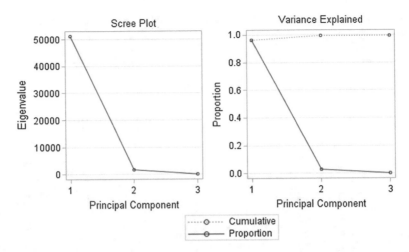

Fig. 10 Number of Principal Component versus Eigenvalue of CPI for the all EU Member Countries. Data period from 1994 to 2015. *Source* Own processing base on OECD data

the individual components. The first principal component reflects each of the country evenly since the first eigenvector shows approximately equal loadings on all variables except Bulgaria on which it acquires higher positive loadings. In addition, the second eigenvector has the higher positive loadings on Croatia and Romania. The third principal component has the highest negative loadings on Romania. Apart from this it seems to reflect the Baltic countries (Estonia, Latvia and Lithuania).

The interdependence between the analyzed countries and the principal components can be seen on the Fig. 12. We found that there is either a strong or a moderate correlation between the second component and the countries.

Figure 13 displays components scores and points out the potential outlying years, which are 1994 and 1997. Behind this result is the crisis in Bulgaria and the high rates of inflation in Croatia and Romania at the end of the past century. The countries in central Europe during their economic transformation had high unstable levels of inflation.

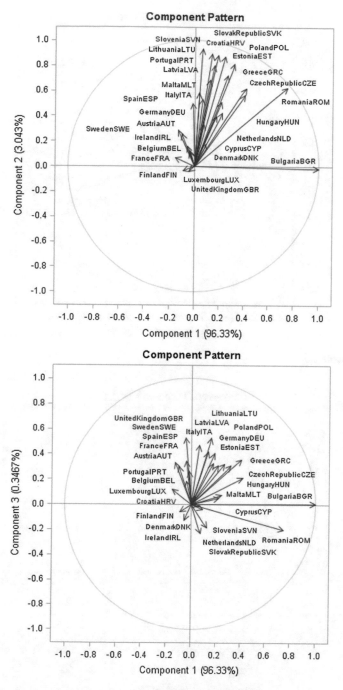

Fig. 11 Pairwise Component Pattern of CPI for the all EU Member Countries. Data period from 1994 to 2015. *Source* Own processing base on OECD data

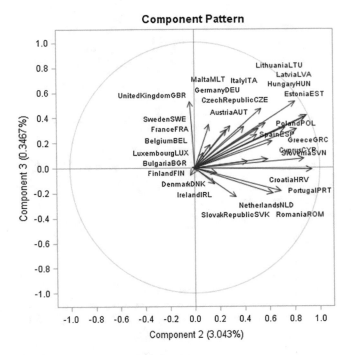

Fig. 11 (continued)

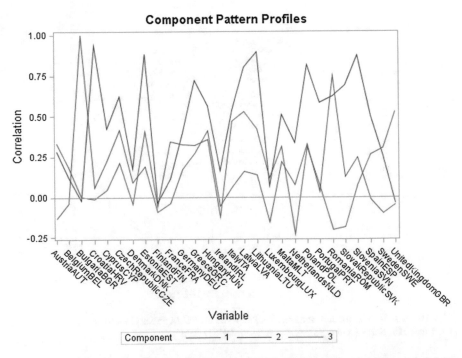

Fig. 12 Component Pattern Profiles of CPI for the all EU Member Countries. Data period from 1994 to 2015. *Source* Own processing base on OECD data

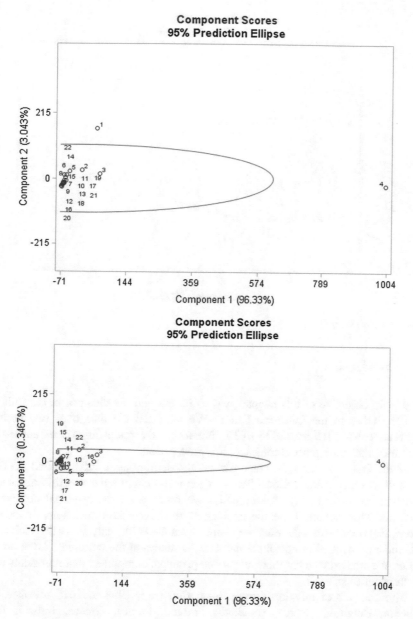

Fig. 13 Component Scores of CPI for the all EU Member Countries. Data period from 1994 to 2015. *Source* Own processing base on OECD data

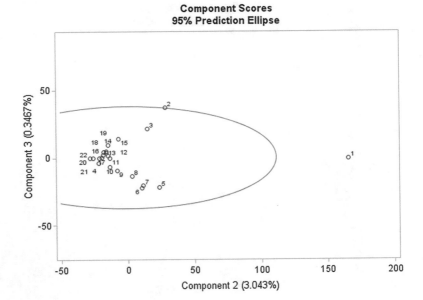

Fig. 13 (continued)

5 Conclusion

The main objective of this chapter was to investigate the changes in the inflation and the GDP in the European Union. We analyzed the data over two periods, between 1967–2015 and 1994–2015. We found the main differences among the countries that took place during the analyzed periods.

In the first subchapter we demonstrated the definitions of inflation and the GDP as well as their characteristics. The chapter also contains the possible impacts of inflation on an economy. Additionally, we focused on the principal component analysis. This statistical method enables us to reduce the dimensions of multidimensional data. After obtaining the results from the PCA analysis we determined an adequate number of principal components to represent the countries' inflation and GDP. We studied whether there exists one principal component that represents most of the countries.

The analysis of inflation during the period from 1994 to 2015 revealed that Austria, Belgium, Cyprus, Denmark, Finland, France, Greece, Ireland, Italy, Luxembourg, Malta, Netherlands, Portugal, Spain and Sweden had lower average inflation and standard deviation in contrary to the analysis over a longer period (1967–2015). The reason behind this could be that the countries, who introduced the euro as their official currency had to meet the Maastricht convergence criteria to be able to entry the euro area and subsequently the common monetary policy.

During the period which stems from 1967 to 2015 we found one principal component explaining 80.36% of the variability in the inflation. This suggests that the inflation in the original countries together with the countries that joined the EU before 1995 was more diverse. These countries formed a more independent system, as we needed four principal components to represent 95.38% of the variability in the data. After comparing the results of the long term (1967–2015) with the shorter-term analyses (1994–2015) we pointed out the dependence of the socialist regime's succession states from the countries of Western Europe. In the first analysis (1994–2015), the first principal component explained more variability in the data than in the second one (1967–2015).

In the case of GDP both analyses indicate that only one principal component is needed to explain more than 95% of the overall variability in the data. This implies that in the EU there is a strong interdependence between the countries. The first principal component maintained the greatest variability of the original data. Additionally, this principal component the most reflected the GDP in France and the United Kingdom.

The composition of the first principal component was not the same in the two analyses, even though in both the southern countries were dominant in the first principal component. In the longer period the first principal component reflects Italy's and some of the southern countries' (Portugal, Greece, Spain) inflation who entered the EU before 1980s. According to the values of the first eigenvector, in the shorter period (1994–2015) a former socialist country, Bulgaria became prevailing in the first principal component. In addition, the PCA results revealed a few problematic years, the hyperinflations in Bulgaria and Yugoslavia in the last century and the global economic and financial crisis in 2009.

In the practical part we explained in more detail the relationship between the principal components, which represent the analyzed countries' inflation/GDP, and France and Luxembourg. From the conclusions of each regression analysis for inflation it was clear that between the principal components and the two countries is a strong linear relationship. Both models explained more than 90% of the changes in the inflation of these countries. The variability in their price growth was caused by fluctuations of inflation in the countries that joined the EU earlier. The remaining variability may had been caused by various other factors, for example by the changes in fiscal policy, unemployment rates, tax policy or financial market regulations.

The results of the regression analysis of Luxembourg's GDP revealed that, not all the assumptions about the random error component were met. The residuals did not have a normal probability distribution and had a systematic tendency. In contrary, the scattering analyses suggested that each of the regression model is statistically significant. The slope coefficients were likewise statistically significant.

Based on the values of the Adj. R squares the regression models for France are better as they explain more variability in the countries' data.

The established regression models provide several viewpoints that can be considered during the monetary and fiscal policy implementations. With relatively high accuracy they can be also applied to predict the GDP in France and Luxembourg.

References

1. Bec F, Mogliani M (2015) Nowcasting French GDP in real-time with surveys and "blocked" regressions: combining forecasts or pooling information? Int J Forecast 31(4):1021–1042. https://doi.org/10.1016/j.ijforecast.2014.11.006
2. Bjørnland HC, Ravazzolo F, Thorsrud LA (2017) Forecasting GDP with global components: this time is different. Int J Forecast 33(1):153–173. https://doi.org/10.1016/j.ijforecast.2016.02.004
3. Blanchard O, Johnson DR (2013) Macroeconomics. Pearson Education, Harlow
4. Colander DC, Gamber EN (2002) Macroeconomics. Pearson Education, New Jersey
5. Čaplánová A, Martincová M (2014) Inflation, unemployment and human capital from macroeconomic of view. Wolters Kluwer, Bratislava
6. Gerdesmeier D (2011) Price stability: why is it important to you? Frankfurt am Main: European Central Bank, 25–26. Available at: https://www.ecb.europa.eu/pub/pdf/other/price_stability_web_2011sk.pdf?3da7d27d233bb8ac51bdfc1a53e96822
7. Gregus M, Kryvinska N (2015) Service orientation of enterprises—aspects, dimensions, technologies. Comenius University in Bratislava, ISBN: 9788022339780
8. Holmes MH (2016) Introduction to scientific computing and data analysis. Springer International Publishing Switzerland (2016)
9. Kaczor S, Kryvinska N (2013) It is all about services—fundamentals, drivers, and business models. The society of service science. J Serv Sci Res 5(2):125–154 (Springer)
10. Kardaun OJWF (2005) Classical methods of statistics. Springer, Berlin
11. Kotlebová J, Sobek O (2007) Monetary policy: strategies, institutions and instruments. Iura Edition, Bratislava
12. Kryvinska N (2012) Building consistent formal specification for the service enterprise agility foundation. The society of service science. J Serv Sci Res 4(2):235–269 (Springer)
13. Kryvinska N, Gregus M (2014) SOA and its business value in requirements, features, practices and methodologies. Comenius University in Bratislava, ISBN: 9788022337649
14. Mankiw NG (2004) Principles of economics. Thomson South-Western Mason, Ohio
15. Ray M, Anderson D (2011) Krugman's economics for AP. Worth Publishers, New York
16. Rusnák M (2016) Nowcasting Czech GDP in real time. Econ Model 54(4):26–39. https://doi.org/10.1016/j.econmod.2015.12.010
17. Schumacher C (2005) Forecasting German GDP using alternative factor models based on large datasets. Discussion Paper Series 1: Economic Studies No 24
18. Schumacher Ch, Breitung J (2008) Real-time forecasting of German GDP based on a large factor model with monthly and quarterly data. Int J Forecast 24(3):386–398
19. Simionescu M, Dobeš K, Brezina I, Gaal A (2016) GDP rate in the European Union: simulations based on panel data models. J Int Stud 9(3):191–202. https://doi.org/10.14254/2071-8330.2016/9-3/15
20. Soares MJ, Aguiar-Conraria L (2014) Inflation rate dynamics convergence within the Euro. In: Murgante B et al (eds) Computational science and its applications—ICCSA 2014. ICCSA 2014. Lecture Notes in Computer Science, vol 8579. Springer, Cham

21. Sorić P (2018) Consumer confidence as a GDP determinant in New EU Member States: a view from a time-varying perspective. Empirica 45(2):261–282. https://doi.org/10.1007/s10663-016-9360-4

22. Stoshikj M, Kryvinska N, Strauss C (2013) Project management as a service. In: The 15th international conference on information integration and web-based applications & services (iiWAS2013), 2–4 December 2013. ACM, Vienna, Austria, pp 220–228

23. Stoshikj M, Kryvinska N, Strauss C (2014) Efficient managing of complex programs with project management services. Global Journal of Flexible Systems Management, Special Issue on Flexible Complexity Management and Engineering by Innovative Services, vol 15, Issue 1. Springer, Berlin, pp 25–38

24. Taş N, Hepsen A, Önder E (2013) Analyzing macroeconomic indicators of economic growth using panel data. J Finance Investment Anal 2(3):41–53. Available at SSRN: https://ssrn.com/abstract=2264388 or http://dx.doi.org/10.2139/ssrn.2264388

25. WorldBank (2016) WorldBank data [Online] www.data.worldbank.org

Office 365 as a Tool for Effective Internal Communication in Corporate Environment

Martin Krajčík

Abstract Email communication as one of the most widespread computer applications today, is in general very popular among its users. As it is used for both—either a personal or a professional communication, there occur also drawbacks regarding email usage: an increasing number of messages that overwhelm users, systems and become extremely time consuming. Despite the massive expansion of social media sites, email communication keeps its position and importance within organizational communication process and tends to increase. However, email communication can also result inefficient and can become a cause of a poor work efficiency. The aim of this chapter is a description and analysis of Office 365 and its components with a focus on SharePoint Online in corporate environment. We will offer general recommendations and guidelines for a specific organization, which will be applicable in practice to other organizations using the system, or for organizations that are considering the implementation. The chapter also describes the partial implementation of services within the selected workplaces. Office 365 components, components of SharePoint Online are implemented to improve an effective communication within the organization and to reduce the number of emails sent or received only to the necessary minimum.

1 Introduction

Email communication is still one of the most popular means of electronic communication within or outside the organization. In the era of social networks such as Facebook, YouTube, LinkedIn, Twitter or Instagram, and many more, people are still using the "old fashioned" email. However, several mainly corporate-focused platforms and applications are taking the power of the content sharing and collaboration to the next level in corporate sphere to improve communication and productivity of the employees [1].

M. Krajčík (✉)
Comenius University in Bratislava, Bratislava, Slovakia
e-mail: martin.krajcik@fm.uniba.sk

© Springer International Publishing AG, part of Springer Nature 2019
N. Kryvinska and M. Greguš (eds.), *Data-Centric Business and Applications*,
Lecture Notes on Data Engineering and Communications Technologies 20,
https://doi.org/10.1007/978-3-319-94117-2_12

The Trend journal reports, frequent email checking increases the stress, lowers IQ and it is more difficult to resist it than to resist the smoking [2]. Most people do not avoid the email at all, and according to a survey of 1100 British employees, more than half of them is responding to a message immediately, as soon as possible after receiving the message. Almost 21% of them do not have a problem to interrupt the meeting to answer the email [3].

Procure Plus, the British company, with around 40 employees undertook a daring attempt, when the management of the company decided that internal communication will be carried out by any means, just not via the e-mail. Employees were distracted by the corporate regulations, bordering with sociological research at first, as the BBC wrote, but soon they got used to it. The experiment lasted only one week, but in the words of Mike Brogan, president of the consortium, they have seen very impressive results. "E-mails have been reduced by 50% and at the offices could be seen more life," he said. Satisfaction with the results has meant that the company has remained the regulations [4].

Office 365 is the brand name of Microsoft Corporation group of services and software that provides productivity software such as Microsoft Word, Excel or PowerPoint, cloud storage service and software, OneDrive, e-mail service, Outlook, social networking services Yammer, SharePoint and other. Office 365 is available in different variations depending on which type of users are going to work with Office 365.

In this chapter, we are describing the possibilities that come with the implementation of Office 365 service. This implementation will take place at the Faculty of Management Comenius University in Bratislava, Slovakia. Nevertheless, it is also applicable in corporate sphere since the faculty environment is similar to corporate environment, in terms of number of users, departments and hierarchy. We will also consider and describe all possibilities that can help improve the communication inside the faculty or can stop wasting the time of the employees on irrelevant processes such as reading unnecessary emails or replying to one email multiple times to different users.

The problem is that there are almost no functional processes at the Faculty of Management in the moment. This also means that even existing processes are not working properly, and employees are wasting their time on meaningless paperwork and tasks or by writing the emails to senior staff members what they are supposed to do. By implementation of the features that come with Office 365 into the employee's work-lives, we would like to set some standards on how to do things more effectively. After the implementation of these features and tools, we will also set the basics for future development of processes in the organization. These processes could be later revised, standardized and used to improve the quality of employee's experience and the organization can benefit from that in many ways.

2 Internal Communication and Its Importance for an Organization

Communication through Internet changed the corporate environment dramatically. Company managers do not have to be in place to lead their subordinates, the contracts can be signed in minutes thanks to electronic signatures and the employees could stay informed about company news at any time and at any place. Internet communication provides us with lot of benefits, but this type of communication has also many downsides, as we will discuss later.

Even though communication outside the organization is very important it is well known, that communication inside the organization could be important as well.

Company leaders often think that employees are the only tool to achieve their goals. Employees do not necessarily need to know everything about the situation in the organization or what is organization planning to the future. The result is that employees could be depressed, frustrated and their work performance is low. High quality internal communication does not affects only working performance or employee satisfaction, but it could also lead to increased reputation of the organization. Employees are also part of the brand image of the organization and it is very important how they behave. Not only PR manager, but also the receptionists answering the phones could have a significant impact on the brand awareness or on the brand image of the organization.

"Internal communication is one of the key factors that affect the performance of the organization and thus their profitability. The primary objective of the survey is to provide a useful picture of internal communication in our region and help organizations set priorities and strategic objectives for its further development", explains Martin Onofre, founder and president of AICO [5].

The survey focused on communication tools, measuring the effectiveness, funding and state of development of internal communication. The survey focused also on the perception of the effectiveness of internal communication in specific areas, such as support in the fight against fraud and corruption of staff.

The survey showed that in most of Slovak organizations competence of internal communication remains the domain of the human resources but compared to the survey of 2012 it could also be seen increased momentum for its implementation in the Communications Department (Fig. 1).

The biggest obstacle to development is the lack of personnel, time, incorrectly adjusted processes and limited budgets. The results also revealed that the majority of Slovak companies in the sector does not measure anything. "If you do not measure performance, efficiency, and return on investment—no matter whether the capacity, time, or other reasons—then top management do not have enough arguments, why invest in the creation or strengthening of teams, their training in advanced communications technologies and other. Thus, our internal communication is chasing its own tail", adds Onofrej [5] (Fig. 2).

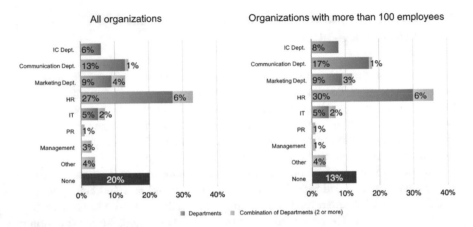

Fig. 1 Department responsible for internal communication

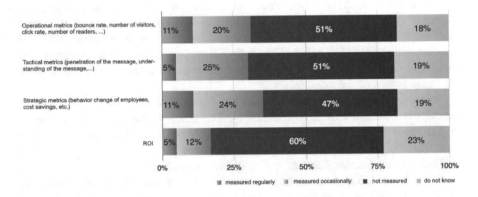

Fig. 2 Measurements in internal communications

According to respondents, the most effective internal communication role is to inform employees. On the other hand, the least effective is in the implementation of internal social networks (and support operating with them), even though this should be the crucial task of the companies of the 21st century. Organizations can streamline the communication and cooperation, promote exchange of knowledge and experience, shorten response times and eliminate organizational silo-effects [5] (Fig. 3).

Respondents also assessed that the internal communication is most effective in promoting awareness of occupational safety and health. On the contrary, the weakest is communicating the sensitive issues (dismissal, benefit and cost reduction). There are also gaps in supporting the fight against fraud and corruption of employees, while the survey revealed that respondents from the top management positions in this area believe and are perceived significantly more positive about the fact [5] (Fig. 4).

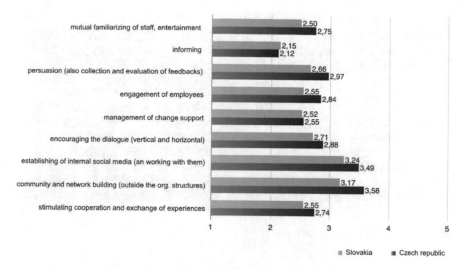

Fig. 3 Comparison of internal communication competences

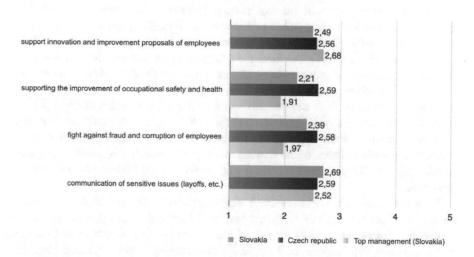

Fig. 4 Assessment of perception of effectivity in specific areas

It is no surprise that email is our dominant communication tool. In the survey did not appear any pioneering organization that has decided to eliminate email from the company's life. Slovakia has remained weak dynamics of the development of modern communication tools, especially social networking sites and in-house videos [5] (Fig. 5).

Budgets are stagnating. Exceptions are intranets and communication training for managers and internal communicators. Intranets are also the only instrument where

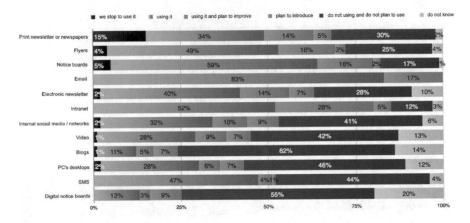

Fig. 5 Usage of communication tools

the companies are not planning the reduction of the budgets. Strengthening the position of employees with competence of internal communication is negligible.

"For the development of internal communication, it is necessary to open a dialogue with the top management of enterprises. Therefore, we are pleased with the increased participation of respondents in top management positions. This indicates that the internal communication begins to be among the top areas of concern. Leaders are starting to show an active interest in discussion on strengthening it. This will open the debate with leaders of major Slovak organizations at the professional event on February 26, 2015 in Bratislava", concludes Onofrej [5].

Association for Internal Communications (AICO) is a non-profit professional association whose primary mission is to enhance the internal communication in an organization with a focus on improving their performance, competitiveness and profitability. AICO members are local and multinational companies as well as government.

Well described and implemented business processes accessible to employees through intranet applications are germ purposeful gathering and processing information that forms the greatest wealth of the company—the wealth of knowledge and experience [6]. Experience has shown that reusing knowledge and accurate information is and will be the main driver of innovation. The main task managers will therefore know exactly what processes in the enterprise, not only in terms of economy, but especially in terms of their management style and how produce value. Among the cornerstones include improving the process of direct management of people and teams (task assignment and control), continuous improvement of systems of indirect management (corporate directives and guidelines), as well as coordination and communication with employees at various levels of management (senior, operational, project management, education, etc.).

Intranet provides all the necessary functions—communication, coordination and cooperation. The main benefit is gathering information in a manner that ensures their repeated use, and that the highest possible number of workers. It creates a knowledge base from which benefit managers in the design of organizational changes or new product development. Considerable contribution is a shortening of the adaptation process.

2.1 Internal Communication Tools

Several internal communication tools can be provided within the organization. Each of the tools have their advantages and disadvantages or as you may know "pros and cons".

Email. Email is still an important piece of the pie and is difficult to eliminate, but there are more powerful internal communication tools you can be using to increase communication, collaboration and efficiency in your organization.

So, what could possibly go wrong with email? Well, there is a list of things that could possibly go wrong, but we will focus on duplicity of emails. Duplicity is, from our point of view, the biggest problem and disadvantage of the emails at the faculty, and probably in most of the organizations. As we were able to see at the Faculty of Management, the duplicity is a big problem there. Students were often getting one information from many sources. For example, students in the final year are getting sample questions for their state examinations. These sample questions are sending Vice Dean for Undergraduate and Graduate Study to all students directly to their faculty mailboxes. A few moments later, Head of Study Department is sending the same information. Finally, most of the faculty departments are sending this information to the student's mailboxes. This also applies to information about the bachelor and master thesis, information about course schedules and more.

Because we do not have the opportunity to evaluate and get the correct number of the amount of duplicate emails, we went through the average student and employee mailbox. Our research shows that approximately 18% of their emails were duplicates (the same emails sent from various sources or forwarded) and another 8% of these emails were similar emails (the same emails with some marginal modification or edited text). We can assume that by implementing the tools that Office 365 offers we can potentially decrease the amount of sent and received emails within the organization up to 26%, which represents one quarter of all email communication at the faculty.

We can see the number of email accounts and users on the Fig. 6 [5]. The number of email accounts has increasing tendency even if the number of active users of social network is increasing from year to year.

"Put it in the mail, so I do not forget about it". Sentence that we use too often at the time of smartphones and computers. There are hundreds of messages coming to the mailboxes every day. Many of them are important, but the vast majority is just useless spam. This is not a problem if we had not checked the mailbox every few minutes. It disturbs our concentration, productivity and wellbeing at work and as researches say it also decreases our IQ.

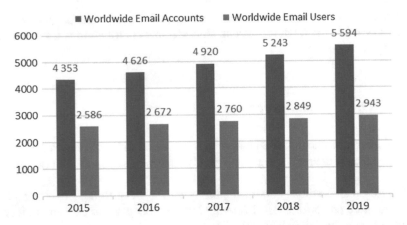

Fig. 6 Worldwide email accounts and user forecast, 2015—2019 (*in millions)

According to a study of consulting company McKinsey, email is the second biggest time consumer at the work. If the communication work more efficient, for example through social networks, an employee could leave work two hours earlier, it would save 25–30% of their working time (Fig. 7).

An average employee in a big corporation spends 680 h a year on email, which is in the normal eight-hour shifts 85 days or four months working (if you have 20 working days a month) [7]. These calculations did former employee of IBM Luis Suarez. In February 2008 he got a revolutionary idea to get rid of the mail completely, even though he works in an international corporation IBM, where email is a major communication channel.

Suarez still owns his email address, but he decided to stop responding. He was receiving to his mailbox event notifications and many information from inside the

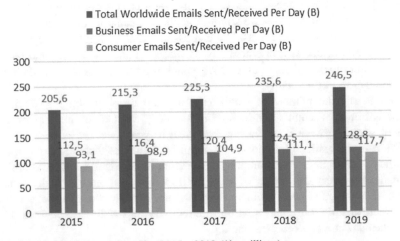

Fig. 7 Worldwide daily email traffic, 2015—2019 (*in millions)

company, which indeed he checked, but refused to answer, even though his boss worked in the USA and Suarez lived and worked in the Spanish Gran Canaria. Colleagues thought that he will last only two weeks, but he lasted until his retirement a few months ago. He worked 17 years for IBM [7].

In the past Suarez was receiving from 30 to 40 emails a day, now he receives 16 per week. At first glance, the Spaniard opted for extreme solution, but in fact he preferred only other services for communication—internal social networking (Connections) and private social networks such as Facebook or Twitter. He said that he increased efficiency and he is less disturbed at work. "It's complicated at the first sight in that you will replace one tool—the e-mail—with several other tools. Nevertheless, once you have it and your brain get used to it you will become more productive", he said recently in Hyde Park CT24 [7]. He also wanted to measure how efficient it was. He installed an application on his computer and measured the time spent on social networks for one year. The research showed that he spent 35 days on the PC compared to 85 days for the period of active use of e-mails.

In fact, he never planned to bury the e-mail, only more modern and faster technology. It was a transition from one form of communication to another. He argues that people should not respond via e-mail but choose a different way to respond. The more you respond to the emails the more emails you will receive. "When you answer them via Twitter they will start to do the same", he says [7]. Suarez is also giving advises to people to get their emails organized into the categories (in his case it was around 40 categories) as meetings, notifications, invitations, photo sharing and so on. The point is to generate one category and determine the service that would have done better for them. For example, you do not have to use email for file sharing instead you can use Dropbox or OneDrive. If you want an immediate response you can make a phone call and if you want to brainstorm a text, upload into Google Documents or you can use Microsoft Office Online.

The idea of Suarez has inspired some other companies. Three years ago, Atos, one of the largest IT companies, has banned internal communication via e-mail [8]. Each employee of Atos received from colleagues 100 emails a day on average. Only 15% of them were yet useful, according to the Atos study. CEO of the company is former French Minister of Finance Thierry Breton, who hates electronic communications. It has been eight years since he sent an e-mail and he does not intend to change anything. "If you people want to contact me they can come and visit me in person or call me," said Breton [8]. Since year 2014 employees are prohibited from sending the emails to each other. Instead of this they can use another type of communication—a corporate social network similar to Facebook.

In 2011, the automaker Volkswagen restricted emails in Germany. Blackberry business phones have been equipped with software that blocks email sending half an hour after the employer changes and unlocks it 30 min before the next work period. This action was the result of increasing criticism of internal emails, which employees were processing all day long.

Deutsche Telekom officially allowed to employees to not respond on work emails and calls after work. Based on the German operator experience, the German

Ministry of Labor issued an order for their employees, that none of the employee, who owns a mobile phone or Internet connection may not use it for working agenda outside the individual working hours.

Lot of changes are coming also from France where the government is preparing laws and regulations about working email and calls after working hours. These emails and calls will be prohibited, or the employees cannot be punished if they will not be responding after working hours.

Intranet. Intranet [9] is well-known tool in lot of organizations. In this case, we will present Intranet as an effective communication tool. Companies use intranet on everyday basis to keep their employees informed about important events, meetings or everyday tasks. You can also use the Intranet as a tool to collaborate, coordinate the work or you can use internal explainer videos, as we will describe below. You can easily implement other tools or plugins to the intranet to store and share documents, create a blog, wiki page or just create information channel. Everything depends on the organization needs and requirements.

We can expect and demand from intranet exactly those features, which are provided to control the exchange of information between managers and employees, but not only in the form of electronic mail.

Email is "too democratic". It is lagging behind the capabilities of current computer technology. The benefit from the PC is almost hundred percent when the employee is writing the letter using the most recent word processor software, but almost zero when the letter must be approved by the supervisor or when the employee wants to find the exact letter after half a year again. The problem is that the maximum level of communication and processing of correspondence represents, in most cases, only email.

Intranet's real value should be examined from the perspective of the value chain, that is, its overall contribution to the business strategy and management culture of the company according to its features and services [10]. It turns out that minimizing the costs is not the main driving force for the development of business activities. Constant pressure on shortening the innovation cycle has resulted into requirements on the managers [11]. It is required from managers to have flexible management methods, to coordinate projects and human resources, make use of their accumulated knowledge and experience. And here is where intranet comes.

When the managers have the access to the reports of the organization economics and its figures and overall value on the market with "intranet technologies" data from the ERP system or data warehouse could be available to them just in time. In this case, we can speak about management of information system. Therefore, the awareness of the top management is utter, and the managers can make better and faster decisions. But how these decisions will be applied and transferred to humans? Which management and control processes will be started? He will call a meeting, distribute the tasks, motivate the teams to solve the urgent problems, maybe he will find a professional who works somewhere in a branch. But the question is how? The same way that he obtained the information for the initial decision. Well, probably not, unless the intranet will be seen only as a tool to publish static corporate directives or phone book, which is often declared by utility of the intranet.

One step closer to our notion of the definition of intranet are managers that introduce a system of assignments, planning meetings, storing and sending out the minutes, manages the project meetings, and much more, because this is the beginning of communication which intranet as a medium can easily handle. The added value of this solution is not only to save some paper, but also that the manager can act immediately, directly contact members of the team as soon as he or she needs to materialize his or her ideas into tasks and have them under control even after some time.

To make exact calculations of application that intranet should solve is impossible. Intranet is not a commodity as an accounting software. In time, there appeared variety of intranet applications made exactly for this type of platform and we think that this trend will continue.

The simplest and most widely used applications are those with publication character. Organizations publish their organizational structures, phone books, annual reports or list of courses. The information is updated by the authorized department or employee using some of the tools for "painting" the web sites. The administrator then uploads these web pages on intranet. It is functional, but the only benefit is the concentration of information in one place and their availability through a computer. Nothing more and nothing less.

Another typical way of using the intranet is accessing data in the economic system through a web browser. Effective and reliable way for those managers who are willing and able to enter information request through the web browser UI.

Another application is administrative system, where can be found the utility value for manager. Document production according to prescribed standards of graphic content, workflow and managed access to documents (comments and approvals).

Most sophisticated intranet applications are based on deep research and experience of business processes. As an example, we can mention staff training, the starting package to increase the productivity. Educational applications in the intranet serves the HR department to compile individual training programs for various professions. Employees can order the intranet training, HR records and publishes achieved levels of staff qualifications. The result? For any senior executive would not be a problem at any time to find staff with the requisite qualifications for a short-term project.

Intranet undergoes the product development cycles it forms into the commodity as a database system, accounting programs or electronic mail few years ago. No one doubts about these systems anymore. The only thing that we can criticize in articles on the web is that intranet is often presented as just simple "painting" in-house web site, claiming that just a few programmers in Java or workers trained in the use of the relevant Microsoft products can create the intranet.

A quality intranet like any complex information system requires training and analysis, often more complex than in the case of database applications. It is required the process analysis in implementation phase, project management approach and good adaptation process of employees, because all employees with PC or mobile phone are connected to the intranet. This fact must be considered, when we are developing and implementing the application (security, governance cycle of documents, audit of the undertaken steps, distribution system, etc.).

Highly debated issue, but sometimes quite underestimated, is the quality of GUI that browsers and applications presented in it provides. While in conventional application GUI, ergonomics is strictly subordinate to labor productivity of user, in the WWW environment there are some limitations. Some manufacturers know about this problem and provide a more user-friendly environment (such as Lotus Notes).

We should also include Java programming language into the intranet's terminology, which is currently considered a panacea for developers or operational problems. Developers and programmers still have much to do to overcome all barriers (for example, so much discussed safety of Java), so major producers can release the solution suitable for large enterprises.

Live Chat. Live Chat or Instant Messaging is commonly used by lot of users of modern information systems (in Office 365 it is Lync or Skype for Business). In education environment, we can use it as a communication tool between students and teachers, when the student is sick and he or she needs to consult something discussed in the class, bachelor or master thesis or just get some information. This type of communication is better than by email, because the discussion is almost instant, and you have possibility to have face-to-face communication (web camera).

In many cases, email is misunderstood. To get the response to your message immediately you should use chat instead of email. This technology is there for a long time and has a strong place in our personal lives. We can mention the services such as WhatsApp, Viber, Skype, Facebook Messenger, Hangouts from Google or iMessage from Apple. Even though this technology wins in our personal lives, it is not common in business. Using the live chat, you are saving yours and your partner's time and you do not have unnecessary additional emails in your mailbox.

We can see on the Fig. 8 how many active users each chat platform has. From the graph, it is obvious that user, mostly from developed countries, are using social

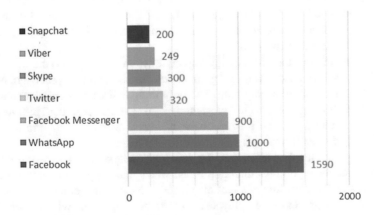

Fig. 8 Number of active users (*in millions)

networks in their private and business time very often. The connection to the internet is almost everywhere. You can connect at home on your WiFi, at the office on the work network, in the cafe or even in the train or in the bus on the way home. The number of active users shows us not only that people are using social networks actively, but also that they got used to it and they prefer to communicate through the social media, maybe rather than any other communication channel. This could help the organizations to decide whether or not implement the corporate social media in their organization.

Corporate Social Media. Social media became important in our lives. Most people are using social media such as Facebook, LinkedIn, Twitter and YouTube can be considered as a social media channel. Corporations know that people know how to use social media and people are keen on to use them also in their job.

Employee Self Service (ESS). Employee Self Service or ESS is used to give employees to access to their personal data, to edit them and view them and to give them access to the electronic payroll where they can see what their wage was, how much from this wage are taxes and which taxes do they pay, and more. It is also possible to create ESS directly on SharePoint site, but this will need further development and implementation of additional software or additional tools to connect the SharePoint to HR software. At the Faculty of Management, they are using SAP (Fig. 9).

There are tools to connecting SAP and SharePoint, which allows us to create the electronic payroll. However, we will not create ESS at the faculty, as it is system that is more complex and cannot be covered in this chapter.

Internal Explanatory Videos. Internal explainer videos or tutorials are one of the most popular means of sharing information. One of the most popular video sharing services, YouTube, uses more than 1 billion people [12]. More than 300 h

Fig. 9 An example of ESS on sharepoint

of video are uploaded on YouTube each day and over 6 billion hours of video are watched per month [12] (Fig. 10).

The uses for video are quite diverse as well [13].

- Brand Promotion. Videos can be used internally (and externally) to promote a company's brand and reinforce core values.
- Major Organizational Initiatives. Change is hard but ensuring employee is crucial to the success of any major organizational initiative. Clear messaging and effective communication is key.
- Training and Orientation. Internal explainer videos have a powerful place in the HR world. Hiring, training, and on boarding greatly benefit with the addition of a library of training videos that offer new employees the chance to learn at their own pace.

Internal video content ranges from webcasts to live-streamed events. Webcasts can include a mixture of video content and slides or just slideshows with an audio track. Examples of employee video communications include CEO messages; employee introductions; training; featured departments; product announcements; and marketing messages [14]. Fifty one percent of communicators measure the effectiveness of their video communications using video tracking software and 76% noted that audiovisual content improved employee communications [14]. Fifty percent of large organizations said that video was "very important to their employee communications" and nine of ten respondents said that video engages employees [14]. Some key video metrics used by employee communicators included video views, time spent watching video, the number of times the video was shared across the organization, and employee comments [14]. Thirty one percent of

Fig. 10 Office 365 Video UI

communicators said they do not measure the effectiveness of employee video campaigns due to not liking the analytics tools available or not having access to video analytics [14].

Desktop Internal Communication Systems. Desktop Internal Communication Systems is a specific communication tool. This tool is very effective in eliminating the number of sent or received emails inside an organization. You can share an information right on the screen of the employee's monitors.

Not only you do not need to send any email, but you can also track the users if they read the notifications or they ignored it—a fully complex statistics is available to a competent person. Based on the analysis of these data you can create latest updates for those users that ignored the first notification. If you notify your users regularly, you can customize the day or hour when the notification will pop up based on analyzed data.

As we can see on the picture, the system enables notification directly on the screen of any PC or laptop. This gives an operator the option to inform employees immediately about any situation that occurs in an organization. This system gives any operator or administrator several advantages.

Measurable. You can report on who has or has not opened the full message contained in a desktop alert and clicked on any related hyperlinks. This is a useful feature if you have to show compliance to any internal policies or legal regulations.

Flexible. Desktop alerts include "read now" or "read later" options, as well as the option to display the full message immediately. This is a useful option for emergency alert notifications. You can preset the size, position and prominence of the desktop alert window for each message.

Location Independent. Desktop alerts reach all staff, regardless of their location or network. Target alert messages to entire locations, individuals, groups, roles, remote and mobile workers.

Organizations are always trying to improve their image in a public, but still more organizations are trying to improve their internal processes, internal perception of the company and improve the overall working environment, so that employees feel more comfortable and that the organization is supporting them in their effort and encourages them in better performance.

Another way to use this system is a screensaver. It is not common for organizations to use screensaver as a place to promote any information. Even though screensaver is a suitable place to promote company surveys, information about new software, event updates and more. On the other hand, screensaver is not intended to be the main tool to promote information. This technology can be used as an additional source of promotion.

3 Goals, Subgoals and Research Methods

The main goal of the research is to describe the possibilities of Office 365, recognize the advantages and disadvantages and to provide better and more effective communication tools mainly inside the organization. We will also create the manual

("Best Practices") for further development of the implemented changes to the Office 365. This manual can help the organization, or any other, to make changes in the environment and involve more departments to the Office 365 without any problem.

To meet these objectives, we set subgoals. First subgoal is to create functional SharePoint site for the chosen department. We will set up the site, fill them with the information and we will share this site with all employees of the specific department that will be, in this scenario, Center of Information Technology at the Faculty of Management Comenius University in Bratislava. The second subgoal is to reduce the email communication whether it is a sent or received email communication. To evaluate this, we will process the data obtained from CIT on the specific object. This object will be the assistant at the Department of Information Systems.

To achieve these objectives, we have set the following hypotheses:

Hypothesis 1 *Office 365 has the potential to replace selected features of the email in most cases and reduce the ineffective email communication.*

Through several years of personal research and study of the cloud-based applications and email services we can assume that the Office 365 has the potential to replace selected features of the email. Email is used to not only send and receive electronic letters, as it is supposed to be, but email is also replacing social networks, instant messaging and also document library, or cloud-based storage. We assume that we will be able to replace some of the unpleasant habits of the employees at the faculty they learned in past years when email was the most preferred and the only communication option for them. To achieve this, we will create the SharePoint sites and tutorials on how to use these sites properly. This procedure has to be also sustainable and we need to think further to the future, as the faculty will continue using the SharePoint Online and other Office 365 features. In order to meet these requirements, we will collaborate and coordinate our activities with the employees of Centre of Information Technologies at the Faculty of Management Comenius University in Bratislava.

Hypothesis 2 *We can reduce the email communication of the selected object by 7% in the first month of implementation.*

We have selected one person at the faculty that we will cooperate with and monitor the mailbox of this person. We will focus on the previous habits of this person to better understand the workflow and the information flow. By continuous monitoring and tracking of the mailbox, we will be able to select the unnecessary emails and sort them into the groups or categories that we will set up. This will help us to identify which of the feature is represented in that particular email or the group of emails and which feature, or features, of the Office 365 is the best one to replace it. For the first sight, the 7% milestone is not one of the highest numbers, but we need to consider that most of the faculty staff is using email on everyday basis for several years. Most of the employees are in the category "45 years and more", which means that they are not the best category group to introduce to modern technologies and features. It takes time to teach them and to explain them why we want them to use more than just an email. After creating the SharePoint sites and after we revise them, we will focus on implementing this technology to the employees work lives.

3.1 Methodology of Research

The initial phase of the research is based on the study of the possibilities of Office 365 environment and its implementation to the organization. Through conversations with employees of the faculty and literature study, we can better understand the requirements according to the future administrators of the system (employees at CIT) and requirements of the employees in the organization in general on the other hand. Then we can process this obtained data for further analysis and it will help us in the empirical framework where we will implement specific solution.

We will collect materials and information from the Faculty of Management Comenius University in close collaboration of the Center of Information Technology. These data are up-to-date, accurate and suitable for further research. This data will be processed, and the results will be published in this research in form or graphs and other figures.

In the empirical framework, as we mentioned above, we will be implementing specific solution for the organization. We will create two SharePoint sites. One site will be created for the CIT as a base for their further development of the department and its services. The second site will be created for the Department of Information Systems, which is one of the main departments at the faculty. This site will be focusing on the employees of the department and their need to collaborate and share their ideas, work, and studies with other colleagues not only from the same department. The other challenge will be to provide them with the opportunity to use the SharePoint site to help them with the administrative tasks to find forms they need, to get the right information about all rights and obligations and to keep them informed about the events of the department.

3.2 Data Processing Technique

The data we will process will be gained from the resources of the CIT. Firstly, we will process and analyze raw data in form of logs. This data has to be exported from the servers then processed into the form readable for humans and finally we will analyze and further process these data into to the graphs and figures.

Another resource for our data will be information directly from the user's mailboxes. These data will be anonymous to preserve the anonymity of each user and not to harm their personal information. For obtaining this type of data, we will use Office 365 Admin Panel, which is providing basic analytical tools. These tools will show us the data in form of numbers (number of sent emails, number of received emails, number of spam), basic ratio between the received emails and spams (or any other) and we will be able to see graphs made automatically from this data. These graphs are processed by the Microsoft's service Office 365 directly.

In the specific case where we will cooperate with the Department of Information Systems, we will track the mailbox of one member. All data that we will gain are also in raw form and no personal details will be processed neither the content of any email. We will observe only the number of received/sent emails, number of emails received from/sent to a one repeating email address, number of emails received/sent with the same or similar subject. This data will serve us to evaluate if our hypotheses are correct, what happened differently than we expected, and this data will also provide us with the overall picture of the situation at the Faculty of Management regarding to email usage.

4   Office 365 and SharePoint Online User Interface

To better understand the potential of Office 365 and creating the SharePoint sites, intranet and implementing some of the Office 365 features, we need to see how it works. The user interface is very important in terms of better communication. If the user interface will not be simple, easy to use and well-structured users can get confused and the implementation will get complicated and we could easily fail. Good or excellent user interface is interface where the designer does not need to explain it and make complicated and long-lasting tutorials for the users.

The biggest advantage of Office 365 is that all services are available in one place. In the upper left corner, you have a menu, where you can choose which app you want to launch. You can access all the Office 365 apps (Outlook, Yammer, SharePoint, Office Online and more), you can add additional apps such as Power BI, Microsoft Dynamics CRM or if your company has its own applications, you can use these and offer them to your employees or customers connected to your Office 365 environment.

4.1   Outlook on the Web

Outlook on the web is a web browser version of Microsoft Outlook that is used by businesses and organizations. Outlook on the web can only be used to access Office 365 for business and other accounts that are hosted on a server that is running Microsoft Exchange Server.

Outlook on the web is available to everyone who is connected to the Internet. To access your mailbox, you can use most popular web browsers. Also, you do not need to worry if the users are using the same operating system (OS) as your organization. The web application works fine with all OS if there is a web browser

Fig. 11 Outlook on the web user interface

installed. That also means that you and your employees do not need to install the MS Office suite to get access to their mailboxes (Fig. 11).

We can say that the Outlook on the web and Microsoft's application MS Outlook have very similar UI. From the left, we can see folders (Navigation Pane), list of received messages (Inbox) and reading pane.

4.2 SharePoint

Cloud-based service hosted by Microsoft Corporation that is designed for businesses of all sizes. Instead of local installation and deployment of SharePoint Server can subscribe to every business plan Office 365 or a separate SharePoint Online. Employees can thus create sites to share documents and information with colleagues, partners and customers (Fig. 12).

SharePoint Online is much more structured and complicated application, but its UI is complex and sophisticated.

The UI is fully editable by an administrator. It is possible to make only some changes that are natively supported. You can easily change the main menu, sidebar menu or add/remove applications from the main page of the site.

In addition, you can change the look of your intranet sites. If you want to change the look of SharePoint site, you can modify the source code or create a modern design using HTML, CSS and JavaScript. Then, these files will be transformed into the readable code for SharePoint and will be uploaded to SharePoint directory.

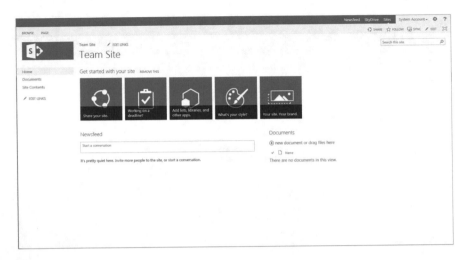

Fig. 12 Default sharepoint team site user interface

4.3 OneDrive for Business

OneDrive for Business (OD4B) is an integral part of Office 365 or SharePoint Server and provides space in the cloud, where you can store, share and synchronize your work. Files can be updated and shared from any device that uses OneDrive for Business. You can even work on Office documents simultaneously with other users. There are three main ways to access the OneDrive for business.

The first way is to access the OD4B is by installing the OneDrive for Business application directly to your computer. You can install the standalone application, or this application could be part of the MS Office bundle. The biggest advantage of installed application is that you can work with the files offline. If your job requires to travel a lot, it is more comfortable to edit documents offline and when you connect to the Internet all your changes will be uploaded in the cloud to your Office 365 environment. A disadvantage could be that while the OneDrive is uploading or downloading larger files you can notice some performance issues. If you are synchronizing the OneDrive for the first time with the server (Office 365) and you already have files in the OneDrive, we recommend running the synchronization in time when you will not be using the computer. When the synchronization is done, you can use your computer as always. All changes to the documents will be synchronized immediately and with almost no reduction in the performance.

The second way is to access the OD4B directly from the web interface. Web interface offers the same features as the installed application except that you have to be connected to the Internet. The web interface can be comfortable for organizations, because they do not need to install any application on computers/workstations (Fig. 13).

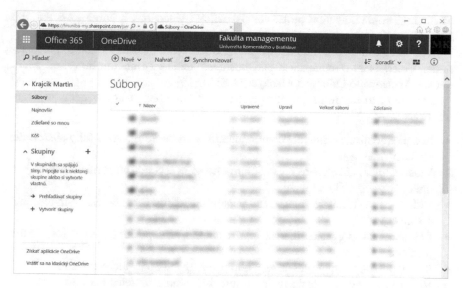

Fig. 13 OneDrive for Business - Web Application

The third option is to access the OD4B from your mobile device. Whether it is a smartphone or tablet or any other mobile device you can download the application. This could be an advantage if the employees are using their mobile devices to access the mailbox, documents or any other service provided by organization. OD4B for mobile devices offers access to online and offline documents, so if you cannot connect to Internet you can easily edit the documents offline and upload them after connecting to Internet.

4.4 Lync and Skype for Business

With Skype for Business, you can stay connected with colleagues and business partners. In one package, you will be able to communicate with them via instant messaging, online meetings, calls, and videoconferences and use the sharing and collaboration features.

4.5 Video

Video is a service similar to YouTube. You can upload videos into your intranet and share them with your colleagues, bosses or subordinates. This service could have an enormous potential in internal communication of the organization. Video is one of the most popular means of content sharing. According to one research, 93%

of internal communication professionals believe video has become essential and
54% of employees expect to see video in the workplace [15].

"No more long, cumbersome marketing documents—no more boring presentations. Video is how companies and business should communicate", said Mark
Leaser, Worldwide Offerings Manager at IBM [15].

According to the research video can provide number of key benefits to organizations [15]:

- Recipients of video internal communications are more likely to fully absorb the
  intended message.
- Video is more effective as it can communicate in seconds what might have taken
  multiple paragraphs to write.
- Video can eliminate many in person meetings, handouts, e-mails and
  documents.
- Research has suggested that people react to video positively as it feels more
  personal.
- It can spread training messages without costly sessions and seminars.
- Video is always available in the future for people to refer back to—email is
  often deleted.
- Video is accessible from a variety of devices from desktop to mobile.
- Video offers the chance to create emotive content that increases engagement.

4.6 Sway

Sway was primarily intended to create and share interactive reports, presentations,
private messages and other files. Sway was designed to replace or at least to extend
the possibilities of conventional presentations made in MS PowerPoint. Not only
you can add pictures, videos and text, but you can also add content from third-party
sites via embed codes. Thus, Sway is an integrated part of Office 365 you can
access all files stored in OD4B (Fig. 14).

4.7 Delve

With Delve application, you can connect and collaborate with other people, as well
as to identify and organize the information that are the most important to you in
Office 365. Delve displays only information you have access. Other users will not
see your private documents. We call Delve a hub that provides us with the benefits
of Office 365 and its collaboration features. If you are working on a project with
your colleagues, in Delve you can see which documents were edited recently and
who edited them (Fig. 15).

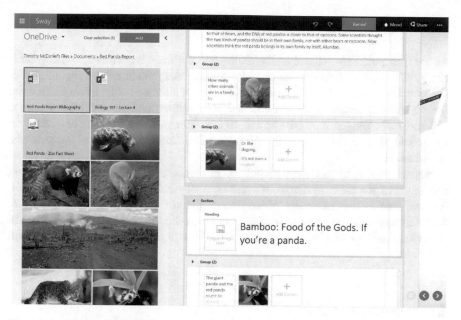

Fig. 14 Sway application interface

Fig. 15 Delve personal site

5 Creating a SharePoint Site

As we mentioned above, one of the Office 365 services we will be focusing on is SharePoint Online. This service provides all necessary features to boost the effective communication. We will create basic Intranet sites and subsites in SharePoint Online and implement some other Office 365 features and services. Intranet sites will serve to users, so they will be able to find all necessary information there and we would like to prevent them from sending unnecessary emails within the organization.

Before we start to create any SharePoint site, it is necessary to focus on planning a structure of the entire system to prevent an unorganized formation of the intranet. We will create basic hierarchy of the SharePoint for the Faculty of Management. This will help us from creating sites that are not well structured and implemented in the system and it will give us instructions on how to proceed when a new site will be created.

5.1 Structure Planning

Planning is one of the most important managerial skills. In this case, we will set the goals and orientation of the Intranet for the future development. These rules where and how to create the sites will lead to better structured and well-organized intranet portal.

The basic idea of intranet for the Faculty of Management is to create a space where everyone (employees, students, and administrative staff) can come and find all necessary information for their work or study. We would like to create a space for sharing the ideas, collaboration and space where the faculty can strengthen its image and organizational culture in depth. We would also like to prevent employees from sending unnecessary emails to their colleagues and thus decrease the overall amount of emails in the organization. This will give the employees an opportunity to use this technology and chance to not only reduce the number of emails in their mailboxes, but also save time.

SharePoint, as we can see on the Fig. 16 is a web application that is based on applications. Blogs, various lists, document libraries—all these are application installed on the SharePoint Online and sites or subsites created on the SharePoint.

The Top-level site is the main site (your domain by default; www.yourdomain. sharepoint.com) is the site from where you will begin to develop the entire system. From there you can create sites and subsites according to your plan (Fig. 17).

As we can see on the Fig. 20. We will divide the main (Top-level) site into several subsites. First, we will create subsite for the departments. There are five departments at the faculty, thus we will create one main subsite called "Departments" under which we will create five additional subsites according to the departments. Each subsite will have their own administrator of the site. We can

Fig. 16 Structure of a site collection in sharepoint 2013

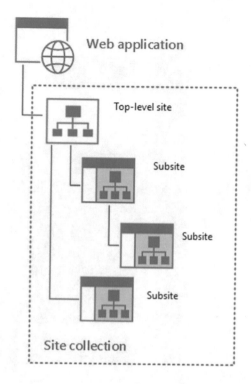

consider the head of the department and assistant of the head of the department as the best option. We will also need to give access to this site to members of the department. These members will not have access as the administrator of the site (or subsite), but they will be listed here as members of the subsite. This role will have permission to add, delete or edit documents, see various lists, calendars and libraries of the department's site.

Another subsite will be site for CIT. As we mentioned before, this department has four employees. These employees will be also administrators of the site. However, we need to give access also to all employees and students to this site, as they will be looking for help and answers on IT topics on this site. Therefore, we will grant them access only to some parts of the CIT site. This department will also need a blog where they can inform the users about modern technologies, techniques and present some tips and tricks for software used at the faculty.

5.2 SharePoint Site for Centre of Information Technology

Center of Information Technology (CIT) is a specific department at faculty. This department provides and supports multiple services for users. CIT is not only administrating the whole information system at the faculty, but also managing and

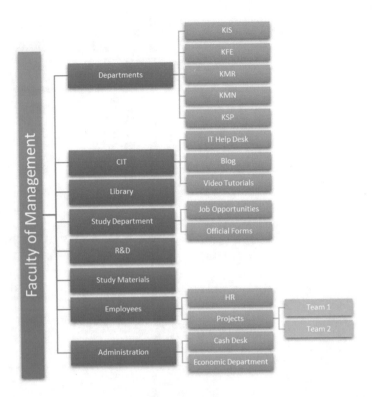

Fig. 17 The proposed structure of the sharepoint sites

supporting Office 365 and its features. This means that there are various requests to the department to be solved and all of them have to be solved on time (because as we know, every user is the most important one). CIT is receiving three requests per day on average, not counting the phone requests. After we analyzed all data, we were able to categorize these requests as follow (Fig. 18):

Password reset/blocked emails. These are the most frequented requests. Users forgot their passwords, or they were not responding to administrator's email, so their email address was blocked. These blocked accounts are mostly students that exceed the length of the study (typically 3 years for bachelor and 2 years for master).

Purchase of goods, services and licenses. CIT also provides installation of the hardware and the software at the faculty. That means all purchased goods, services or licenses must be processed by the IT technicians at CIT. Therefore, all requests are dealt through CIT employees, which means they are dealing with lot of emails.

Reasonable service and configuration requests. These are typical requests from users such as the service of computers or printers, software installation, software and hardware configuration and other. Also, all malfunctioning printers, pc's or other devices must be checked by the technicians and if it is not in

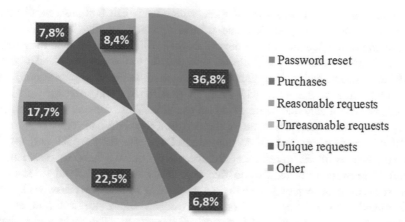

Fig. 18 Categorization of received requests

competence of the technician to repair it or change the device, they have to send a request to the external organization.

Unreasonable requests. We can also call them mails that are sent to the wrong recipient. These are request to reset password to Academic Information System (ASI2), which is under administration of rectorate and not under faculty's departments. Nevertheless, these requests must be closed and employees at CIT have to answer them and send contact details to rectorate. We can prevent this by posting the information to the website that is accessible to public. Even though CIT can receive similar requests, because as we have learnt in the organization not all users are reading the manuals or tutorials (this applies to most users of information technologies).

Unique or disposable requests. These requests are requests that are not repeating regularly. Example of such a request could be turning the internet off in the PC room at the faculty on particular day at particular time. This type of request appears from time to time depending on period of the school year. These requests cannot be planned and automated, because they are appearing as a current need of teachers and it is changing from year to year and from semester to semester.

Other. These other emails are emails that do not contain any requests. This could be an information email or an email from automatic email systems.

This categorization will help us to better understand the structure of received emails, which we will try to reduce. Also, according to this categorization, we will create each part of the SharePoint site for the CIT.

While we are creating the site, we must consider that one part of the site for CIT has to be visible for public audience, in this case for the employees of faculty. The second part will be accessible only by CIT employees for their internal needs. As we mentioned above, this department is administrating the whole information system at faculty. This means that the site has to be well structured and easy to navigate, so users will be able to search answers for their questions and find all necessary information. Thanks to the site where users can find answers to most of their questions, we expect decrease of the received emails. This will lead to timesaving of the department and the employees will be able to spend their time on more important tasks (they will not need to answer to the same or similar requests of different users, such as configuration of the email on the smartphone, etc.). In the case that some of the users will send an email, requesting the answers to the question that is mentioned or answered on the SharePoint site employees can easily send a link to this user where he or she can find the right answer.

In this case, we cannot create only SharePoint site for the CIT and use it as primary source of the information for the users, but we have to create new or maintain existing website. Due to fact that users are not allowed to reset their passwords on their own they have to contact the CIT to reset the password for them. CIT already has existing and functional website where we can find all basic information about the Office 365 services at the faculty. The main topics that are covered on the website are about mobile device configuration, MS Outlook configuration, possibility to get MS Office for free as a part of license, contact information and more. Still, there is no Frequently Asked Questions (FAQ) section on the website, but we would like to create such a section on the SharePoint site. This site will be visible only to the faculty users and to public audience outside the organization. We want to prevent the audience from the outside of the organization to access this data, because some of the information mentioned in the FAQ, or any other place on the intranet, might be described as a sensitive data for the organization or sensitive data that we do not want to share with a public audience.

Document Library. Document library is one of the most suitable feature to reduce the number of emails. Most of the documents have an attachment with one or more document files. These documents can be easily uploaded and stored in the cloud using the OneDrive or these files can be accessed through SharePoint site. In addition, lot of employees in many organizations are using their mailbox as a file storage. They create new message, attach a file and save it as a draft or even worst they send this kind of email to themselves. This means that one copy of the email is in Sent folder and the other copy is in their Inbox. This could lead to unnecessary overload of the mailbox and additional costs due to storage limitations (you need to buy additional disk drive in the case you are using Office 365 on the premise or additional space in the cloud for Office 365 online users).

We will create two document libraries. First document library will be managed by Centre of Information Technology. This library will contain all documents necessary to share among the employees of CIT. These documents could be documents on property records, logos of the organization, procedures and guidance for solving various problems, publications and books (literature, tutorials, etc.), phonebook, drivers for various hardware, presentations or any other documents.

The second document library will be open to public. This library will be filled with documents that can help users and can instruct them how to use given hardware (PCs and notebooks) or will guide them to correctly use the installed software.

Blog. Blog is a very good opportunity to communicate with your customers. In this case our customers are employees of the Faculty of Management. We would like to keep them informed what is happening in the CIT, what are the technologies that they are using, or they are implementing right now. We would also like to post some "Tips and Tricks" blogposts, which are very popular nowadays. Thanks to a blog that we will create, we would like to improve the technical skills of our users and involve them into the process of implementation of modern technologies. We are expecting also less questions from the users and better awareness.

You can see the basic setup for the blog in SharePoint on Fig. 22. You can find a list of categories on the left side. You can edit this list easily by adding, deleting or renaming the items in the list. This list will be also visible for the visitors of the blog, so they can more easily and in more convenient way access the blog post they want to see.

Every new entry, a blog post, will be categorize based on the category that we have created, or we can create a new category within a few clicks. Thanks to categories, we will be able to sort the blog posts by the specific criteria and to group posts that are related to similar or the same topic. It is better manageable, and our blog will not be full of various blog posts that are not related to each other. In addition, users will be able to read related blog posts to that one they are reading in the moment.

After publishing the blog post, we can share this blog post via email. Even if we want to reduce the number of emails, we need to start by using the email. All employees at the faculty are using only the email as their communication tool. If we will be able to teach the employees regularly visit this blog, we can reduce the number of notification emails to minimum or zero or these emails could be replaced by a newsletter that will be sent to their mailboxes once in two weeks. Employees will be able to find all necessary information and data in this newsletter; thus, the emails reduction will be visible for everyone.

Video Tutorials. We can use SharePoint as a knowledge base, so we can provide employees with tons of tutorials on how to connect to Intranet via mobile devices, tutorials according to MS Office (Excel, Word, PowerPoint, etc.), so we can make employee's life easier, etc.

We can also share these videos with students, colleagues and external participants. In fact, most people can learn more and be able to stay informed by watching the short videos rather than by reading the posts.

The average consumer with an Internet connection watches roughly 206 videos per month, and Nielsen claims 64% of marketers expect video to dominate their strategies in the near future [16]. Video is processed easily with almost no effort. The audiovisual content had better influence on us and our brain is more likely to remember the given information in the video.

CIT is already using the YouTube channel for uploading their video tutorials on how to install MS Office or how-to setup your email on your smartphone. Because CIT is dealing with the information system at the faculty, it will be more convenient and safe to upload these videos into the SharePoint Video. We can set the permission for watch video to control which video will be visible to whom.

Security/Permissions. The security could be very complex in the SharePoint yet still easy to setup. We will be setting the permission to the CIT site. We can easily set the global permission for whole SharePoint, partially for sites, subsites, but also for document libraries or any other site app individually. We will create 3 main groups at the department's SharePoint site.

- CIT—members
- CIT—visitors
- CIT—owners

It is also possible to use predefined groups that SharePoint is offering. These groups have set permissions and all you need to do is to delegate these permissions or add members to the particular group.

In the group "CIT—members" we will add all employees of the CIT, the head of the department and other employees that are working at CIT (part-time jobs, evening shift, etc.). Members will be able to read the content and to access the document library (add, edit and remove documents). They will also basic permission to add, edit and remove content from the site, write a blog, post a message or a link to the group.

"CIT—visitors" will be created for those users who work outside the department, but CIT wants them to access their data and information. These users could be students or employees, or both groups.

The last group, "CIT—owners" will be only 3 out of 4 employees at CIT. These members will have the administrator permission and will be responsible for the content published on the CIT site. They will have other permissions too, such as adding or deleting the members of each group (also the owners group), creating new subsites and many more.

5.3   SharePoint Site for Department of Information Systems

This is an another type of SharePoint site. The difference between this site and the site of the CIT are the users. In CIT, we have four employees, but the Department of Information systems have 20+ employees and also their interests and information requests are different.

Contact List. Contact list is a basic feature but can be very useful. While CIT do not need contact list for only four employees, it is a necessity at Department of IS. The number of internal and external employees and internal PhD. students are around fifty. We can easily set up this list by adding employees and students manually or automatically. We will describe the manual procedure, as the automatic procedure requires some system and settings changes in Active Directory used at faculty.

Assistant at Department of IS receives approximately 26 emails a month and 1 call a day average on requesting the contact details of employees at the department. The average time spent on reply to these emails is 4.5 min (receiving the email, going through the email, answering the email and searching for contact details, sending the email and receiving the feedback or confirmation email). This can lead to loss of 126 min a month of working time on emails.

Document Library. Document library for this department will be different from the one at CIT. It is due to fact that employees of this department are using documents differently. Employees need to share documents that are necessary for their administration duties.

Firstly, we set up basic site under the main site at faculty's SharePoint. As we stated above, we have created the site using the defined model. We have created site "Department of Information Systems" (../KIS). Further, we need to know what apps will be added to this site.

Tasks. Tasks are very useful feature. You can use tasks in your MS Outlook, Outlook on the web or even in the SharePoint Online. To create tasks section, we will install app from the selection. In the site content, we will find app called "Tasks", click on the app and enter the name of the app. In this case, we will call it "Tasks during Semester". In this app, we will add tasks like "write a report for a master thesis" or "update personal details for HR department". Besides the tasks for all employees, we can also add individual tasks by assigning one or more employees to a task. This feature can be helpful if the department will create a team, which has to cooperate and coordinate their steps to achieve the goals.

In the tasks application we can see also the timeline, progress bar, with an option to add one or more subtasks. Everyone in team can mark their own progress or they can see the overall progress of the whole team. In addition, the history of tasks is available to the team members. They can use it as an advantage and if any similar task they were dealing with before appears, they can go back to that task, review it and use the data from that previous task or improve the procedure to improve the final results.

Self-Service Portal. In coordination with HR Department, we will create simple Employee Self-Service site that can be extended with some other features in the future. We will call this ESS simply Self-Service Portal. It is possible to connect SAP, which is used at the faculty, to SharePoint, so every employee will be able to see his or her monthly earnings, monthly balance of days off, personal details and many more.

After brainstorming with HR department at faculty, we decided to create Self-Service Portal to share documents that are important for other departments at the faculty. HR department is dealing not only with contracting, but also has to manage the attendance records, personal details of all employees and many more.

The current situation is that HR department is sending all documents via email. This causes two main problems. First problem is that the amount of emails sent by the HR department may confuse employees and they can keep working with the old documents. If the department is sending all updates through the email and employees working with these documents do not categorize or prioritize their emails it is possible to overlook the most recent email with an updated document in attachment. The search in the mailbox is very inconvenient according to finding the right attachment, thus the searching for the right document can take longer time.

The second situation is obvious the department is sending nonproductive emails. If the HR department wants to correct some mistakes or update just some part or information in the document the department has to send another email.

Firstly, we create a site under the top-level site called "HR". This site will temporary contain only general information about the HR department, working hours, employees of this department and a faculty phone book. In the future, we are planning to extend this site with more functionalities.

Then, we will create the Self-Service Portal subsite. This subsite will be extended document library, because we will add also general information about the service, upload tutorials on how to use the library and give employees an example on how to fill in the documents. Afterwards, we will start to upload the predefined documents. Users can easily upload multiple documents using the drag-and-drop feature or if this feature is not available, you can still upload the documents through upload window.

The main advantage of this solution is that HR department do not has to send any documents to the employee's mailboxes. If there is a change in the document, it is possible to make this change in the Office online right in the document without even downloading it.

The other option is to reupload the corrected document. HR department can send hyperlink, or URL, to documents via email. The problem that can occur is that the link could get broken if the document shared by the link is moved, renamed or modified. In this case, the department do not have to worry about the broken hyperlinks to their documents. SharePoint Online not only offers versioning of the documents, but it is also possible to reupload, edit or even rename the documents and keep the same hyperlink to that document. This feature is very helpful. HR department can send only one email with link to the document and employees can access this document at time, even after the document has been changed.

6 Conclusion

Intranet is growing fast and can easily replace some of the feature of the email. There are still areas where intranet can be improved, and the developers should focus on these areas, but overall, we consider intranet as a great advantage for the companies and their employees. Intranet is not only helping employees to collaborate, share their ideas, experiences and files with others or to stay informed no matter where or when they connect to the site, but intranet also improves the employee's work experience, save their time by decreasing the number of received emails or it can improve the internal communication in an organization [17–19].

Office 365 offers all necessary features and services that give the administrators in the organizations tools for boosting the internal communication while reducing the number of email communication. As we proved in this chapter, email communication can be time consuming, ineffective and in some case, it can frustrate the employees and cause poor work efficiency.

Analyzing the Faculty of Management communication channels and tolls they were using we ended up with complex data, which were processed to gain necessary information. In the analysis, we were able to identify the main problems of the communication at the faculty and we have provided possible solutions that could be implemented.

As we proved, the internal communication is very important for the organizations even though many organizations do not pay attention to internal communication. Internal communication can improve also the external communication in many ways. The employees are informed about the company's status, so they can respond to the external flows more efficiently, professionally and more confidentially as they have all necessary information few clicks away. Managers have access to Intranet where organization can store the latest data. Managers could process these data whenever they need to, so they will have the latest possible information about the company, budget or personal details of each employee [20, 21].

In the empirical framework, we were focusing on implementation of the SharePoint Online feature to reduce the number of sent or received emails at the Department of Information Systems. We have deeply analyzed the mailboxes at the Centre of Information Technology and assistant at the Department of Information Systems. We were able to divide emails into several categories. These categories gave us a picture what tools we should implement to potentially reduce the number of received or sent emails.

To mention some of the features we were implementing can mention document libraries at first. Document libraries should be essential for file sharing as it is more convenient way to share the documents and to collaborate on them. We have uploaded all documents that are typically sent by the email to the SharePoint site of the Department of Information System. The second step was to set the permissions for members of the department, so they will be able to access data specified by the assistant or head of the department.

Video is one of the most popular means of communication nowadays. SharePoint Online contains a service called "Video" which can be fully implemented to the intranet. Organizations can share the educational and explanatory videos within the organization. Introducing the video portal to the employee's lives can save managers time, because they will be able to answers some of the question easily through explanatory video. Administrators can create video tutorials to teach employees how to use the software tools installed on their computers and thus improve their IT skills and probably reduce the number of requests related to user's settings of some applications.

Implementation of these solutions resulted into reduction of emails by 8% in assistant's mailbox at Department of Information Systems. This is 1% over our expectations. This result was achieved one month after full implementation of the solution. We are expecting to develop the system more in depth and to reduce the emails in the whole organization, in this case Faculty of Management Comenius University in Bratislava.

308   M. Krajčík

References

1. Molnár E, Molnár R, Kryvinska N, Greguš M (2014) Web Intelligence in practice. The Society of Service Science. J Ser Sci Res Springer 6(1):149–172
2. Brejčák P Časté kontrolovanie e-mailov zvyšuje stres, znižuje IQ a odoláva sa muťažšie ako cigarete. [Online] 06 25, 2014. [Cited: 02 12, 2016.] http://www.etrend.sk/technologie/caste-cekovanie-e-mailov-zvysuje-stres-znizuje-iq-a-odolava-sa-mu-tazsie-ako-alkoholu.html
3. BBC News 'Infomania' worse than marijuana. [Online] 04 22, 2005. [Cited: 02 12, 2016.] http://news.bbc.co.uk/2/hi/uk_news/4471607.stm
4. Garone E Email backlash at the office? [Online] BBC, 01 20, 2014. [Cited: 02 12, 2016.] http://www.bbc.com/capital/story/20140117-email-backlash-at-the-office
5. AICO V slovenských firmách je potenciál internej komunikácie nevyužitý. [Online] AICO, 01 2015. [Cited: 02 12, 2016.] http://www.internalcommunication.eu/sk/novinky/v-slovenskych-firmach-je-potencial-internej-komunikacie-nevyuzity
6. Gregus M, Kryvinska N (2015) Service orientation of enterprises—aspects, dimensions, technologies. Comenius University in Bratislava. ISBN: 9788022339780
7. McMillan R Wired.com. IBM Gives Birth to Amazing E-mail-less Man. [Online] 01 16, 2012. [Cited: 02 12, 2016.] http://www.wired.com/2012/01/luis-suarez/
8. ATOS Zero email: insights and stories. [Online] 05-08 2014. [Cited: 02 12, 2016.] http://atos.net/en-us/home/we-are/zero-email/zero-email-insights-and-stories.html
9. Schade A, Caya P, Nielsen J (2010) Intranet design annual 2010: the year's 10 Best Intranets. [Online] 01 2010. [Cited: 02 12, 2016.] http://isqa.ru/wp-content/uploads/2013/06/intranet_design_annual_2010.pdf
10. Kryvinska N (2012) Building consistent formal specification for the service enterprise agility foundation. Soc Ser Sci J Ser Sci Res Springer 4(2):235–269
11. Kryvinska N, Gregus M (2014) "SOA and its business value in requirements, features, practices and methodologies. Comenius University in Bratislava, ISBN: 9788022337649
12. Smith C expandedramblings.com. By the numbers: 125 + Amazing YouTube Statistics. [Online] 04 29, 2016. [Cited: 04 29, 2016.] http://expandedramblings.com/index.php/youtube-statistics/
13. Herzog C lightanimations.com. Three Internal Communications Tools You Need To Be Using. [Online] 08 31, 2015. [Cited: 02 12, 2016.] http://www.lightanimations.com/2015/08/31/three-internal-communications-tools-you-need-to-be-using/
14. O'Meara J flimp.net. 72 percent of Employee Communicators Plan to Increase Video Use. [Online] 05 15, 2013. [Cited: 02 12, 2016.] http://www.flimp.net/blog/72-percent-of-employee-communicators-plan-to-increase-video-use/
15. ANDY Improving internal communication with video. [Online] enginecreative.co.uk, 2012. [Cited: 02 12, 2016.] http://www.enginecreative.co.uk/insights/improving-internal-communication-video
16. Margalit L Did video kill text content marketing? entrepreneur.com. [Online] 16. 04 2015. [Dátum: 12. 02 2016.] https://www.entrepreneur.com/article/245003
17. Charfaoui E (2007) Klasické vzdelávanie s použitím prvkov e-vzdelávania pri sprostredkovaní nemeckej ekonomickej terminológie. Odborný jazyk na vysokých školách 3 [elektronický zdroj]. - Praha: Česká zemědelská univerzita v Praze, 74–76. ISBN 978-80-213-1615-7
18. Urikova O, Ivanochko I, Kryvinska N, Strauss C, Zinterhof P (2013) Consideration of aspects affecting the evolvement of collaborative ebusiness in service organizations. Inderscience Publishers. International Journal of Services. Economics and Management (IJSEM). Transform Inn Business Ser Models Modelling 5(1/2):72–92

19. Stoshikj M, Kryvinska N, Strauss C (2014) Efficient managing of complex programs with project management services. Springer, Global Journal of Flexible Systems Management, Special Issue on Flexible Complexity Management and Engineering by Innovative Services. 15(1):25–38, https://doi.org/10.1007/s40171-013-0051-8
20. Charfaoui E (2007) Výuka odborného jazyka s využitím informačných systémov. Kompetence v cizích jazycích jako důležitá součást profilu absolventa vysoké školy. Brno, CJP UO. 127–131. ISBN 978-807231-261-0
21. Kaczor S, Kryvinska N (2013) It is all about services—fundamentals, drivers, and business models. Soc Ser Sci J Ser Sci Res Springer 5(2):125–154

Simulating and Reengineering Stress Management System—Analysis of Undesirable Deviations

Oleg Kuzmin, Mykhailo Honchar, Volodymyr Zhezhukha
and Vadym Ovcharuk

Abstract The article proposes a method for diagnosing the level of criticality of an undesirable deviation in the supply-production-marketing chain based on simulating the impact of a potential or actual incident that causes such a deviation on each link in the chain using a number of representative parameters (reliability of the supply chain, inventory management system, negative response from external marketing stakeholders, competitiveness of the marketing complex, quality of the production process, flexibility of the production process, level of planned processes violation). Given that in terms of their content the absolute majority of parameters for simulating the impact of a potential or actual incident that causes undesirable deviation for each link in the supply-production-marketing chain are fuzzy, it is feasible and necessary to use fuzzy set tools to solve the specified problem. Also, according to the results of the performed research, a set of heuristic rules for diagnosing incidents in the supply-production-marketing chain was proposed.

Keywords Reengineering · Stress management · Stress management system
Supply-production-marketing chain · Undesirable deviation

O. Kuzmin · M. Honchar · V. Zhezhukha (✉) · V. Ovcharuk
Lviv Polytechnic National University, Bandera Street, 12, Lviv, Ukraine
e-mail: volodymyr.y.zhezhukha@lpnu.ua

O. Kuzmin
e-mail: oleh.y.kuzmin@lpnu.ua

M. Honchar
e-mail: mykhailo.f.honchar@lpnu.ua

V. Ovcharuk
e-mail: vadym.v.ovcharuk@lpnu.ua

© Springer International Publishing AG, part of Springer Nature 2019
N. Kryvinska and M. Greguš (eds.), *Data-Centric Business and Applications*,
Lecture Notes on Data Engineering and Communications Technologies 20,
https://doi.org/10.1007/978-3-319-94117-2_13

311

1 Introduction

One of the key conditions for a timely response to critical undesirable deviations that are substantial, extreme and exert significant negative impact on the enterprises operation is the availability of an effective stress management system. The multi-dimensional and diverse nature of these deviations necessitates the appropriate theoretical and practical training of managers engaged in such processes.

The policy for managing personal, group or corporate stress within the stress management systems at the business entities should ensure the implementation of complex key tasks such as:

- Identification and determination of potential vulnerabilities in the production and economic activity of the enterprise from the point of view of emergence of critical undesirable deviations;
- Development, implementation and formalization of the key elements of the stress management system;
- Development of a list of potential critical undesirable deviations for each key business process of the company;
- Development of a complex of potential measures for influence on critical undesirable deviations;
- Establishment of criteria for the interpretation of critical undesirable deviations in such a way as to necessitate the use of managerial measures to influence them.

Critical undesirable deviations in the activities of business entities may take different forms determined by a number of parameters, namely: duration of the individual stages of deviations, the reasons behind their occurrence, level of influence on business processes, tools for managing them, etc.

The study of theory and practice makes it possible to conclude that the low level of process management within the stress management systems determines not only the increase in critical undesirable deviations, but also their negative impact on the activities of business entities. Therefore, it is important to have effective tools for simulating and reengineering such processes.

To build stress management systems at enterprises, it is expedient to use the simulation tools. It makes it possible to clearly identify the functional boundaries of each manager within this process, recipients and disseminators of information, timing for the implementation of the relevant work, level of responsibility and authority, etc. Simulation also provides for transparency of stress management systems since it allows you to understand the directions of documentation, infor-mation, communications, etc.

In addition, since in terms of their content the absolute majority of parameters for simulating the impact of a potential or actual incident that causes undesirable deviation for each link in the supply-production-marketing chain are fuzzy, it is feasible and necessary to use fuzzy set tools to solve the specified problem.

2 Paper Preparation

2.1 Features of Incident Criticality Simulation in the Stress Management Systems

In the analyzed context, taking into account the statement by A. I. Hizun [1, p. 48], it should be noted that the concept of critical undesirable deviation, which is essential, extreme and has a significant negative impact on the enterprise's operation, is related to the concept of the incident. According to the author, the reason behind any crisis situation is "an incident with the highest level of criticality determined by the level of losses, the number of victims and other characteristics, that is, the incident/potential crisis situation". Simulation of the incidents criticality in the stress management systems should be carried out in the supply-production-management chain. To formalize the above, we introduce a set of potential incidents P_{nv}:

$$P_{nv} = \left\{ \bigcup_{i=1}^{n} P_{nv_i} \right\} = \{ P_{nv_1} + P_{nv_2}, \ldots, P_{nv_n} \}, (i = \overline{1, n}), \tag{1}$$

where n is the number of potential incidents in the supply-production-marketing chain, each of which can be represented as a function similar to [2–4]:

$$P_{nv_i} = \langle P_{nv_i}, P_i, E_i, Z_i, R_i, L_{ks} \rangle, \tag{2}$$

where

P_{nv_i} is the designation of the i-th incident in the supply-production-marketing chain which may lead to a critical undesirable deviation;

P_i a subset of parameters used to identify the i-th incident (or its prediction);

E_i a subset of all possible fuzzy (linguistic) values of the i-th incident;

Z_i a subset of actual values of the i-th incident at a certain time and space;

R_i a subset of heuristic rules formed on the basis of fuzzy logic for each individual i-th incident;

L_{ks} the level of criticality of the undesirable deviation in the supply-production-marketing chain caused by the i-th incident.

Taking into account formula (2) and the information described in the work [1, p. 49], the level of undesirable deviation in the stress management system can be considered critical, provided that $L_{ks} \geq L_{ks_c}$ where L_{ks_c} is the average level of criticality of the undesirable deviation. Provided that $L_{ks} < L_{ks_c}$ (the low level of criticality of the undesirable deviation caused by the i-th incident in the supply-production-marketing chain), the enterprise may decide on the influence on it using conventional methods or ignoring it as not causing significant damage to the production and economic activity of the business entity.

Considering the designation of the i-th incident in the supply-production-marketing chain which could lead to a critical undesirable deviation (P_{nv_i}), it should be noted that such incidents are proposed to be considered within the supply, production, and marketing processes of each individual entity. Given this, the set of identifiers for i-th incidents in the supply-production-marketing chain with $n = 3$ should be displayed as follows:

$$P_{nv} = \left\{ \bigcup_{i=1}^{3} P_{nv_i} \right\} = \{P_{nv_1} + P_{nv_2} + P_{nv_3}\} = P_{nv_{post}} + P_{nv_{vyr}} + P_{nv_{zb}}, \qquad (3)$$

where $P_{nv_1} = P_{nv_{post}}$, $P_{nv_2} = P_{nv_{vyr}}$, $P_{nv_3} = P_{nv_{zb}}$ are the identifiers of incident groups within the supply, production and marketing processes of the enterprise, respectively.

Considering a subset of parameters used to identify the i-th incident in the supply-production-marketing chain (P_i), it should be noted that this subset must include key parameters P. Moreover, it is fair to assert that $P_i \subseteq P$ and

$$P = \left\{ \bigcup_{j=1}^{m} P_j \right\} = \{P_1, P_2, \ldots, P_m\}, j = \overline{1, m}, \qquad (4)$$

where m is the total number of diagnostic parameters used to identify the i-th incident in the supply-production-marketing chain.

2.2 Parameters for Simulating the Impact of a Potential or Actual Incident on the Links of the Supply-Production-Marketing Chain

Study of theory and practice, as well as the results of the research carried out, makes it possible to identify the following main parameters of simulating the effect of a potential or actual incident which causes undesirable deviation on each link in the supply-production-marketing chain:

1. Reliability of the supply chain (N_{lp}) $(j = 1)$. The parameter, based on the review and generalization of literary sources, is one of the key indicators in the supply logistics and, to some extent, generalizes its effectiveness. Taking into account the results of research by V. Ermoshyn [5, p. 56], it should be noted that the reliability parameter of the supply chain indicates the effectiveness of supplying "the right products to the right place, at the right time, in the right condition and packaging, with the right documentation". In a way, this parameter allows identifying both the flexibility of supply systems, the time to respond to changes in factors of the internal and external environment, production necessity

planning, procurement plans development, ensuring operational schedule of supply, acceptance of raw materials, materials and components, checking their quality at the enterprise, etc.

The enterprise supply system is considered effective subject to absolute reliability of the supply chain. Within the scope of the research, we will use the indicator of the level of supply chain reliability, known in theory and practice (defined as the ratio of exactly completed deliveries by all parameters to their total required quantity) as the measure of the parameter.

2. Inventory management system (S_{uz}) $(j = 2)$. The problems in the field of supply can be evidenced by the untenable and inefficient inventory management system at the enterprise. Its typical features may, in particular, include constant shortage of stocks, on the one hand, and their surplus—on the other. Thus, the stock shortages are known to cause interruptions in the production process, and, consequently, decrease in sales volumes and decrease in revenue from the sales of products. In turn, the surplus of stocks also leads to a number of problems such as increase in costs for the storage of raw materials, materials or components, physical and moral aging of the latter at the enterprise, etc.

As the theory and practice studies show, today there are many methods, approaches and criteria in the field of optimizing inventory management processes. They, in particular, allow you to calculate the required and sufficient level of stock, the scope of the optimal order, etc., thus forming an integrated inventory management system. Within the analyzed aspect, it is expedient to highlight technological complexes for inventory management common in practice (for example, with fixed order scope, with its fixed frequency or with the established frequency of replenishment of stocks to a sufficient level), different models and methods of such management (for example, heuristic, technical and economic calculations, economic and mathematical calculations, etc.), etc. The level of the inventory management system development should be considered as the measure of this parameter. It is proposed to be established by an expert method taking into account the current trends in these processes.

3. Negative reaction from external sales stakeholders $(R_{zs})(j = 3)$. This parameter is suitable primarily for simulating the impact of incidents in the field of sales. As the study of theory and practice suggests, a significant indicator of the future crisis at an enterprise may include negative reaction to cooperation with the company under review primarily on the part of sales intermediaries and end users (however, potential negative issues can also be evidenced by negative assessments from intermediaries, creditors, financial institutions, shareholders, etc.). This negative reaction may be due to various reasons (e.g., deterioration in product quality, default by the company on its contractual obligations, regular changes in strategy, objectives and policies, frequent conflicts with business entities, etc.). In any case, negative reaction from external sales stakeholders means problems in the field of product sales, in particular, a significant decrease

in its volumes, and thus a decrease in the revenue from the sales of products, increase in inventory balances, deterioration of turnover, etc. Within the scope of the study, we will consider the level of negative reaction as a measure of this negative reaction parameter on the part of external sales stakeholders; it is proposed to be calculated as the ratio of the number of external sales stakeholders that negatively assesses the cooperation with the enterprise-producer to the total number of such stakeholders.

4. Competitiveness of the marketing mix (K_{km}) $(j = 4)$. Identification of incidents in the field of sales can be done by identifying problems with the false forecast of demand for products, ineffective positioning of products in the market, inappropriate pricing policies, negative changes in the structure of the orders portfolio, imbalances in marketing policies, inadequate sales promotion, etc. The above should be generalized in the competitiveness parameter of the marketing mix.

The level of competitiveness should be considered as the measure of this marketing mix competitiveness parameter; it is proposed to be established by an expert method taking into account the current trends and industry development.

5. Production process quality (Y_{vp}) $(j = 5)$. This parameter should also be considered generalized during the simulation of the impact of incidents in the industrial sphere of the enterprise. Thus, problems can arise in case of inefficient production technology, deficit or surplus of resources taking into account anticipated expected production volumes, inappropriate level of production capacity, low level of control over individual parts of the production process, inappropriate priorities in the spheres of research and development, etc. It is proposed to consider the above in general within the parameters of the production process quality. Low quality of the production process can result, for example, in the failure of planned production volumes, inappropriate energy consumption, increase in equipment downtime, etc. The measure of this parameter, as in the previous case, should be such quality level, which is also proposed to be established in an organization using an expert method.

6. Production process flexibility (H_{vp}) $(j = 6)$. As theory and practice suggest, production process flexibility is important under the conditions of dynamic environment of the enterprise. It is considered from different aspects in the literature, while most authors hold an opinion that the flexibility indicates the ability of the production process to be promptly readjusted to manufacture new or modified products as a result of changing market requirements. Such a reorganization may include solving various problems, in particular, reequipping of technological processes, changes in the structure of production process or integrating new elements into it, introducing parallel technological channels, etc. It is obvious that the measure of the production process flexibility parameter should be the time to adjust it for the manufacture of a new or improved product (expert estimates should apply).

7. The level of planned processes violation (P_{zp}) $(j = 7)$. This parameter is suitable for simulating the impact of incidents in the field of supply, as well as in the field of production and marketing. Any deviation from the planned parameters may be promptly eliminated and significantly violate the established rules, mechanisms, policies, regulations, etc. The higher the level of violations, the higher the negative effects of crisis phenomena. This parameter must be both a measure and assessed by an expert method at the enterprise.

Each of the abovementioned parameters characterizes certain side of the incidents impact simulation in the supply-production-marketing chain. Obviously, under certain conditions, such parameters can complement each other and thereby increase the negative effect. Understanding how these parameters affect the entity's activities allows you to identify potential problems, make predictions, take preventive measures, etc.

Each business entity has its priority in the identification of incidents in the supply-production-marketing chain within the framework of stress management systems. In many respects, such priority is determined by the activity of the enterprise, its stage of the life cycle, management parameters, etc. On the other hand, it should be borne in mind that different companies will react differently to existing or potential problems in the supply-production-marketing chain.

2.3 Justification of Feasibility of Using the Fuzzy Sets Theory for Diagnosing the Undesirable Deviation Criticality Level

In terms of their content the absolute majority of parameters for simulating the impact of a potential or actual incident that causes undesirable deviation for each link in the supply-production-marketing chain are fuzzy, it is feasible and necessary to use fuzzy set tools for further research. Based on the review and generalization of literary sources [6–10], the theory of fuzzy sets is widely used in solving various economic problems.

Given the above, a subset of parameters used to diagnose the i-th incident (or its prediction) within the framework of stress management systems P_i should be formed on the basis of a set of parameters P, while:

$$\left\{ \bigcup_{i=1}^{n} P_i \right\} = \{P_1, \ldots, P_n\}, \ P_i \subseteq P; \tag{5}$$

$$P_i = \left\{ \bigcup_{j=1}^{k_i} P_{ij} \right\} = \{P_{i1}, \ldots, P_{ik_i}\}, \tag{6}$$

where

n is the number of potential incidents in the supply-production-marketing chain;

k_i the number of parameters within the stress management systems associated with the i-th incident.

Thus, taking into account the results presented in the papers [1, 3, 11, 12], we get:

$$\left\{\bigcup_{i=1}^{n} P_i\right\} = \left\{\bigcup_{i=1}^{n}\left\{\bigcup_{j=1}^{k_i} P_{ij}\right\}\right\} = \{\{P_{11}, \ldots, P_{1k_i}\}, \ldots, \{P_{n1}, \ldots, P_{nk_n}\}\}. \quad (7)$$

In our case, taking into account the above parameters in the supply-production-marketing chain (that is, $n = 3$, $k_1 = k_2 = k_3 = 3$) within the framework of stress management systems, we obtain the following: $P_{11} = P_1, P_{12} = P_2, P_{13} = P_7, P_{21} = P_5, P_{22} = P_6, P_{23} = P_7,$ $P_{31} = P_3, P_{32} = P_4, P_{33} = P_7$. Thus, formula (7) will be as follows:

$$\left\{\bigcup_{i=1}^{3} P_i\right\} = \left\{\bigcup_{i=1}^{3}\left\{\bigcup_{j=1}^{k_i} P_{ij}\right\}\right\} = \{\{P_{11}, P_{12}, P_{13}\}, \{P_{21}, P_{22}, P_{23}\}, \{P_{31}, P_{32}, P_{33}\}\}$$

$$= \{\{N_{lp}, S_{uz}, P_{zp}\}, \{Y_{vp}, H_{vp}, P_{zp}\}, \{R_{zs}, K_{km}, P_{zp}\}\}, \quad (8)$$

where

$P_{11} = N_{lp}, P_{12} = S_{uz}, P_{13} = P_{zp}$ appropriate parameters for identifying incidents within the supply section P_{nv_1};

$P_{21} = Y_{vp}, P_{22} = H_{vp}, P_{23} = P_{zp}$ appropriate parameters for identifying incidents within the production section;

$P_{31} = R_{zs}, P_{32} = K_{km}, P_{33} = P_{zp}$ appropriate parameters for identifying incidents within the marketing section.

Summarizing the above, identification of incidents in the supply-production-marketing chain involves simulating their impact on the corresponding element of the chain. If necessary, the impact of incidents on the entire chain can also be analyzed. Moreover, in each individual case, we should take into account appropriateness of parameters for identifying incidents in terms of sections (in other words, each individual section in the supply-production-marketing chain corresponds to its own set of parameters), as shown in Table 1.

2.4 Description from the Position of Fuzzy Sets of a Subset of All Possible Fuzzy (Linguistic) Values of Incidents

Then, it is necessary to describe a subset of all possible fuzzy (linguistic) values of the i-th incident, as well as a subset of heuristic rules formed on the basis of fuzzy logistics for each such incident from the position of fuzzy sets. This will eventually

Table 1 Parameters for diagnosing incidents in the supply-production-marketing chain

Parameters	Sections		
	Supply section	Production section	Marketing section
Reliability of the supply chain (N_{lp})	X		
Inventory management system (S_{uz})	X		
Negative reaction on the part of external sales stakeholders (R_{zs})			X
Competitiveness of the marketing mix (K_{km})			X
Production process quality (Y_{vp})		X	
Production process flexibility (H_{vp})		X	
The level of planned processes violation (P_{zp})	X	X	X

Note developed by authors

identify the level of undesirable deviation in the supply-production-marketing chain caused by the i-th incident.

Considering a subset of all possible fuzzy (linguistic) values of the i-th incident, it is appropriate to note that, in terms of their content, critical undesirable deviations within stress management systems are poorly-structures concepts with high levels of relativity. Therefore, it is proposed to use the combination of expert methods and fuzzy logic to solve the above problem. In particular, it is expedient to use the trapezoidal membership function known in the theory and practice from among the available $L - R$ functions (Fig. 1).

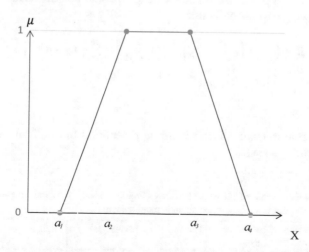

Fig. 1 Trapezoidal membership function in the theory of fuzzy sets *Note* provided on the basis of [6, 13–15]

Thus, description of a subset of all possible fuzzy (linguistic) values of the i-th incident within the framework of stress management systems should be made taking into account the following equations [6, 13–15]:

$$
\mu_A(x) = \begin{cases} 0, x < a_1; \\ \dfrac{x - a_1}{a_2 - a_1} & a_1 \le x < a_2; \\ 1 & a_2 \le x \le a_3; \\ \dfrac{a_4 - x}{a_4 - a_3} & a_3 < x \le a_4, \end{cases} \tag{9}
$$

where a_1, a_2, a_3, a_4—are the numerical parameters (actual values) of the trapezoidal $L - R$ function, in turn $a_1 \le a_2 \le a_3 \le a_4$; a_1—are the left zero value of the function; a_4—right zero value of the function; a_2 i a_3—points where the value of the membership function is maximum (trapezium vertex) (i.e. the certainty of the allegation of affiliation to a certain term-set).

Taking into account the peculiarities of the tasks solved in the context of simulating critical undesirable deviations in the supply-production-marketing chain, we should stress the appropriateness of using the following approach to the reflection of linguistic variables within the limits given in Fig. 1 of the trapezoidal membership function, i.e.: a_1—pessimistic assessment; a_4—optimistic assessment; a_2, a_3—the most realistic assessment.

Thus, trapezoidal scale of all possible fuzzy (linguistic) values of the diagnosis of the i-th incident will involve a combination of linguistic variables and trapezoidal assessments (Table 2).

Considering the subset of heuristic rules generated by an expert method based on the fuzzy logic for a particular i-th incident R_i, it should be noted that for each such rule the following expression is true:

$$
\begin{aligned}
R &= \bigcup_{i=1}^{n} \left\{ \bigcup_{p=1}^{R_i} R_{ip} \right\} = \left\{ \bigcup_{i=1}^{n} \left\{ \bigcup_{p=1}^{R_i} L_{ip} \to I_{ip} \right\} \right\} \\
&= \left\{ \bigcup_{i=1}^{n} \left\{ \bigcup_{p=1}^{R_i} R_{ip} = (L_{ip} \to I_{ip}) \right\} \right\},
\end{aligned} \tag{10}
$$

where

R_{ip} is p-th rule for the diagnosis of a potential i-th incident within stress management systems;

Table 2 Trapezoidal scale of all possible fuzzy (linguistic) values of the diagnosis of the i-th incident within the framework of stress management systems

Linguistic variables	Very low	Low	Average	High	Very high
Trapezoidal scales of assessments	(0; 0; 0.1; 0.3)	(0.1; 0.3; 0.3; 0.5)	(0.3; 0.5; 0.5; 0.7)	(0.5; 0.7; 0.7; 0.9)	(0.7; 0.9; 1.0; 1.0)

Note provided on the basis of [6, 13, 16, 17]

L a unique state identifier for each i-th incident (and this state takes into account the number of diagnostic parameters of the i-th incident and the number of terms);

I_{ip} element of the set of linguistic variables of the i-th incident.

Thus, taking into account the above parameters of diagnosing incidents in the supply-production-marketing chain, it is necessary to indicate the need for the formation of a set of specified rules for each individual section of the chain. Using the logic-linguistic connections within the fuzzy set theory and taking into account that I_{ip} can acquire the values "very low", "low", "average", "high" and "very high", and also denoting the value of the output parameters as H—low, C—average, B—high and D—very high, we give a set of heuristic rules R_{ip} for the diagnosis of incidents in the supply-production-marketing chain (Table 3).

Table 3 A set of heuristic rules R_{ip} for diagnosing incidents in the supply-production-marketing chain

Rule number p	Indicators in the supply-production-marketing chain			Result
	$N_{lp}/Y_{vp}/R_{zs}$	$S_{uz}/H_{vp}/K_{km}$	$P_{zp}/P_{zp}/P_{zp}$	
1	VL	VL	VL	H
2	VL	VL	LW	H
3	VL	VL	AR	H
4	VL	VL	HG	C
5	VL	VL	VH	C
6	VL	LW	VL	H
7	VL	LW	LW	H
8	VL	LW	AR	H
9	VL	LW	HG	C
10	VL	LW	VH	C
11	VL	AR	VL	H
12	VL	AR	LW	H
13	VL	AR	AR	C
14	VL	AR	HG	C
15	VL	AR	VH	C
16	VL	HG	VL	C
17	VL	HG	LW	C
18	VL	HG	AR	C
19	VL	HG	HG	B
20	VL	HG	VH	B
21	VL	VH	VL	C
22	VL	VH	LW	C
23	VL	VH	AR	C
24	VL	VH	HG	B

(continued)

Table 3 (continued)

Rule number p	Indicators in the supply-production-marketing chain			Result
	$N_{lp}/Y_{vp}/R_{zs}$	$S_{uz}/H_{vp}/K_{km}$	$P_{zp}/P_{zp}/P_{zp}$	
25	VL	VH	VH	B
26	LW	VL	VL	H
27	LW	VL	LW	H
28	LW	VL	AR	C
29	LW	VL	HG	C
30	LW	VL	VH	C
31	LW	LW	VL	H
32	LW	LW	LW	H
33	LW	LW	AR	H
34	LW	LW	HG	C
35	LW	LW	VH	C
36	LW	AR	VL	H
37	LW	AR	LW	H
38	LW	AR	AR	C
39	LW	AR	HG	C
40	LW	AR	VH	C
41	LW	HG	VL	H
42	LW	HG	LW	H
43	LW	HG	AR	C
44	LW	HG	HG	C
45	LW	HG	VH	B
46	LW	VH	VL	C
47	LW	VH	LW	C
48	LW	VH	AR	C
49	LW	VH	HG	B
50	LW	VH	VH	B
51	AR	VL	VL	H
52	AR	VL	LW	H
53	AR	VL	AR	C
54	AR	VL	HG	C
55	AR	VL	VH	C
56	AR	LW	VL	H
57	AR	LW	LW	H
58	AR	LW	AR	C
59	AR	LW	HG	C
60	AR	LW	VH	C
61	AR	AR	VL	C
62	AR	AR	LW	C
63	AR	AR	AR	C

(continued)

Table 3 (continued)

Rule number p	Indicators in the supply-production-marketing chain			Result
	$N_{lp}/Y_{vp}/R_{zs}$	$S_{uz}/H_{vp}/K_{km}$	$P_{zp}/P_{zp}/P_{zp}$	
64	AR	AR	HG	C
65	AR	AR	VH	B
66	AR	HG	VL	C
67	AR	HG	LW	C
68	AR	HG	AR	C
69	AR	HG	HG	B
70	AR	HG	VH	B
71	AR	VH	VL	C
72	AR	VH	LW	C
73	AR	VH	AR	C
74	AR	VH	HG	B
75	AR	VH	VH	D
76	HG	VL	VL	H
77	HG	VL	LW	H
78	HG	VL	AR	C
79	HG	VL	HG	C
80	HG	VL	VH	B
81	HG	LW	VL	H
82	HG	LW	LW	H
83	HG	LW	AR	C
84	HG	LW	HG	B
85	HG	LW	VH	B
86	HG	AR	VL	C
87	HG	AR	LW	C
88	HG	AR	AR	C
89	HG	AR	HG	B
90	HG	AR	VH	B
91	HG	HG	VL	C
92	HG	HG	LW	B
93	HG	HG	AR	B
94	HG	HG	HG	B
95	HG	HG	VH	B
96	HG	VH	VL	B
97	HG	VH	LW	B
98	HG	VH	AR	B
99	HG	VH	HG	B
100	HG	VH	VH	D
101	VH	VL	VL	H
102	VH	VL	LW	H

(continued)

Table 3 (continued)

Rule number p	Indicators in the supply-production-marketing chain			Result
	$N_{lp}/Y_{vp}/R_{zs}$	$S_{uz}/H_{vp}/K_{km}$	$P_{zp}/P_{zp}/P_{zp}$	
103	VH	VL	AR	C
104	VH	VL	HG	C
105	VH	VL	VH	B
106	VH	LW	VL	H
107	VH	LW	LW	H
108	VH	LW	AR	C
109	VH	LW	HG	B
110	VH	LW	VH	B
111	VH	AR	VL	C
112	VH	AR	LW	C
113	VH	AR	AR	C
114	VH	AR	HG	B
115	VH	AR	VH	D
116	VH	HG	VL	B
117	VH	HG	LW	B
118	VH	HG	AR	B
119	VH	HG	HG	B
120	VH	HG	VH	D
121	VH	VH	VL	B
122	VH	VH	LW	B
123	VH	VH	AR	D
124	VH	VH	HG	D
125	VH	VH	VH	D

Note developed by the authors

Moreover, it should be noted that due to the same number of diagnostic parameters of these incidents, as well as the same formed set I_{ip} ("very low", "low", "average", "high" and "very high") and the same approach to displaying the output parameters (H—low, C—average, B—high and D—very high), the number and contents of the heuristic rules R_{ip} for the diagnosis of incidents in the supply-production-marketing chain will be the same irrespective of the elements of this chain, which somehow simplifies further calculations in this area within of stress management systems at the enterprise.

As shown in information in Table 3, in the analyzed context, we should mention the existence of 125 heuristic rules R_{ip} for the diagnosis of incidents in the supply-production-marketing chain.

2.5 Integrated Models for Incident Groups in the Supply-Production-Marketing Chain

Summarizing the above, in stress management systems, an integrated model for incident groups within the supply chain will be as follows:

$$
\begin{aligned}
P_{nv_1} &= \langle P_{nv_1}, P_1, E_1, Z_1, R_1, L_{ks_1} \rangle \\
&= \langle P_{nv_{post}}, \{P_{11}, P_{12}, P_{13}\}, \{\{E_{111}, E_{112}, E_{113}, E_{114}, E_{115}\}, \\
&\quad \{E_{121}, E_{122}, E_{123}, E_{124}, E_{125}\}, \{E_{131}, E_{132}, E_{133}, E_{134}, E_{135}\}\}, \\
&\quad \{Z_{11}, Z_{12}, Z_{13}\}, \{R_{11}, R_{12}, R_{13}, ..., R_{11}, \}, L_{ks_1} \rangle \\
&= \left\langle P_{nv_{post}}, \{N_{lp}, S_{uz}, P_{zp}\}, \left\{ \left\{ E_{P_{nv_{post}}}^{N_{lp1}}, E_{P_{nv_{post}}}^{N_{lp2}}, E_{P_{nv_{post}}}^{N_{lp3}}, E_{P_{nv_{post}}}^{N_{lp4}}, E_{P_{nv_{post}}}^{N_{lp5}} \right\}, \right. \right. \\
&\quad \left\{ E_{P_{nv_{post}}}^{S_{uz1}}, E_{P_{nv_{post}}}^{S_{uz2}}, E_{P_{nv_{post}}}^{S_{uz3}}, E_{P_{nv_{post}}}^{S_{uz4}}, E_{P_{nv_{post}}}^{S_{uz5}} \right\}, \left\{ E_{P_{nv_{post}}}^{P_{zp1}}, E_{P_{nv_{post}}}^{P_{zp2}}, E_{P_{nv_{post}}}^{P_{zp3}}, E_{P_{nv_{post}}}^{P_{zp4}}, E_{P_{nv_{post}}}^{P_{zp5}} \right\} \right\}, \\
&\quad \left. \{Z_{N_{lp}}, Z_{S_{uz}}, Z_{P_{zp}}\}, \{R_{11}, R_{12}, R_{13}, ..., R_{1125}, \}, L_{ks_1} \right\rangle.
\end{aligned}
\tag{11}
$$

Similarly, an integrated model for the groups of incidents within the production processes will be as follows:

$$
\begin{aligned}
P_{nv_2} &= \langle P_{nv_2}, P_2, E_2, Z_2, R_2, L_{ks_2} \rangle \\
&= \langle P_{nv_{vyr}}, \{P_{21}, P_{22}, P_{23}\}, \{\{E_{211}, E_{212}, E_{213}, E_{214}, E_{215}\}, \\
&\quad \{E_{221}, E_{222}, E_{223}, E_{224}, E_{225}\}, \{E_{231}, E_{232}, E_{233}, E_{234}, E_{235}\}\}, \\
&\quad \{Z_{21}, Z_{22}, Z_{23}\}, \{R_{21}, R_{22}, R_{23}, ..., R_{11}, \}, L_{ks_2} \rangle \\
&= \left\langle P_{nv_{vyr}}, \{Y_{vp}, H_{vp}, P_{zp}\}, \left\{ \left\{ E_{P_{nv_{vyr}}}^{Y_{vp1}}, E_{P_{nv_{vyr}}}^{Y_{vp2}}, E_{P_{nv_{vyr}}}^{Y_{vp3}}, E_{P_{nv_{vyr}}}^{Y_{vp4}}, E_{P_{nv_{vyr}}}^{Y_{vp5}} \right\}, \right. \right. \\
&\quad \left\{ E_{P_{nv_{vyr}}}^{H_{vp1}}, E_{P_{nv_{vyr}}}^{H_{vp2}}, E_{P_{nv_{vyr}}}^{H_{vp3}}, E_{P_{nv_{vyr}}}^{H_{vp4}}, E_{P_{nv_{vyr}}}^{H_{vp5}} \right\}, \left\{ E_{P_{nv_{vyr}}}^{P_{zp1}}, E_{P_{nv_{vyr}}}^{P_{zp2}}, E_{P_{nv_{vyr}}}^{P_{zp3}}, E_{P_{nv_{vyr}}}^{P_{zp4}}, E_{P_{nv_{vyr}}}^{P_{zp5}} \right\} \right\}, \\
&\quad \left. \{Y_{vp}, H_{vp}, P_{zp}\}, \{R_{21}, R_{22}, R_{23}, ..., R_{2125}, \}, L_{ks_2} \right\rangle.
\end{aligned}
\tag{12}
$$

An integrated model for the groups of incidents within the marketing processes will be as follows:

$$
\begin{aligned}
P_{nv_3} &= \langle P_{nv_3}, P_3, E_3, Z_3, R_3, L_{ks_3} \rangle \\
&= \langle P_{nv_{zb}}, \{P_{31}, P_{32}, P_{33}\}, \{\{E_{311}, E_{312}, E_{313}, E_{314}, E_{315}\}, \\
&\quad \{E_{321}, E_{322}, E_{323}, E_{324}, E_{325}\}, \{E_{331}, E_{332}, E_{333}, E_{334}, E_{335}\}\}, \\
&\quad \{Z_{31}, Z_{32}, Z_{33}\}, \{R_{31}, R_{32}, R_{33}, ..., R_{31}, \}, L_{ks_3} \rangle \\
&= \left\langle P_{nv_{zb}}, \{R_{zs}, K_{km}, P_{zp}\}, \left\{ \left\{ E_{P_{nv_{zb}}}^{R_{zs1}}, E_{P_{nv_{zb}}}^{R_{zs2}}, E_{P_{nv_{zb}}}^{R_{zs3}}, E_{P_{nv_{zb}}}^{R_{zs4}}, E_{P_{nv_{zb}}}^{R_{zs5}}, \right. \right. \right. \\
&\quad \left\{ E_{P_{nv_{zb}}}^{K_{km1}}, E_{P_{nv_{zb}}}^{K_{km2}}, E_{P_{nv_{zb}}}^{K_{km3}}, E_{P_{nv_{zb}}}^{K_{km4}}, E_{P_{nv_{zb}}}^{K_{km5}} \right\}, \left\{ E_{P_{nv_{zb}}}^{P_{zp1}}, E_{P_{nv_{zb}}}^{P_{zp2}}, E_{P_{nv_{zb}}}^{P_{zp3}}, E_{P_{nv_{zb}}}^{P_{zp4}}, E_{P_{nv_{zb}}}^{P_{zp5}} \right\} \right\}, \\
&\quad \left. \{R_{zs}, K_{km}, P_{zp}\}, \{R_{31}, R_{32}, R_{33}, ..., R_{3125}, \}, L_{ks_3} \right\rangle.
\end{aligned}
\tag{13}
$$

Simulation of stress management systems in the supply-production-marketing chain enables us to identify the level of influence of potential incidents on each of these areas. If necessary, it is also possible to determine the level of influence of these incidents on the chain as a whole. Therefore, under these conditions, the set of heuristic rules will include the following components:

$$P_{nv_1} = P_{nv_{post}}, P_{nv_2} = P_{nv_{vyr}}, P_{nv_3} = P_{nv_{zb}}. \tag{14}$$

The proposed method for diagnosing the level of criticality of the undesirable deviation in the supply-production-marketing chain, based on simulating the effect of a potential or actual incident that causes such a deviation on each link in a given chain using a number of representative parameters, has been applied in a number of domestic companies. For example, at PJSC "Lviv Locomotive Repair Plant", certain actual incidents in Q4 2017 in the supply and marketing chain were diagnosed as having caused critical undesirable deviations (Table 4).

Thus, the results of diagnosing the criticality of certain actual incidents at PJSC "Lviv Locomotive Repair Plant" in Q4 2017 make it possible to conclude that it is expedient to interpret delays in the supply of brake equipment, as well as negative feedback on the quality of wheel pairs repair as critical. The foregoing suggests the possibility of practical use of the proposed method for diagnosing the level of criticality of undesirable deviation in the supply-production-marketing chain.

It should be noted that PJSC "Lviv Locomotive Repair Plant" is one of the leading enterprises of Ukrainian railway, and this justifies its selection as an object for diagnosing the possibility of implementation the method of undesirable deviation criticality level diagnostic in the supply-production-marketing chain. Today, this is one of the leading enterprise in Ukraine in repairing locomotives. The Company is equipped with the current techniques and the breakthrough technologies are being widely used. The plant carries out the medium repairs and overhaul of the trunk electric locomotives, both direct and alternating current, repairs of industrial railway transport and the quarry electric locomotives traction units, as well as repairs the traction motors, auxiliary electrical machines, wheel sets, etc.

PJSC "Lviv Locomotive Repair Plant" is a successful modern enterprise, which is developing dynamically and efficiently provides its services in the rolling stock repairs and modernization of the whole post-Soviet area railways. For the consistently high quality and benefits achieved in the Ukrainian business of the repair and maintenance of the railway rolling stock, the plant was repeatedly awarded with the "Star of Quality" and as the "Enterprise of the Year" of the National Rating of Goods and Services Quality of Ukraine, as well as with the International Economic Rating "League of the Bests". In recent years, the plant won the winner's fame of the "Face of the City" contest in the category "Lviv Industrial".

At PJSC "Lviv Locomotive Repair Plant", the certification center TUV SUD Management Service GmbH has realized the certified audit of the quality management system for compliance with the requirements of ISO 9001:2015. Based on the audit results (report number 707084328), a positive decision according the certificate issue has been taken. The Certificate is in force from 01/05/2018 to 01/04/2021.

Table 4 Results of diagnosing criticality of some actual incidents at PJSC "Lviv Locomotive Repair Plant" in Q4 2017

Links of the supply-production-marketing chain	Incidents	Fuzzy (linguistic) values of influence			Level of criticality (identified by a set of heuristic rules R_{ip})
Supply	Delays in the supply of brake equipment	Reliability of the supply chain (N_{lp}) High	Inventory management system (S_{uz}) High	The level of planned processes violation (P_{zp}) High	High
	Noncompliance with the standards of purchased contactors of the MK series	Reliability of the supply chain (N_{lp}) Average	Inventory management system (S_{uz}) Average	The level of planned processes violation (P_{zp}) High	Average
Marketing	Different decrease in demand for the thermal treatment of shafts by external customers	Negative reaction on the part of external sales stakeholders (R_{zs}) Low	Competitiveness of the marketing mix (K_{km}) Average	The level of planned processes violation (P_{zp}) Low	Low
	Negative feedback on the quality of wheeled pairs repair	Negative reaction on the part of external sales stakeholders (R_{zs}) High	Competitiveness of the marketing mix (K_{km}) High	The level of planned processes violation (P_{zp}) Average	High

Note established by the authors

A more thorough analysis of the undesirable deviations criticality in the supply-production-marketing chain at PJSC "Lviv Locomotive Repair Plant" made it possible to single out the following factors of risk:

- the market conditions change;
- failure to provide the plant activities with the sufficient amount of raw materials, materials and energy;
- a lack in training the plant employees to produce new types of competitive products;
- deterioration of the general economic situation in Ukraine, the tax law instability.

Key problems that make a significant influence on the PJSC "Lviv Locomotive Repair Plant" activities from the position of critical undesirable deviations are the problems of political, economic and production-technological character. In particular, the political instability in the country and management change of the branch made a negative impact on the stability and sustainability of production due to the insufficient planning and financing the orders volumes of Ukrzaliznytsia. Too harsh pricing policy of the monopolist-customer has a harmful impact on the production and solvency; because of the lack of working capital, the timely manufacture kitting-up gets broken, the time schedules and technologies of repairs works deform, and this leads to quality deterioration and the production costs increase. Besides, in such conditions the company is not able to renew the fixed assets according the plan and fully, to absorb new technologies and new types of overhaul and modernization. The financial plan of PJSC "Lviv Locomotive Repair Plant" has proposed a complex of measures for reducing possible risks, which would allow to minimize these risks, as the financial plan is based on the calculations and on the company's abilities to improve the financial, organization and production situation. At this stage the Company requires the support of the Government of Ukraine in matters of the plant reconstruction by supporting the potential investors.

At the same time, it should be mentioned that the production and services in electric locomotives overhaul at PJSC "Lviv Locomotive Repair Plant" is profitable and have perspectives of the further demand growth on the market of Ukraine and CIS countries. Business operations financing is realized according to the principle of planned self-financing. The funds receipted from the sold production are being allocated by the enterprise on its own to cover the items of production costs, wage fund, taxes, contributions to the social security funds, losses, etc. The Company, in accordance with the agreements terms, must receive cash for the performed products (services) within a period not exceeding 10 days after the works are completed. Most often it happens that the customers postpone their payments. That is why the Issuer is making every effort to replenish the working capital in order to provide payments of wages, taxes, compulsory payments, and so on. To do this, the Issuer uses the banking services to get the short-term credits. It is also provided for the Company to use the funds to purchase the fixed assets, intangible assets and for the other purposes that contribute to the production development, improve social

Table 5 Information about the product sales of PJSC "Lviv Locomotive Repair Plant" in the 1st quarter of 2018

#	The main types of the enterprise products	Output volume		Sales volume	
		In monetary terms, thousand UAH	In percent to the total output (%)	In monetary terms, thousand UAH	In percent to the total sales volume (%)
1	Services in the rolling stock repair and modernization	14,397	54.13	20,475	67.69
2	Services in the wheel sets repair	8234	30.96	7275	24.05
3	Services in the traction motors repair	2549	9.58	1674	5.53
4	Other services	1218	4.58	824	2.72

Note constructed by the authors on the basis of PJSC "Lviv Locomotive Repair Plant" financial statements

conditions of the Company's employees and their material welfare. Renovation of the production fixed assets, implementation of the new breakthrough technologies, mastering the new types of repairs require the commitment of significant financial resources in the form of profitable investment projects. The separate investment projects of production diversification by means of electric locomotives modernization development are promising.

The implementation results of the proposed method of diagnosing the level of undesirable deviation criticality in the supply-production-marketing chain allowed PJSC "Lviv Locomotive Repair Plant" not only to simulate the influence of the potential or current incident, which stipulates for such a deviation, to the each link of the mentioned chain using the range of representative parameters, but in so doing to increase the volumes of sales and improve the financial indicators due to the appropriate decision-making (Table 5).

3 Conclusion

The proposed method for diagnosing the level of criticality of an undesirable deviation in the supply-production-marketing chain enables the analysts to simulate the impact of a potential or actual incident that causes such a deviation on each link in the chain using a number of representative parameters (reliability of the supply chain, inventory management system, negative response from external marketing stakeholders, competitiveness of the marketing complex, quality of the production process, flexibility of the production process, level of planned processes violation).

Simulation of stress management systems in the supply-production-marketing chain enables us to identify the level of influence of potential incidents on each of these areas. If necessary, it is also possible to determine the level of influence of these incidents on the chain as a whole.

References

1. Hizun AI (2015) Metody ta zasoby otsinky parametriv bezpeky dlya vyyavlennya kryzovykh sytuatsiy u informatsiyniy sferi. Kandydat nauk. Natsional'nyy aviatsiynyy universytet
2. Korchenko AH, Ivachenko YEV. y Kazmyrchuk SV (2011) Yntehryrovannoe predstavlenye parametrov ryska. Zakhyst informatsiyi, 1, 96–101
3. Korchenko AO (2014) Metod a-rivnevoyi nominalizatsiyi nechetkykh chysel dlya system vyyavlennya vtorhnen'. Zakhyst informatsiyi 16(4):304–311
4. Stoshikj M, Kryvinska N, Strauss C (2014) Efficient managing of complex programs with project management services. Springer, Glob J Flex Syst Manag Spec Issue Flex Complex Manag Eng Innovative Serv (March 2014) 15(1):25–38, https://doi.org/10.1007/s40171-013-0051-8
5. Ermoshyn VA (2015) Klyuchovi pokaznyky efektyvnosti yak instrument upravlinnya lohistykoyu postachannya MTR. Matematyka ta informatsiyni tekhnolohiyi v naftohazovomu kompleksi 2:53–70
6. Didyk AM (2016) Sotsial'no-ekonomichni vazheli zabezpechennya polivektornoho ro-zvytku pidpryyemstv. Doktor nauk. Natsional'nyy universytet «L'vivs'ka politekhnika»
7. Dubrovyn VI (2009) Metod otrymannya vektora priorytetiv yikh nechetkykh matryts' poparnykh porivnyan'. Shtuchnyy intelekt 3:464–470
8. Kahan Ye S (2012) Vykorystannya metodu analizu khrererkhiyi ta teoriyi nechetkykh mnozhyn dlya otsinky skladnykh sotsial'no-ekonomichnykh yavyshch. Yzvestyya Altays'koho derzhavnoho universytetu, 1-1, pp 160–163
9. Rybak VA (2010) Vykorystannya teoriyi nechetkykh mnozhyn dlya otsinky ekoloho-ekonomichnoyi efektyvnosti. Novosty nauky y tekhnolohyy 1:21–29
10. Kryvinska N (2012) Building consistent formal specification for the service enterprise agility foundation. Soc Serv Sci J Serv Sci Res, Springer 4(2):235–269
11. Urikova O, Ivanochko I, Kryvinska N, Strauss C, Zinterhof P (2013) Consideration of aspects affecting the evolvement of collaborative eBusiness in service organizations. Inderscience Publishers, Int J Serv Econ Manag (IJSEM), Spec Issue Transform Innovations Bus Serv Models Model 5(1/2):72–92
12. Gregus M, Kryvinska N (2015) Service orientation of enterprises—aspects, dimensions, technologies. Comenius University, Bratislava. ISBN 9788022339780
13. An'shyn VM, Demkin IV, Tsar'kov IN, Nykonov YM (2008) Prymenenye teoryy nechětkykh mnozhestv k zadache formyrovanyya portfelya proektov. Problemy analyza ryska 3/5:8–21
14. Kaczor S, Kryvinska N (2013) It is all about services—fundamentals, drivers, and business models. Soc Serv Sci J Serv Sci Res, Springer 5(2):125–154
15. Kryvinska N, Gregus M (2014) SOA and its business value in requirements, features, practices and methodologies. Comenius University, Bratislava. ISBN 9788022337649
16. Molokanova VM (2012) Otsinyuvannya yakisnykh pokaznykiv portfelya proektiv za dopomohoyu teoriyi nechitkykh mnozhyn. Upravlinnya proektamy ta rozvytok vyrobnytstva 3:106–114
17. Molnár E, Molnár R, Kryvinska N, Greguš M (2014) Web Intelligence in practice. The Society of Service Science. J Serv Sci Res, Springer 6(1):149–172

Printed in the United States
By Bookmasters